T0348708

Renewable Energy Conversion, Transmission and Storage

Other books by the author:

Hydrogen and Fuel Cells. 2005
Renewable Energy, 3rd ed. 2004
Life-cycle analysis of energy systems (with Kuemmel and Nielsen). 1997
Blegdamsvej 17. 1989
Superstrenge. 1987
Fred og frihed. 1985
Fundamentals of Energy Storage (with Jensen). 1984
Energi for fremtiden (with Hvelplund, Illum, Jensen, Meyer and Nørgård).
     1983
Energikriser og Udviklingsperspektiver (with Danielsen). 1983
Renewable Energy. First edition, 1979; Second edition 2000
Skitse til alternativ energiplan for Danmark (with Blegaa, Hvelplund, Jen-
     sen, Josephsen, Linderoth, Meyer and Balling). 1976

More information about the author at http://energy.ruc.dk

Renewable Energy Conversion, Transmission and Storage

Bent Sørensen

Roskilde University
Energy, Environment and Climate Group,
Department of Environmental, Social and Spatial Change
Universitetsvej 1, P. O. Box 260
DK-4000 Roskilde, Denmark

AMSTERDAM • BOSTON • HEIDELBERG • LONDON
NEW YORK • OXFORD • PARIS • SAN DIEGO
SAN FRANCISCO • SINGAPORE • SYDNEY • TOKYO

SEVIER

Academic Press is an imprint of Elsevier

Academic Press is an imprint of Elsevier

30 Corporate Drive, Suite 400, Burlington, MA 01803, USA
525 B Street, Suite 1900, San Diego, California 92101-4495, USA
84 Theobald's Road, London WC1X 8RR, UK

This book is printed on acid-free paper. ∞

Copyright © 2007, Elsevier Inc. All rights reserved.

No part of this publication may be reproduced or transmitted in any form or by any means, electronic or mechanical, including photocopy, recording, or any information
tion
storage and retrieval system, without permission in writing from the publisher.

Permissions may be sought directly from Elsevier's Science & Technology Rights Department in Oxford, UK: phone: (+44) 1865 843830, fax: (+44) 1865 853333, E-mail: permissions@elsevier.com. You may also complete your request on-line via the Elsevier homepage (http://elsevier.com), by selecting "Support & Contact" then "Copyright and Permission" and then "Obtaining Permissions."

Library of Congress Cataloging-in-Publication Data
Application Submitted

British Library Cataloguing-in-Publication Data
A catalogue record for this book is available from the British Library.

ISBN: 978-0-12-374262-9

For information on all Academic Press publications
visit our Web site at www.books.elsevier.com

Layout and print-ready electronic-medium manuscript by author
Printed and bound by CPI Group (UK) Ltd, Croydon, CR0 4YY
Transferred to Digital Print 2011

Working together to grow
libraries in developing countries

www.elsevier.com | www.bookaid.org | www.sabre.org

ELSEVIER   BOOK AID
              International   Sabre Foundation

Preface

It is increasingly becoming accepted that renewable energy has a decisive place in the future energy system and that the "future" may not be very far away, considering not just issues of greenhouse gas emissions and the finiteness of fossil and nuclear resources, but also their uneven distribution over the Earth and the increasing political instability of precisely those regions most endowed with the remaining non-renewable resources.

Renewable energy sources have been the backbone of our energy system during most of human history, interrupted by a brief interval of cheap fuels that could be used for a few hundred years in a highly unsustainable way. Unfortunately, this interval has also weakened our sensibility over wasteful uses of energy. For a long time, energy was so cheap that most people did not think it worthwhile to improve the efficiency of energy use, even if there was money to save. Recent analysis has shown that a number of efficiency improvements that would use already existing technology could have been introduced at a cost lower than that of the energy saved, even at the prevailing low prices. We now know that any renewal of our energy supply-system would probably be more (although not necessarily a lot more) expensive than the present cost of energy, and although this book is about the prospects for filling our future energy needs with a range of renewable technologies, it must still be emphasised that carrying though all efficiency improvements in our conversion system, that can be made at lower cost than the new system, should be done first, and thereby buying us more time to make the supply transition unfold smoothly.

This book is based on the energy conversion, transmission and storage parts of the author's *Renewable Energy*, the book that in 1979 placed the topic on the academic agenda and actually got the term "renewable energy" accepted. While *Renewable Energy* (now in its third edition) deals with the physical, technical, social, economic and environmental aspects of renewable energy, the present book concentrates on the engineering aspects, in order to provide a suitable textbook for the many engineering courses in renewable energy coming on-line, and hopefully at the same time providing a handy primer for people working in this important field.

Gilleleje, June 2007, *Bent Sørensen*

Contents

Preface   v
Contents   vi
Units and Conversion factors   viii

    Chapter 1. Introduction   1

I. General principles
    Chapter 2. Basic principles of energy conversion   3
    Chapter 3. Thermodynamic engine cycles   13

II. Heat energy conversion processes
    Chapter 4. Direct thermoelectric conversion   17
    Chapter 5. Engine conversion of solar energy   22
    Chapter 6. Heat pumps   26
    Chapter 7. Geothermal and ocean-thermal energy conversion   30

III. Mechanical energy conversion processes
    Chapter 8. Basic description of flow-driven converters   34
    Chapter 9. Propeller-type converters   43
    Chapter 10. Cross-wind and other alternative converter concepts   64
    Chapter 11. Hydro and tidal energy conversion   76
    Chapter 12. Magneto hydrodynamic converters   81
    Chapter 13. Wave energy converters   83

IV. Solar radiation conversion processes
    Chapter 14. Photovoltaic conversion   94
    Chapter 15. Photo-electrochemical conversion   127
    Chapter 16. Solar thermal conversion   137
    Chapter 17. Solar thermal electricity generators   159
    Chapter 18. Solar cooling and other applications   163

V. Electrochemical energy conversion processes
    Chapter 19. Fuel cells   169
    Chapter 20. Other electrochemical energy conversion   179

VI. Bioenergy conversion processes
    Chapter 21. Combustion   186
    Chapter 22. Biological conversion into gaseous fuels   195
    Chapter 23. Biological conversion into liquid fuels   207
    Chapter 24. Thermochemical conversion to gaseous and other fuels   214

CONTENTS

VII. Energy Transmission
   Chapter 25. Heat transmission   224
   Chapter 26. Power transmission   228
   Chapter 27. Fuel transmission   232

VIII. Heat storage
   Chapter 28. Heat capacity storage   234
   Chapter 29. Latent heat and chemical transformation storage   250

IX. High-quality energy storage
   Chapter 30. Pumped hydro storage   261
   Chapter 31. Flywheels   267
   Chapter 32. Compressed gas storage   275
   Chapter 33. Battery storage   288
   Chapter 34. Other storage forms   295

Mini-projects and exercises   299

References   307

Index   323

Units and conversion factors

Powers of 10$^{\square}$

Prefix	Symbol	Value	Prefix	Symbol	Value
atto	a	10^{-18}	kilo	k	10^3
femto	f	10^{-15}	mega	M	10^6
pico	p	10^{-12}	giga	G	10^9
nano	n	10^{-9}	tera	T	10^{12}
micro	μ	10^{-6}	peta	P	10^{15}
milli	m	10^{-3}	exa	E	10^{18}

SI units

Basic unit	Name	Symbol
length	metre	m
mass	kilogram	kg
time	second	s
electric current	ampere	A
temperature	Kelvin	K
luminous intensity	candela	cd
plane angle	radian	rad
solid angle	steradian	sr
amount$^{\#}$	mole	mol

Derived unit	Name	Symbol	Definition
energy	joule	J	$kg\,m^2\,s^{-2}$
power	watt	W	$J\,s^{-1}$
force	newton	N	$J\,m^{-1}$
electric charge	coulomb	C	$A\,s$
potential difference	volt	V	$J\,A^{-1}\,s^{-1}$
pressure	pascal	Pa	$N\,m^{-2}$
electric resistance	ohm	Ω	$V\,A^{-1}$
electric capacitance	farad	F	$A\,s\,V^{-1}$
magnetic flux	weber	Wb	$V\,s$
inductance	henry	H	$V\,s\,A^{-1}$
magnetic flux density	tesla	T	$V\,s\,m^{-2}$
luminous flux	lumen	lm	$cd\,sr$
illumination	lux	lx	$cd\,sr\,m^{-2}$
frequency	hertz	Hz	$cycle\,s^{-1}$

$^{\square}$ G, T, P, E are called milliard, billion, billiard, trillion in Europe, but billion, trillion, quadrillion, quintillion in the USA. M as million is universal.
$^{\#}$ The amount containing as many particles as there are atoms in 0.012 kg ^{12}C.

UNITS AND CONVERSION FACTORS

Conversion factors

Type	Name	Symbol	Approximate value
energy	electon volt	eV	1.6021×10^{-19} J
energy	erg	erg	10^{-7} J (exact)
energy	calorie (thermochemical)	cal	4.184 J
energy	British thermal unit	Btu	1055.06 J
energy	Q	Q	10^{18} Btu (exact)
energy	quad	q	10^{15} Btu (exact)
energy	tons oil equivalent	toe	4.19×10^{10} J
energy	barrels oil equivalent	bbl	5.74×10^{9} J
energy	tons coal equivalent	tce	2.93×10^{10} J
energy	m^3 of natural gas		3.4×10^{7} J
energy	kg of methane		6.13×10^{7} J
energy	m^3 of biogas		2.3×10^{7} J
energy	litre of gasoline		3.29×10^{7} J
energy	kg of gasoline		4.38×10^{7} J
energy	litre of diesel oil		3.59×10^{7} J
energy	kg of diesel oil/gasoil		4.27×10^{7} J
energy	m^3 of hydrogen at 1 atm		1.0×10^{7} J
energy	kg of hydrogen		1.2×10^{8} J
energy	kilowatt hour	kWh	3.6×10^{6} J
power	horsepower	hp	745.7 W
power	kWh per year	kWh/y	0.114 W
radioactivity	curie	Ci	3.7×10^{8} s^{-1}
radioactivity	becqerel	Bq	$1\ s^{-1}$
radiation dose	rad	rad	10^{-2} J kg^{-1}
radiation dose	gray	Gy	J kg^{-1}
dose equivalent	rem	rem	10^{-2} J kg^{-1}
dose equivalent	sievert	Sv	J kg^{-1}
temperature	degree Celsius	°C	K − 273.15
temperature	degree Fahrenheit	°F	9/5 C + 32
time	minute	min	60 s (exact)
time	hour	h	3600 s (exact)
time	year	y	8760 h

continued next page

Type	Name	Symbol	Approximate value
pressure	atmosphere	atm	1.013×10^5 Pa
pressure	bar	bar	10^5 Pa
pressure	pounds per square inch	psi	6890 Pa
mass	ton (metric)	t	10^3 kg
mass	pound	lb	0.4536 kg
mass	ounce	oz	0.02835 kg
length	Ångström	Å	10^{-10} m
length	inch	in	0.0254 m
length	foot	ft	0.3048 m
length	mile (statute)	mi	1609 m
volume	litre	l	10^{-3} m^3
volume	gallon (US)		3.785×10^{-3} m^3

INTRODUCTION

The structure of this book is to start with general principles of energy conversion and then move on to more specific types of conversion suitable for different classes of renewable energy such as wind, hydro and wave energy, solar radiation used for heat or power generation, secondary conversions in fuel cell or battery operation, and a range of conversions related to biomass, from traditional combustion to advanced ways of producing liquid or gaseous biofuels.

Because some of the renewable energy sources are fundamentally intermittent, and sometimes beyond what can be remedied by regional trade of energy (counting on the variability being different in different geographical regimes), energy storage must also be treated as an important partner to many renewable energy systems. This is done in the final chapters, after a discussion of transmission or transport of the forms of energy available in a renewable energy system. In total, the book constitutes an introduction to all the technical issues to consider in designing renewable energy systems. The complementary issues of economy, environmental impacts and planning procedures, as well as a basic physical-astronomical explanation of where the renewable energy sources come from and how they are distributed, may be found in the bulkier treatise of Sørensen (2004).

If used for energy courses, the teacher may find the "mini-projects and exercises" attached at the end useful. They comprise simple problems but in most cases can be used as mini-projects, which are issues discussed by individual students or groups of students for a period of one to a couple of weeks, and completed by submission of a project report of some 5-25 pages for evaluation and grading. These mini-projects may involve small computer models made by the students for getting quantitative results to the problems posed.

General principles do not wear with time, and the reference list contains many quite old references, reflecting a preference for quoting those who first discussed a given issue rather than the most recent marginal improvement.

I. GENERAL PRINCIPLES

CHAPTER

2

BASIC PRINCIPLES OF ENERGY CONVERSION

A large number of energy conversion processes take place in nature. Man is capable of performing a number of additional energy conversion processes by means of various devices invented during the history of man. Such devices may be classified according to the type of construction used, according to the underlying physical or chemical principle, or according to the forms of energy appearing before and after the action of the device. In this chapter, a survey of conversion methods, which may be suitable for the conversion of renewable energy flows or stored energy, will be given. A discussion of general conversion principles will be made below, followed by an outline of engineering design details for specific energy conversion devices, ordered according to the energy form being converted and the energy form obtained. The collection is necessarily incomplete and involves judgment about the importance of various devices.

2.1 Conversion between energy forms

For a number of energy forms, Table 2.1 lists some examples of energy conversion processes or devices currently in use or contemplated, organised according to the energy form emerging after the conversion. In several cases more than one energy form will emerge as a result of the action of the device, e.g. heat in addition to one of the other energy forms listed. Many devices also perform a number of energy conversion steps, rather than the single ones given in the table. A power plant, for example, may perform the conversion process chain between energy forms: chemical → heat → mechanical → electrical. Diagonal transformations are also possible, such as conversion of mechanical energy into mechanical energy (potential energy of elevated fluid → kinetic energy of flowing fluid → rotational energy of turbine) or of heat into heat at a lower temperature (convection, conduction). The second law of thermodynamics forbids a process in which the only change is that heat is transferred from a lower to a higher temperature. Such transfer can be established if at the same time some high-quality energy is degraded, e.g. by a heat pump (which is listed as a converter of electrical into heat energy in Table 2.1, but is discussed further in Chapter 6).

Initial energy form	Converted energy form				
	Chemical	Radiant	Electrical	Mechanical	Heat
Nuclear					Reactor
Chemical			Fuel cell, battery discharge		Burner, boiler
Radiant	Photolysis		Photovoltaic cell		Absorber
Electrical	Electrolysis, battery charging	Lamp, laser		Electric motor	Resistance, heat pump
Mechanical			Electric generator, MHD	Turbines	Friction, churning
Heat			Thermionic & thermoelectric generators	Thermodynamic engines	Convector, radiator, heat pipe

Table 2.1. Examples of energy conversion processes listed according to the initial energy form and one particular converted energy form (the one primarily wanted).

The efficiency with which a given conversion process can be carried out, i.e. the ratio between the output of the desired energy form and the energy input, depends on the physical and chemical laws governing the process. For the heat engines, which convert heat into work or vice versa, the description of thermodynamic theory may be used in order to avoid the complication of a microscopic description on the molecular level (which is, of course, possible, e.g. on the basis of statistical assumptions). According to thermodynamic theory (again the "second law"), no heat engine can have an efficiency higher than that of a reversible Carnot process, which is depicted in Fig. 2.1, in terms of different sets of thermodynamic state variables,

(P, V) = (pressure, volume),
(T, S) = (absolute temperature, entropy),

and

(H, S) = (enthalpy, entropy).

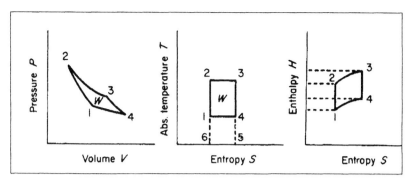

Figure 2.1. The cyclic Carnot process in different representations. Traversing the cycle in the direction $1\rightarrow 2\rightarrow 3\rightarrow 4$ leads to the conversion of a certain amount of heat into work (see text for details).

The change of the entropy S during a process (e.g. an energy conversion process), which brings the system from a state 1 to a state 2, is defined by

$$\Delta S = \int_{T_1}^{T_2} T^{-1}\, dQ, \tag{2.1}$$

where the integral is over successive infinitesimal and reversible process steps (not necessarily related to the real process, which may not be reversible), during which an amount of heat dQ is transferred from a reservoir of temperature T to the system. The imagined reservoirs may not exist in the real process, but the initial and final states of the system must have well-defined temperatures T_1 and T_2 in order for (2.1) to be applicable. The entropy may contain an arbitrary common constant fixed by the third law of thermodynamics (Nernst's law), which states that S may be taken as zero at zero absolute temperature ($T = 0$).

The enthalpy H is defined by

$$H = U + PV,$$

in terms of P, V and the internal energy U of the system. According to the first law of thermodynamics, U is a state variable given by

$$\Delta U = \int dQ + \int dW, \tag{2.2}$$

in terms of the amounts of heat and work added to the system [Q and W are not state variables, and the individual integrals in (2.2) depend on the paths of integration]. The equation (2.2) determines U up to an arbitrary constant, the zero point of the energy scale. Using the definition (2.1),

$$dQ = T\, dS$$

and

$$dW = -P\,dV,$$

both of which are valid only for reversible processes The following relations are found among the differentials:

$$dU = T\,dS - P\,dV,$$
$$dH = T\,dS + V\,dP. \tag{2.3}$$

These relations are often assumed to have general validity.

If chemical reactions occur in the system, additional terms $\mu_i\,dn_i$ should be added on the right-hand side of both relations (2.3), in terms of the chemical potentials μ_i (see e.g. Maron and Prutton, 1959).

For a cyclic process such as the one shown in Fig. 2.1, $\int dU = 0$ upon returning to the initial locus in one of the diagrams, and thus according to (2.3) $\int T\,dS = \int P\,dV$. This means that the area enclosed by the path of the cyclic process in either the (P, V) or the (T, S) diagram equals the work $-W$ performed by the system during one cycle (in the direction of increasing numbers on Fig. 2.1).

The amount of heat added to the system during the isothermal process 2-3 is $\Delta Q_{23} = T(S_3 - S_2)$, if the constant temperature is denoted T. The heat added in the other isothermal process, 4-1, at a temperature T_{ref} is $\Delta Q_{41} = -T_{ref}(S_3 - S_2)$. It follows from the (T, S) diagram that $\Delta Q_{23} + \Delta Q_{41} = -W$. The efficiency by which the Carnot process converts heat available at temperature T into work, when a reference temperature of T_{ref} is available, is then

$$\eta = \frac{-W}{\Delta Q_{23}} = \frac{T - T_{ref}}{T}. \tag{2.4}$$

The Carnot cycle (Fig. 2.1) consists of four steps: 1-2, adiabatic compression (no heat exchange with the surroundings, i.e. $dQ = 0$ and $dS = 0$); 2-3, heat drawn reversibly from the surroundings at constant temperature (the amount of heat transfer ΔQ_{23} is given by the area enclosed by the path 2-3-5-6-2 in the (T, S)-diagram); 3-4, adiabatic expansion; and 4-1, heat given away to the surroundings by a reversible process at constant temperature [$|\Delta Q_{41}|$ equal to the area of the path 4-5-6-1-4 in the (T, S)-diagram].

The (H, S)-diagram is an example of a representation in which energy differences can be read directly on the ordinate, rather than being represented by an area.

It requires long periods of time to perform the steps involved in the Carnot cycle in a way that approaches reversibility. As time is important for man (the goal of the energy conversion process being power rather than just an amount of energy), irreversible processes are deliberately introduced into the thermodynamic cycles of actual conversion devices. The thermodynamics of irreversible processes are described below using a practical approximation, which will be referred to in several of the examples to follow. Readers without special interest in the thermodynamic description may go lightly over the formulae (unless such readers are up for an exam!).

2.2 Irreversible thermodynamics

The degree of irreversibility is measured in terms of the rate of energy dissipation,

$$D = T \, dS/dt, \tag{2.5}$$

where dS/dt is the entropy production of the system while held at the constant temperature T (i.e. T may be thought of as the temperature of a large heat reservoir, with which the system is in contact). In order to describe the nature of the dissipation process, the concept of free energy may be introduced (cf. E.G. Callen, 1960).

The free energy of a system, G, is defined as the maximum work that can be drawn from the system under conditions where the exchange of work is the only interaction between the system and its surroundings. A system of this kind is said to be in thermodynamic equilibrium if its free energy is zero.

Consider now a system divided into two subsystems, a small one with extensive variables (i.e. variables proportional to the size of the system) U, S, V, etc. and a large one with intensive variables T_{ref}, P_{ref}, etc., which is initially in thermodynamic equilibrium. The terms "small system" and "large system" are meant to imply that the intensive variables of the large system (but not its extensive variables U_{ref}, S_{ref}, etc.) can be regarded as constant, regardless of the processes by which the entire system approaches equilibrium.

This implies that the intensive variables of the small system, which may not even be defined during the process, approach those of the large system when the combined system approaches equilibrium. The free energy, or maximum work, is found by considering a reversible process between the initial state and the equilibrium. It equals the difference between the initial internal energy, $U_{init} = U + U_{ref}$, and the final internal energy, U_{eq}, or it may be written (all in terms of initial state variables) as

$$G = U - T_{ref} S + P_{ref} V, \tag{2.6}$$

plus terms of the form $\Sigma \mu_{i,ref} n_i$ if chemical reactions are involved, and similar generalisations in case of e.g. electromagnetic interactions.

If the entire system is closed, it develops spontaneously towards equilibrium through internal, irreversible processes, with a rate of free energy change

$$\frac{dG}{dt} = \frac{d}{dt}(U_{init} - U_{eq}(t)) = \left(\frac{\partial}{\partial S(t)}U_{eq}(t)\right)\frac{dS(t)}{dt},$$

assuming that the entropy is the only variable. $S(t)$ is the entropy at time t of the entire system, and $U_{eq}(t)$ is the internal energy that would be possessed by a hypothetical equilibrium state defined by the actual state variables at time t, that is $S(t)$ etc. For any of these equilibrium states, $\partial U_{eq}(t)/\partial S(t)$ equals T_{ref} according to (2.3), and by comparison with (2.5) it is seen that the rate of dissipation can be identified with the loss of free energy, as well as with the increase in entropy,

$$D = -dG/dt = T_{ref}\, dS(t)/dt. \tag{2.7}$$

For systems met in practice, there will often be constraints preventing the system from reaching the absolute equilibrium state of zero free energy. For instance, the small system considered above may be separated from the large one by walls keeping the volume V constant. In such cases the available free energy (i.e. the maximum amount of useful work that can be extracted) becomes the absolute amount of free energy, (2.6), minus the free energy of the relative equilibrium, which the combined system can be made to approach in the presence of the constraint. If the extensive variables in the constrained equilibrium state are denoted U^0, S^0, V^0, etc., then the available free energy becomes

$$\Delta G = (U - U^0) - T_{ref}(S - S^0) + P_{ref}(V - V^0), \tag{2.8}$$

eventually with the additions involving chemical potentials. In the form (2.6) or (2.8), G is called the Gibbs potential. If the small system is constrained by walls, so that the volume cannot be changed, the free energy reduces to the Helmholtz potential $U - TS$, and if the small system is constrained so that it is incapable of exchanging heat, the free energy reduces to the enthalpy H. The corresponding forms of (2.8) give the maximum work that can be obtained from a thermodynamic system with the given constraints.

A description of the course of an actual process as a function of time requires knowledge of "equations of motion" for the extensive variables, i.e. equations that relate the currents such as

$J_s = dS/dt$ (entropy flow rate) or $J_Q = dQ/dt$ (heat flow rate),

$J_m = dm/dt$ (mass flow rate) or $J_\theta = d\theta/dt$ (angular velocity), $\qquad$ (2.9)

$J_q = dq/dt = I$ (charge flow rate or electrical current), etc.

to the (generalised) forces of the system. As a first approximation, the relation between the currents and the forces may be taken as linear (Onsager, 1931),

$$J_i = \sum_j L_{ij} F_j. \tag{2.10}$$

The direction of each flow component is J_i / J_i. The arbitrariness in choosing the generalised forces is reduced by requiring, as did Onsager, that the dissipation be given by

$$D = -dG/dt = \sum_i J_i \cdot F_i. \tag{2.11}$$

Examples of the linear relationships (2.10) are Ohm's law, stating that the electric current J_q is proportional to the gradient of the electric potential ($F_q \propto$ grad ϕ), and Fourier's law for heat conduction or diffusion, stating that the heat flow rate $E^{sens} = J_Q$ is proportional to the gradient of the temperature.

Considering the isothermal expansion process required in the Carnot cycle (Fig. 2.1), heat must be flowing to the system at a rate $J_Q = dQ/dt$, with $J_Q = LF_Q$ according to (2.10) in its simplest form. Using (2.11), the energy dissipation takes the form

$$D = T\,dS/dt = J_Q F_Q = L^{-1} J_Q^2.$$

For a finite time Δt, the entropy increase becomes

$$\Delta S = (dS/dt)\,\Delta t = (LT)^{-1} J_Q^2 \Delta t = (LT\Delta t)^{-1} (\Delta Q)^2,$$

so that in order to transfer a finite amount of heat ΔQ, the product $\Delta S\,\Delta t$ must equal the quantity $(LT)^{-1} (\Delta Q)^2$. In order that the process approaches reversibility, as the ideal Carnot cycle should, ΔS must approach zero, which is seen to imply that Δt approaches infinity. This qualifies the statement made in the beginning of this subsection that, in order to go through a thermodynamic engine cycle in a finite time, one has to give up reversibility and accept a finite amount of energy dissipation and an efficiency that is smaller than the ideal one (2.4).

2.3 Efficiency of an energy conversion device

A schematic picture of an energy conversion device is shown in Fig. 2.2, sufficiently general to cover most types of converters in practical use (Angrist, 1976; Osterle, 1964). There is a mass flow into the device and another one out from it, as well as an incoming and outgoing heat flow. The work output may be in the form of electric or rotating shaft power.

It may be assumed that the converter is in a steady state, implying that the incoming and outgoing mass flows are identical and that the entropy of the device itself is constant, that is, that all entropy created is being carried away by the outgoing flows.

From the first law of thermodynamics, the power extracted, E, equals the net energy input,

$$E = J_{Q,in} - J_{Q,out} + J_m (w_{in} - w_{out}).$$ (2.12)

The magnitude of the currents is given by (2.9), and their conventional signs may be inferred from Fig. 2.2. The specific energy content of the incoming mass flow, w_{in}, and of the outgoing mass flow, w_{out}, are the sums of potential energy, kinetic energy and enthalpy. The significance of the enthalpy to represent the thermodynamic energy of a stationary flow is established by Bernoulli's theorem (Pippard, 1966). It states that for a stationary flow, if heat conduction can be neglected, the enthalpy is constant along a streamline. For the uniform mass flows assumed for the device in Fig. 2.2, the specific enthalpy, h, thus becomes a property of the flow, in analogy with the kinetic energy of motion and, for example, the geopotential energy,

$$w = w^{pot} + w^{kin} + h.$$ (2.13)

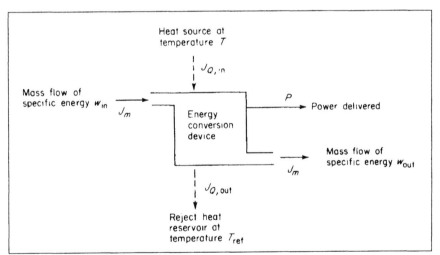

Figure 2.2. Schematic picture of an energy conversion device with a steady–state mass flow. The sign convention is different from the one used in (2.2), where all fluxes into the system were taken as positive.

The power output may be written

$$E = -J_\theta \cdot F_\theta - J_q \cdot F_q,$$ (2.14)

with the magnitude of currents given by (2.9) and the generalised forces given by

$$F_\theta = \int r \times dF_{mech}(r) \qquad \text{(torque)},$$

$$F_q = -\text{grad}(\phi) \qquad \text{(electric field)} \tag{2.15}$$

corresponding to a mechanical torque and an electric potential gradient. The rate of entropy creation, i.e. the rate of entropy increase in the surroundings of the conversion device (as mentioned, the entropy inside the device is constant in the steady-state model), is

$$dS/dt = (T_{ref})^{-1} J_{Q,out} - T^{-1} J_{Q,in} + J_m (s_{m,out} - s_{m,in}),$$

where $s_{m,in}$ is the specific entropy of the mass (fluid, gas, etc.) flowing into the device, and $s_{m,out}$ is the specific entropy of the outgoing mass flow. $J_{Q,out}$ may be eliminated by use of (2.12), and the rate of dissipation obtained from (2.7),

$$D = T_{ref} \, dS/dt =$$
$$J_{Q,in} (1 - T_{ref}/T) + J_m (w_{in} - w_{out} - T_{ref}(s_{m,in} - s_{m,out})) - E = \max(E) - E. \tag{2.16}$$

The maximum possible work (obtained for $dS/dt = 0$) is seen to consist of a Carnot term (closed cycle, i.e. no external flows) plus a term proportional to the mass flow. The dissipation (2.16) is brought in the Onsager form (2.11),

$$D = J_{Q,in} F_{Q,in} + J_m F_m + J_\theta \cdot F_\theta + J_q \cdot F_{q'} \tag{2.17}$$

by defining generalised forces

$$F_{Q,in} = 1 - T_{ref}/T,$$
$$F_m = w_{in} - w_{out} - T_{ref} (s_{m,in} - s_{m,out}) \tag{2.18}$$

in addition to those of (2.15).

The efficiency with which the heat and mass flow into the device is converted to power is, in analogy to (2.4),

$$\eta = \frac{E}{J_{Q,in} + J_m w_{in}}, \tag{2.19}$$

where the expression (2.16) may be inserted for E. This efficiency is sometimes referred to as the "first law" efficiency, because it only deals with the amounts of energy input and output in the desired form and not with the "quality" of the energy input related to that of the energy output.

In order to include reference to the energy quality, in the sense of the second law of thermodynamics, account must be taken of the changes in entropy taking place in connection with the heat and mass flows through the conversion device. This is accomplished by the "second law" efficiency, which for power-generating devices is defined by

$$\eta^{(2.law)} = \frac{E}{\max(E)} = -\frac{\mathbf{J}_\theta \cdot \mathbf{F}_\theta + \mathbf{J}_q \cdot \mathbf{F}_q}{J_{Q,in}F_{Q,in} + J_m F_m},\qquad(2.20)$$

where the second expression is valid specifically for the device considered in Fig. 2.2, while the first expression is of general applicability, when max(E) is taken as the maximum rate of work extraction permitted by the second law of thermodynamics. It should be noted that max(E) depends not only on the system and the controlled energy inputs, but also on the state of the surroundings.

Conversion devices for which the desired energy form is not work may be treated in a way analogous to the example in Fig. 2.2. In the form (2.17), no distinction is made between input and output of the different energy forms. Taking, for example, electrical power as input (sign change), output may be obtained in the form of heat or in the form of a mass stream. The efficiency expressions (2.19) and (2.20) must be altered, placing the actual input terms in the denominator and the actual output terms in the numerator. If the desired output energy form is denoted W, the second law efficiency can be written in the general form

$$\eta^{(2.\,law)} = W / \max(W).\qquad(2.21)$$

For conversion processes based on principles other than those considered in the thermodynamic description of phenomena, alternative efficiencies could be defined by (2.21), with max(W) calculated under consideration of the non-thermodynamic types of constraints. In such cases, the name "second law efficiency" would have to be modified.

CHAPTER 3

THERMODYNAMIC ENGINE CYCLES

A number of thermodynamic cycles, i.e. (closed) paths in a representation of conjugate variables, have been demonstrated in practice. They offer examples of the compromises made in modifying the "prototype" Carnot cycle into a cycle that can be traversed in a finite amount of time. Each cycle can be used to convert heat into work, but in traditional uses the source of heat has mostly been the combustion of fuels, i.e. an initial energy conversion process, by which high-grade chemical energy is degraded to heat at a certain temperature, associated with a certain entropy production.

Figure 3.1 shows a number of engine cycles in (P, V)-, (T, S), and (H, S)-diagrams corresponding to Fig. 2.1.

The working substance of the Brayton cycle is a gas, which is adiabatically compressed in step 1-2 and expanded in step 3-4. The remaining two steps take place at constant pressure (isobars), and heat is added in step 2-3. The useful work is extracted during the adiabatic expansion 3-4, and the simple efficiency is thus equal to the enthalpy difference $H_3 - H_4$ divided by the total input $H_3 - H_1$. Examples of devices operating on the Brayton cycle are gas turbines and jet engines. In these cases, the cycle is usually not closed, since the gas is exhausted at point 4 and step 4-1 is thus absent. The somewhat contradictory name given to such processes is "open cycles".

The Otto cycle, presently used in a large number of automobile engines, differs from the Brayton cycle in that steps 2–3 and 4–1 (if the cycle is closed) are carried out at constant volume (isochores) rather than at constant pressure.

The Diesel cycle (common in ship, lorry/truck and increasingly in passenger car engines) has step 2-3 as isobar and step 4-1 as isochore, while the two remaining steps are approximately adiabates. The actual designs of the machines, involving turbine wheels or piston-holding cylinders, etc., may be found in engineering textbooks (e.g. Hütte, 1954).

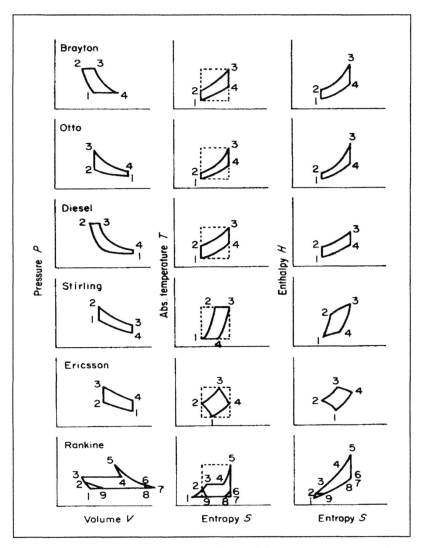

Figure 3.1. Examples of thermodynamic cycles in different representations. For comparison, the Carnot cycle is indicated in the (P, S)-diagram (dashed lines). Further descriptions of the individual cycles are given in the text (cf. also Chapter 5 for an alternative version of the Ericsson cycle).

Closer to the Carnot ideal is the Stirling cycle, involving two isochores (1-2 and 3-4) and two isotherms.

The Ericsson cycle has been developed with the purpose of using hot air as the working fluid. It consists of two isochores (2-3 and 4-1) and two curves somewhere between isotherms and adiabates (cf. e.g. Meinel and Meinel, 1976).

The last cycle depicted in Fig. 3.1 is the Rankine cycle, the appearance of which is more complicated owing to the presence of two phases of the working fluid. Step 1-2-3 describes the heating of the fluid to its boiling point. Step 3-4 corresponds to the evaporation of the fluid, with both fluid and gaseous phases being present. It is an isotherm as well as an isobar. Step 4-5 represents the superheating of the gas, followed by an adiabatic expansion step 5-7. These two steps are sometimes repeated one or more times, with the superheating taking place at gradually lowered pressure, after each step of expansion to saturation. Finally, step 7-1 again involves mixed phases with condensation at constant pressure and temperature. The condensation often does not start until a temperature below that of saturation is reached. Useful work is extracted during the expansion step 5-7, so the simple efficiency equals the enthalpy difference $H_5 - H_7$ divided by the total input $H_6 - H_1$. The second law efficiency is obtained by dividing the simple efficiency by the Carnot value (4.4), for $T = T_5$ and $T_{ref} = T_7$.

Thermodynamic cycles such as those of Figs. 2.1 and 3.1 may be traversed in the opposite direction, thus using the work input to create a low temperature T_{ref} (cooling, refrigeration; T being the temperature of the surroundings) or to create a temperature T higher than that (T_{ref}) of the surroundings (heat pumping). In this case step 7-5 of the Rankine cycle is a compression (8-6-5 if the gas experiences superheating). After cooling (5-4), the gas condenses at the constant temperature T (4-3), and the fluid is expanded, often by passage through a nozzle. The passage through the nozzle is considered to take place at constant enthalpy (2-9), but this step may be preceded by undercooling (3-2). Finally, step 9-8 (or 9-7) corresponds to evaporation at the constant temperature T_{ref}.

For a cooling device the simple efficiency is the ratio of the heat removed from the surroundings, $H_7 - H_9$, and the work input, $H_5 - H_7$, whereas for a heat pump it is the ratio of the heat delivered, $H_5 - H_2$, and the work input. Such efficiencies are often called "coefficients of performance" (COP), and the second law efficiency may be found by dividing the COP by the corresponding quantity ε_{Carnot} for the ideal Carnot cycle (cf. Fig. 2.1),

$$\varepsilon_{Carnot}^{cooling} = \frac{\Delta Q_{14}}{W} = \frac{T_{ref}}{T - T_{ref}}, \tag{3.1}$$

$$\varepsilon_{Carnot}^{heatpump} = \frac{\Delta Q_{32}}{W} = \frac{T}{T - T_{ref}}. \tag{3.2}$$

In practice, the compression work $H_5 - H_7$ (for the Rankine cycle in Fig. 3.1) may be less than the energy input to the compressor, thus further reducing the COP and the second law efficiency, relative to the primary source of high-quality energy.

II. HEAT ENERGY CONVERSION PROCESSES

CHAPTER 4

DIRECT THERMOELECTRIC CONVERSION

If the high-quality energy form desired is electricity, and the initial energy is in the form of heat, there is a possibility of utilising direct conversion processes, rather than first using a thermodynamic engine to create mechanical work and then in a second conversion step using an electricity generator.

4.1 Thermoelectric generators

One direct conversion process makes use of the thermoelectric effect associated with heating the junction of two different conducting materials, e.g. metals or semiconductors. If a stable electric current, I, passes across the junction between the two conductors A and B, in an arrangement of the type depicted in Fig. 4.1, then quantum electron theory requires that the Fermi energy level (which may be regarded as a chemical potential μ_i) is the same in the two materials ($\mu_A = \mu_B$). If the spectrum of electron quantum states is different in the two materials, the crossing of negatively charged electrons or positively charged "holes" (electron vacancies) will not preserve the statistical distribution of electrons around the Fermi level,

$$f(E) = (\exp((E - \mu_i)/kT) + 1)^{-1}, \qquad (4.1)$$

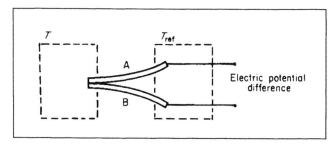

Figure 4.1. Schematic picture of a thermo-electric generator (thermocouple). The rods A and B are made of different materials (metals or better p- and n-type semiconductors).

With E being the electron energy and k being the Boltzmann's constant. The altered distribution may imply a shift towards a lower or a higher temperature, such that the maintenance of the current may require addition or removal of heat. Correspondingly, heating the junction will increase or decrease the electric current. The first case represents a thermoelectric generator, and the voltage across the external connections (Fig. 4.1) receives a term in addition to the ohmic term associated with the internal resistance R_{int} of the rods A and B,

$$\Delta\phi = - IR_{int} + \int_{T_{ref}}^{T} \alpha\, dT'.$$

The coefficient α is called the Seebeck coefficient. It is the sum of the Seebeck coefficients for the two materials A and B, and it may be expressed in terms of the quantum statistical properties of the materials (Angrist, 1976). If α is assumed independent of temperature in the range from T_{ref} to T, then the generalised electrical force (2.15) may be written

$$F_q = R_{int}J_q - \alpha T F_{Q,in}, \tag{4.2}$$

where J_q and $F_{q,in}$ are given in (2.9) and (2.18).

Considering the thermoelectric generator (Fig. 4.1) as a particular example of the conversion device shown in Fig. 3.1, with no mass flows, the dissipation (2.11) may be written

$$D = J_Q F_Q + J_q F_q.$$

In the linear approximation (2.10), the flows are of the form

$$J_Q = L_{QQ} F_Q + L_{Qq} F_q,$$
$$J_q = L_{qQ} F_Q + L_{qq} Fq,$$

with $L_{Qq} = L_{qQ}$ because of microscopic reversibility (Onsager, 1931). Considering F_Q and J_q (Carnot factor and electric current) as the "controllable" variables, one may solve for F_q and J_Q, obtaining F_q in the form (2.24) with $F_Q = F_{Q,in}$ and

$$L_{qq} = (R_{int})^{-1}; \qquad L_{qQ} = L_{Qq} = \alpha T/R_{int}.$$

The equation for J_Q takes the form

$$J_Q = CTF_Q + \alpha T J_q, \tag{4.3}$$

where the conductance C is given by

$$C = (L_{QQ}L_{qq} - L_{Qq}L_{qQ})/(LqqT).$$

Using (4.2) and (4.3), the dissipation may be written

$$D = CTF_Q^2 + R_{int} J_q^2 , \qquad (4.4)$$

and the simple efficiency (2.19) may be written

$$\eta = \frac{-J_q F_q}{J_Q} = \frac{-J_q (R_{int} J_q - \alpha T F_Q)}{CTF_Q + \alpha T J_q} . \qquad (4.5)$$

If the reservoir temperatures T and T_{ref} are maintained at a constant value, F_Q can be regarded as fixed, and the maximum efficiency can be found by variation of J_q. The efficiency (4.5) has an extremum at

$$J_q = \frac{CF_Q}{\alpha} \left(\left(1 + \frac{\alpha^2 T}{R_{int} C} \right)^{1/2} - 1 \right), \qquad (4.6)$$

corresponding to a maximum value

$$\max(\eta) = F_Q \frac{(1 + ZT)^{1/2} - 1}{(1 + ZT)^{1/2} + 1}, \qquad (4.7)$$

with $Z = \alpha^2 (R_{int} C)^{-1}$. Equation (4.7) is accurate only if the linear approximation (4.10) is valid. The maximum second law efficiency is obtained from (4.29) by division by F_Q [cf. (2.20)].

The efficiencies are seen to increase with temperature, as well as with Z. Z is largest for certain materials (A and B in Fig. 4.1) of semiconductor structure and small for metals as well as for insulators. Although R_{int} is small for metals and large for insulators, the same is true for the Seebeck coefficient α, which appears squared. C is larger for metals than for insulators. Together, these features combine to produce a peak in Z in the semiconductor region. Typical values of Z are about 2×10^{-3} (K)$^{-1}$ at $T = 300$ K (Angrist, 1976). The two materials A and B may be taken as a p-type and an n-type semiconductor, which have Seebeck coefficients of opposite signs, so that their contributions add coherently for a configuration of the kind shown in Fig. 4.1.

4.2 Thermionic generators

Thermionic converters consist of two conductor plates separated by vacuum or by a plasma. The plates are maintained at different temperatures. One, the emitter, is at a temperature T large enough to allow a substantial emission of electrons into the space between the plates due to the thermal statistical spread in electron energy (4.1). The electrons (e.g. of a metal emitter) move in a potential field characterised by a barrier at the surface of the plate. The shape of this barrier is usually such that the probability of an electron penetrating it is small until a critical temperature, after which it increases rapidly ("red-glowing" metals). The other plate is maintained at a lower temperature T_{ref}. In order not to have a build-up of space charge between the emitter and the collector, atoms of a substance such as caesium may be introduced in this area. These atoms become ionised near the hot emitter (they give away electrons to make up for the electron deficit in the emitter material), and for a given caesium pressure the positive ions exactly neutralise the space charges of the travelling electrons. At the collector surface, recombination of caesium ions takes place.

The layout of the emitter design must allow the transfer of large quantities of heat to a small area in order to maximise the electron current responsible for creating the electric voltage difference across the emitter–collector system, which may be utilised through an external load circuit. This heat transfer can be accomplished by a so-called "heat pipe" – a fluid-containing pipe that allows the fluid to evaporate in one chamber when heat is added. The vapour then travels to the other end of the pipe, condenses and gives off the latent heat of evaporation to the surroundings, whereafter it returns to the first chamber through capillary channels, under the influence of surface tension forces.

The description of the thermionic generator in terms of the model converter shown in Fig. 2.2 is very similar to that of the thermoelectric generator. With the two temperatures T and T_{ref} defined above, the generalised force F_Q is defined. The electrical output current, J_q, is equal to the emitter current, provided that back-emission from the collector at temperature T_{ref} can be neglected and provided that the positive-ion current in the intermediate space is negligible in comparison with the electron current. If the space charges are saturated, the ratio between ion and electron currents is simply the inverse of the square root of the mass ratio, and the positive-ion current will be a fraction of a percent of the electron current. According to quantum statistics, the emission current (and hence J_q) may be written

$$J_Q = AT^2 \exp(-e\,\phi_e / (kT)), \tag{4.8}$$

where ϕ_e is the electric potential of the emitter, $e\phi_e$ is the potential barrier of the surface in energy units, and A is a constant (Angrist, 1976). Neglecting heat conduction losses in plates and the intermediate space, as well as light emission, the heat $J_{Q,in}$ to be supplied to keep the emitter at the elevated temperature T equals the energy carried away by the electrons emitted,

$$J_{Q,in} = J_q\,(\phi_e + \delta + 2kT / e), \tag{4.9}$$

where the three terms in brackets represent the surface barrier, the barrier effectively seen by an electron due to the space charge in the intermediate space, and the original average kinetic energy of the electrons at temperature T (divided by e), respectively.

Finally, neglecting internal resistance in plates and wires, the generalised electrical force equals the difference between the potential ϕ_e and the corresponding potential for the collector ϕ_c,

$$-F_q = \phi_c - \phi_e \, , \tag{4.10}$$

with insertion of the above expressions (4.30) to (4.32). Alternatively, these expressions may be linearised in the form (2.10) and the efficiency calculated exactly as in the case of the thermoelectric device. It is clear, however, that a linear approximation to (4.8), for example, would be very poor.

CHAPTER 5

ENGINE CONVERSION OF SOLAR ENERGY

The conversion of heat to shaft power or electricity is generally achieved by one of the thermodynamic cycles, examples of which were shown in Fig. 3.1. The cycles may be closed as in Fig. 3.1, or they may be "open", in that the working fluid is not recycled through the cooling step (4-1 in most of the cycles shown in Fig. 3.1). Instead, new fluid is added for the heating or compression stage, and "used" fluid is rejected after the expansion stage.

It should be kept in mind that the thermodynamic cycles convert heat of temperature T into work plus some residual heat of temperature above the reference temperature T_{ref} (in the form of heated cooling fluid or rejected working fluid). Emphasis should therefore be placed on utilising both the work and the "waste heat". This is done, for example, by co-generation of electricity and water for district heating.

The present chapter looks at a thermodynamic engine concept considered particularly suited for conversion of solar energy. Other examples of the use of thermodynamic cycles in the conversion of heat derived from solar collectors into work will be given in Chapters 17 and 18. The dependence of the limiting Carnot efficiency on temperature is shown in Fig. 17.3 for selected values of a parameter describing the concentrating ability of the collector and its short-wavelength absorption to long-wavelength emission ratio. The devices described in Chapter 18 aim at converting solar heat into mechanical work for water pumping, while the devices of interest in Chapter 17 convert heat from a solar concentrator into electricity.

5.1 Ericsson hot-air engine

The engines in the examples mentioned above were based on the Rankine or the Stirling cycle. It is also possible that the Ericsson cycle (which was actually invented for the purpose of solar energy conversion) will prove advantageous in some solar energy applications. It is based on a gas (usually air) as a working fluid and may have the layout shown in Fig. 5.1. In order to describe the cycle depicted in Fig. 3.1, the valves must be closed at definite times and the pistons must be detached from the rotating shaft (in contrast to the situation shown in Fig. 5.1), such that the heat may be supplied at constant volume. In a different mode of operation, the valves open and close as soon as a pressure difference between the air on the two sides begins to develop. In this case, the heat is added at constant pressure, as in the Brayton cycle, and the piston movement is approximately at constant temperature, as in the Stirling cycle (this variant is not shown in Fig. 3.1).

The efficiency can easily be calculated for the latter version of the Ericsson cycle, for which the temperatures T_{up} and T_{low} in the compression and expansion piston cylinders are constant, in a steady situation. This implies that the air enters the heating chamber with temperature T_{low} and leaves it with temperature T_{up}. The heat exchanger equations (cf. section 6.2, but not assuming that T_3 is constant) take the form

$$-J_m^f C_p^f \, dT^f(x)/dx = h' \, (T^f(x) - T^g(x)),$$
$$J_m^g C_p^g \, dT^g(x)/dx = h' \, (T^f(x) - T^g(x)),$$

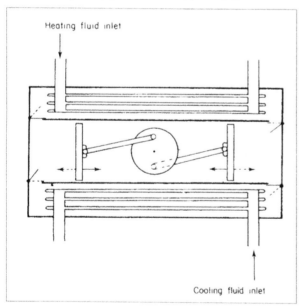

Heating fluid inlet

Cooling fluid inlet

Figure 5.1. Example of an Ericsson hot-air engine.

where the superscript g stands for the gas performing the thermodynamic cycle, f stands for the fluid leading heat to the heating chamber heat exchanger, and x increases from zero at the entrance to the heating chamber to a maximum value at the exit. C_p is a constant-pressure heat capacity per unit mass, and J_m is a mass flow rate. Both these and h', the heat exchange rate per unit length dx, are assumed constant, in which case the equations may be explicitly integrated to give

$$T_{c,in} = T_{c,out} - \frac{J_m^g C_p^g}{J_m^f C_p^f}(T_{up} - T_{low}) = \frac{J_m^g C_p^g (1-H)}{J_m^g C_p^g + J_m^f C_p^f} T_{low} + \left(H + \frac{J_m^f C_p^f (1-H)}{J_m^g C_p^g + J_m^f C_p^f} \right) T_{c,out}.$$

(5.1)

Here $T_{c,out} = T$ is the temperature provided by the solar collector or other heat source, and $T_{c,in}$ is the temperature of the collector fluid when it leaves the heat exchanger of the heating chamber, to be re-circulated to the collector or to a heat storage connected to it. H is given by

$$H = \exp(-h((J_m^f C_p^f)^{-1} + (J_m^g C_p^g)^{-1})),$$

where $h = \int h' \, dx$. Two equations analogous to (5.1) may be written for the heat exchange in the cooling chamber of Fig. 5.1, relating the reject temperature $T_{r,in}$ and the temperature of the coolant at inlet, $T_{r,out} = T_{ref}$, to T_{low} and T_{up}. $T_{c,in}$ may then be eliminated from (5.1) and $T_{r,in}$ from the analogous equation, leaving two equations for determination of T_{up} and T_{low} as functions of known quantities, notably the temperature levels T and T_{ref}. The reason for not having to consider equations for the piston motion in order to determine all relevant temperatures is, of course, that the processes associated with the piston motion have been assumed to be isothermal.

The amounts of heat added, Q_{add}, and rejected, Q_{rej}, per cycle are

$$Q_{add} = mC_p^g (T_{up} - T_{low}) + n\mathscr{R} T_{up} \log(V_{max}/V_{min}),$$

$$Q_{rej} = mC_p^g (T_{up} - T_{low}) + n\mathscr{R} T_{low} \log(V_{max}/V_{min}) + Q'_{rej},$$

(5.2)

where m is the mass of air involved in the cycle and n is the number of moles of air involved. $\mathscr{R}$ is the gas constant, and V_{min} and V_{max} are the minimum and maximum volumes occupied by the gas during the compression or expansion stages (for simplicity the "compression ratio" V_{max}/V_{min} has been assumed to be the same for the two processes, although they take place in different cylinders). The ideal gas law has been assumed in calculating the relation between heat amount and work in (5.2), and Q'_{rej} represents heat losses not contributing to transfer of heat from working gas to coolant flow (piston friction, etc.). The efficiency is

$$\eta = (Q_{add} - Q_{rej})/Q_{add},$$

and the maximum efficiency that can be obtained with this version of the Ericsson engine is obtained for negligible Q'_{rej} and ideal heat exchangers providing $T_{up} = T$ and $T_{low} = T_{ref}$.

$$\max(\eta) = \left(1 - T_{ref}/T\right) \Bigg/ \left(1 + \frac{mC_p^g}{n\mathscr{R}\log(V_{max}/V_{min})}\left(1 - T_{ref}/T\right)\right). \qquad (5.3)$$

The ideal Carnot efficiency may even be approached, if the second term in the denominator can be made small. (However, to make the compression ratio very large implies an increase in the length of time required per cycle, such that the rate of power production may actually go down, as discussed in Chapter 2). The power may be calculated by evaluating (5.2) per unit time instead of per cycle.

CHAPTER

6

HEAT PUMPS

In some cases, it is possible to produce heat at precisely the temperature needed by primary conversion. However, often the initial temperature is lower or higher than required, in the latter case even considering losses in transmission and heat drop across heat exchangers. In such situations, appropriate temperatures are commonly obtained by mixing (if the heat is stored as sensible heat in a fluid such as water, this water may be mixed with colder water, a procedure often used in connection with fossil fuel burners). This procedure is wasteful in the sense of the second law of thermodynamics, since the energy is, in the first place, produced with a higher quality than subsequently needed. In other words, the second law efficiency of conversion (2.20) is low, because there will be other schemes of conversion by which the primary energy can be made to produce a larger quantity of heat at the temperature needed at load. An extreme case of a "detour" is the conversion of heat to heat by first generating electricity by thermal conversion (as it is done today in fossil power plants) and then degrading the electricity to heat of low temperature by passing a current through an ohmic resistance ("electric heaters"). However, there are better ways:

6.1 Heat pump operation

If heat of modest temperature is required, and a high-quality form of energy is available, some device is needed that can derive additional benefits from the high quality of the primary energy source. This can be achieved by using one of the thermodynamic cycles described in Chapter 3, provided that a large reservoir of approximately constant temperature is available. The cycles (cf. Fig. 3.1) must be traversed "anti-clockwise", such that high-quality energy (electricity, mechanical shaft power, fuel combustion at high temperature, etc.) is added, and heat energy is thereby delivered at a temperature T higher than the temperature T_{ref} of the reference reservoir from which it is drawn. Most commonly the Rankine cycle with maximum efficiencies bounded by (3.1) or (3.2) is used (e.g. in an arrangement of the type shown in Fig. 6.1). The fluid of the closed cycle, which should have a liquid and a gaseous phase in the temperature interval of interest, may be a fluoro-chloromethane compound (which needs to be recycled owing to climatic effects caused if it is released to the atmosphere). The external circuits may contain an inexpensive fluid (e.g. water), and they may be omitted if it is practical to circulate the primary working fluid directly to the load area or to the reference reservoir.

The heat pump contains a compressor, which performs step 7-5 in the Rankine cycle depicted in Fig. 3.1, and a nozzle, which performs step 2-9. The intermediate steps are performed in the two heat exchangers, giving the working fluid the temperatures T_{up} and T_{low}, respectively. The equations for determining these temperatures are given below in Section 6.2. There are four such equations, which must be supplemented by equations for the compressor and nozzle performance, in order to allow a determination of all the unknown temperatures indicated in Fig. 6.1, for given T_{ref}, given load and a certain energy expenditure to the compressor. Losses in the compressor are in the form of heat, which in some cases can be credited to the load area.

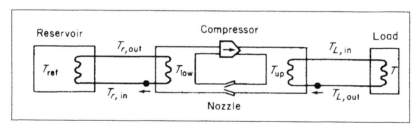

Figure 6.1. Schematic picture of a heat pump.

An indication of the departures from the Carnot limit of the "coefficients of performance", $\varepsilon^{heat\ pump}$, encountered in practice, is given in Fig. 6.2, as a function of the temperature difference $T_{up}-T_{low}$ at the heat pump and for selected values of T_{up}. In the interval of temperature differences covered, the $\varepsilon^{heat\ pump}$ is about 50% of the Carnot limit (3.2), but it falls more and more below the Carnot value as the temperature difference decreases, although the absolute value of the coefficient of performance increases.

Several possibilities exist for the choice of the reference reservoir. Systems in use for space heating or space cooling (achieved by reversing the flow in the compressor and expansion-nozzle circuit) have utilised river, lake and sea water, and air, as well as soil as reservoirs. The temperatures of such reservoirs are not entirely constant, and it must therefore be acceptable that the performance of the heat pump systems will vary with time. Such variations are damped if water or soil reservoirs at sufficient depth are used, because the weather-related temperature variations disappear as one goes just a few metres down into the soil. Alternative types of reservoirs for use with heat pumps include city waste sites, livestock manure, ventilation air from households or from livestock barns (where the rate of air exchange has to be particularly high), and heat storage tanks connected to solar collectors.

In connection with solar heating systems, the heat pump may be connected between the heat store and the load area (whenever the storage temperature is too low for direct circulation), or it may be connected between the heat store and the collector, such that the fluid let into the solar collector is cooled in order to improve the collector performance. Of course, a heat pump operating on its own reservoir (soil, air, etc.) may also provide the auxiliary heat for a solar heating system of capacity below the demand.

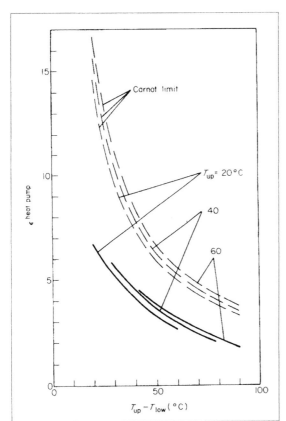

Figure 6.2. Measured coefficient of performance, $\varepsilon^{heat\ pump}$, for a heat pump with a semi-hermetic piston-type compressor (solid lines, based on Trenkowitz, 1969), and corresponding curves for ideal Carnot cycles.

The high-quality energy input to the compressor of a heat pump may also come from a renewable resource, e.g. by wind or solar energy conversion, either directly or via a utility grid carrying electricity generated by renewable energy resources. As for insulation materials, concern has been expressed over the use of CFC gases in the processing or as a working fluid, and substitutes believed to have less negative impacts have been developed.

6.2 Heat exchange

A situation like the one depicted in Fig. 6.3 is often encountered in energy supply systems. A fluid is passing through a reservoir of temperature T_3, thereby changing the fluid temperature from T_1 to T_2. In order to determine T_2 in terms of T_1 and T_3, in a situation where the change in T_3 is much smaller than the change from T_1 to T_2, the incremental temperature change of the fluid by travelling a short distance dx through the pipe system is related to the amount of heat transferred to the reservoir, assumed to depend linearly on the temperature difference,

$$J_m C_p^{fluid} \, dT^{fluid}/dx = h' \, (T_3 - T^{fluid}).$$

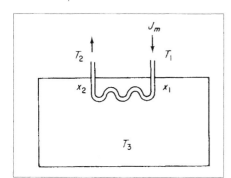

Figure 6.3. Heat exchanger, an idealised example of a well-mixed T_3-reservoir.

Integrating from T_1 at the inlet ($x = x_1$) gives

$$T^{fluid}(x) - T_3 = (T_1 - T_3) \exp(- h'(x - x_1)/(J_m C_p^{fluid})), \qquad (6.1)$$

where h' is the heat transfer per unit time from a unit length of the pipe for a temperature difference of one unit. The heat transfer coefficient for the entire heat exchanger is

$$h = \int_{x_1}^{x_2} h' \, dx,$$

which is sometimes written $h = U_h A_h$, with U_h being the transfer coefficient per unit area of pipe wall and A_h being the effective area of the heat exchanger. For $x = x_2$, (6.1) becomes (upon re-ordering)

$$(T_1 - T_2) = (T_1 - T_3)(1 - \exp(- h/(J_m C_p^{fluid}))). \qquad (6.2)$$

CHAPTER 7

GEOTHERMAL AND OCEAN-THERMAL ENERGY CONVERSION

7.1 Conversion of geothermal heat

Geothermal heat sources have been utilised by means of thermodynamic engines (e.g. Brayton cycles), in cases where the geothermal heat has been in the form of steam (water vapour). In some regions, geothermal sources exist that provide a mixture of water and steam, including suspended soil and rock particles, such that conventional turbines cannot be used. Work has been done on a special "brine screw" that can operate under such conditions (McKay and Sprankle, 1974).

However, in most regions the geothermal resources are in the form of heat-containing rock or sediments, with little possibility of direct use. If an aquifer passes through the region, it may collect heat from the surrounding layers and allow a substantial rate of heat extraction, for example, by drilling two holes from the surface to the aquifer, separated from each other, as indicated in Fig. 7.1a. Hot water (not developing much steam unless the aquifer lies very deep – several kilometres - or its temperature is exceptionally high) is pumped or rises by its own pressure to the surface at one hole and is re-injected through a second hole, in a closed cycle, in order to avoid pollution from various undesired chemical substances often contained in the aquifer water. The heat extracted from a heat exchanger may be used directly (e.g. as district heating; cf. Clot, 1977) or may generate electricity through one of the "low-temperature" thermodynamic cycles considered above in connection with solar collectors (Mock *et al.*, 1997).

If no aquifer is present to establish a "heat exchange surface" in the heat-containing rock, it may be feasible to create suitable fractures artificially (by explosives or induced pressure). An arrangement of this type is illustrated in Fig. 7.1b, counting on the fluid that is pumped down through one drilling hole to make its way through the fractured region of rock to the return drilling hole in such a way that continued heat extraction can be sustained. The heat transfer can only be predicted in highly idealised cases (see e.g. Gringarten *et al.*, 1975), which may not be realised as a result of the fairly uncontrolled methods of rock fractionation available.

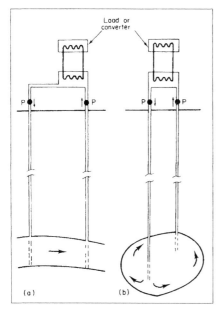

Figure 7.1. Examples of the utilisation of geothermal heat: (a) based on the presence of an aquifer; (b) based on a region of fractured rock.

One important result of the model calculations is that the heat extraction rate deemed necessary for practical applications is often higher than the geothermal flux into the region of extraction, so that the temperature of the extracted heat will be dropping. This non-sustainable use of geothermal energy is apparent in actual installations in New Zealand and Italy (where temperatures of extracted steam are dropping by something like 1°C per year, the number being highly dependent on fracture distribution, rock structure, etc.).

7.2 Conversion of ocean thermal energy

Downward gradients of temperature exist in most oceans, and they are particularly stable (i.e. without variations with time) in the tropical oceans (Sørensen, 2004). The utilisation of such temperature gradients for electricity generation (e.g. by use of a Rankine cycle) has been considered several times since the first suggestions by d'Arsonval (1881).

The temperature differences available over the first 500–1000 m of water depth are only about 25°C. Considering a closed Rankine cycle, with a working fluid (e.g. ammonia) which evaporates and condenses at convenient temperatures, placed near the ocean surface, it will be required to pump colder water through a pipe from the depth to a heat exchanger for condensation of the working fluid. Further, a warm water heat exchanger is required for evaporating the working fluid. If the heat exchange surface is such that, say, 5°C is "lost" at each heat exchanger, the temperature difference available to the thermodynamic cycle is only 15°C, corresponding to a limiting Carnot efficiency of roughly 0.05. For an actual engine, the efficiency is still lower, and from the power generated should be subtracted the power needed to pump hot and cold water through the heat exchangers and to pump cold water from its original depth to the converter level. It is expected that overall efficiencies around 0.02 may be achieved (cf. e.g. McGowan, 1976).

In order to save energy to pump the hot water through the heat exchanger, it has been suggested that these converters be placed in strong currents such as the Gulf Stream (Heronemus, 1975). The possibility of adverse environmental effects from the power extraction from ocean thermal gradients cannot be excluded. Such dangers may be increased if ocean currents are incorporated into the scheme, because of the possible sensitivity of the itinerary of such currents to small perturbations, and because of the dependence of climatic zones on the course of currents such as the Gulf Stream and the Kuro Shio. (Similar worries are discussed in connection with global warming caused by increased anthropogenic injection of greenhouse gases into the atmosphere).

Open thermodynamic cycles have also been suggested for conversion of ocean thermal energy (Claude, 1930; Beck, 1975; Zener and Fetkovich, 1975), for example, based on creating a rising mixture of water and steam bubbles or "foam", which is separated at a height above sea-level, such that the water can be used to drive a turbine rotor.

If viable systems could be developed for conversion of ocean thermal energy, then there would be a number of other applications of such conversion devices in connection with other heat sources of a temperature little higher than that of the surroundings, especially when such heat sources can be regarded as "free". Examples are the reject or "waste" heat flows from the range of other conversion devices operating at higher initial temperature differences, including fuel-based power plants.

III. MECHANICAL ENERGY CONVERSION PROCESSES

BASIC DESCRIPTION OF FLOW-DRIVEN CONVERTERS

CHAPTER 8

A turbine is a device delivering rotational shaft power on the basis of some other type of mechanical energy. If the temperature of the surroundings is regarded as fixed, the simple model in Fig. 2.2 allows the energy dissipation (2.17) to be written

$$D = J_m F_m + J_\theta F_\theta, \tag{8.1}$$

since from (2.18) $F_{Q,in}$ is zero, and no electrical output has been considered in this conversion step. The output variables are the angular velocity of the shaft, J_θ (2.9), and the torque acting on the system, F_θ (2.15), while the input variables are the mass flow rate, J_m (2.9), and the generalised force F_m given in (2.18). The specific energy contents w_{in} and w_{out} are of the form (2.13), corresponding to e.g. the geopotential energy of a given water head,

$$w_{in}^{pot} = g \Delta z, \qquad w_{out}^{pot} = 0, \tag{8.2}$$

the kinetic energy of the working fluid,

$$w_{in}^{kin} = \tfrac{1}{2} u_{in}^2, \qquad w_{out}^{kin} = \tfrac{1}{2} u_{out}^2, \tag{8.3}$$

and the enthalpy connected with the pressure changes,

$$h_{in} = P_{in} / \rho_{in}, \qquad h_{out} = P_{out} / \rho_{out}, \tag{8.4}$$

where the internal energy term U in H, assumed constant, has been left out, and the specific volume has been expressed in terms of the fluid densities ρ_{in} and ρ_{out} at input and output.

If a linear model of the Onsager type (2.10) is adopted for J_m and J_θ and these equations are solved for J_m and F_θ, one obtains

$$J_m = L_{m\theta} J_\theta / L_{\theta\theta} + (L_{mm} - L_{m\theta} L_{\theta m} / L_{\theta\theta}) F_m,$$

$$-F_\theta = -J_\theta / L_{\theta\theta} + L_{\theta n} F_m / L_{\theta\theta}$$ (8.5)

The coefficients may be interpreted as follows: $L_{m\theta} / L_{\theta\theta}$ is the mass of fluid displaced by the turbine during one radian of revolution, $(L_{mm} - L_{m\theta} L_{\theta n} / L_{\theta\theta})$ is a "leakage factor" associated with fluid getting through the turbine without contributing to the shaft power, and finally, $L_{\theta\theta}^{-1}$ represents the friction losses. Insertion into (8.1) gives the linear approximation for the dissipation,

$$D = (L_{mm} - L_{m\theta} L_{\theta n} / L_{\theta\theta}) (F_m)^2 + (J_\theta)^2 / L_{\theta\theta}$$ (8.6)

An ideal conversion process may be approached if no heat is exchanged with the surroundings, in which case (2.19) and (2.12) give the simple efficiency

$$\eta = (w_{in} - w_{out}) / w_{in}.$$ (8.7)

The second law efficiency in this case is, from (2.20), (2.14) and (2.12),

$$\eta^{(2.\,law)} = (w_{in} - w_{out}) \,/\, (w_{in} - w_{out} - T_{ref}(s_{m,in} - s_{m,out})).$$ (8.8)

The second law efficiency becomes unity if no entropy change takes place in the mass stream. The first law efficiency (8.7) may approach unity if only potential energy change of the form (8.2) is involved. In this case $w_{out} = 0$, and the fluid velocity, density and pressure are the same before and after the turbine. Hydroelectric generators approach this limit if working entirely on a static water head. Overshot waterwheels may operate in this way, and so may the more advanced turbine constructions, if the potential to kinetic energy conversion (in penstocks) and pressure build-up (in the nozzle of a Pelton turbine and in the inlet tube of many Francis turbine installations) are regarded as "internal" to the device (cf. Chapter 11). However, if there is a change in velocity or pressure across the converter, the analysis must take this into account, and it is no longer obvious whether the first law efficiency may approach unity.

8.1 Free stream flow turbines

Consider, for example, a free stream flow passing horizontally through a converter. In this case, the potential energy (8.2) does not change and may be left out. The pressure may vary near the converting device, but far behind and far ahead of the device the pressure is the same if the stream flow is free. Thus,

$$w = w^{kin} = \tfrac{1}{2} (u_x^2 + u_y^2 + u_z^2) = \tfrac{1}{2}\, \boldsymbol{u} \cdot \boldsymbol{u},$$

and

$$w_{in} - w_{out} = \tfrac{1}{2} (\boldsymbol{u}_{in} - \boldsymbol{u}_{out}) \cdot (\boldsymbol{u}_{in} + \boldsymbol{u}_{out}).$$ (8.9)

This expression and hence the efficiency would be maximum if u_{out} could be made zero. However, the conservation of the mass flow J_m requires that u_{in} and u_{out} satisfy an additional relationship. For a pure, homogeneous streamline flow along the x-axis, the rate of mass flow is

$$J_m = \rho A_{in} \, u_{x,in} = \rho A_{out} \, u_{x,out},$$ (8.10)

in terms of areas A_{in} and A_{out} enclosing the same streamlines, before and after the passage through the conversion device. In a more general situation, assuming rotational symmetry around the x-axis, there may have been induced a radial as well as a circular flow component by the device. This situation is illustrated in Fig. 8.1. It will be further discussed in Chapter 9, and the only case treated here is the simple one in which the radial and tangential components of the velocity field, u_r and u_t, which may be induced by the conversion device, can be neglected.

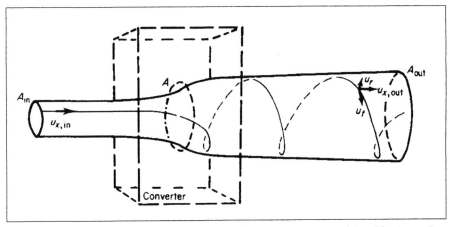

Figure 8.1. Schematic picture of a free stream flow converter or turbine. The incoming flow is a uniform streamline flow in the x-direction, while the outgoing flow is allowed to have a radial and a tangential component. The diagram indicates how a streamline may be transformed into an expanding helix by the device. The effective area of the converter, A, is defined in (8.12).

The axial force ("thrust") acting on the converter equals the momentum change,

$$F_x = J_m \, (u_{x,in} - u_{x,out}).$$ (8.11)

If the flow velocity in the converter is denoted u, an effective area of conversion, A, may be defined by

$$J_m = \rho A u_x,$$ (8.12)

according to the continuity equation (4.42). Dividing (8.10) by ρA, one obtains the specific energy transfer from the mass flow to the converter, within the conversion area A. This should equal the expression (8.9) for the change in specific energy, specialised to the case of homogeneous flows u_{in} and u_{out} along the x-axis,

$$u_x \left(u_{x,in} - u_{x,out} \right) = \tfrac{1}{2} \left(u_{x,in} + u_{x,out} \right) \left(u_{x,in} - u_{x,out} \right)$$

or

$$u_x = \tfrac{1}{2} \left(u_{x,in} + u_{x,out} \right). \tag{8.13}$$

The physical principle behind this equality is simply energy conservation, and the assumptions so far have been the absence of heat exchange [so that the energy change (2.12) becomes proportional to the kinetic energy difference (8.9)] and the absence of induced rotation (so that only x-components of the velocity need to be considered). On both sides of the converter, Bernoulli's equation is valid, stating that the specific energy is constant along a streamline. Far from the converter, the pressures are equal but the velocities are different, while the velocity just in front of or behind the converter may be taken as u_x, implying a pressure drop across the converter,

$$\Delta P = \tfrac{1}{2} \, \rho \left(u_{x,in} + u_{x,out} \right) \left(u_{x,in} - u_{x,out} \right). \tag{8.14}$$

The area enclosing a given streamline field increases in a continuous manner across the converter at the same time as the fluid velocity continuously decreases. The pressure, on the other hand, rises above the ambient pressure in front of the converter, then discontinuously drops to a value below the ambient one, and finally increases towards the ambient pressure again, behind ("in the wake of") the converter.

It is customary (see e.g. Wilson and Lissaman, 1974) to define an "axial interference factor", a, by

$$u_x = u_{x,in} \left(1 - a \right), \tag{8.15}$$

in which case (8.13) implies that $u_{x,out} = u_{x,in} \left(1 - 2a \right)$. With this, the power output of the conversion device can be written

$$E = J_m \left(w_{in} - w_{out} \right) = \rho A \left(u_{x,in} \right)^3 2a \left(1 - a \right)^2, \tag{8.16}$$

and the efficiency can be written

$$\eta = E / \left(J_m w_{in} \right) = 4a \left(1 - a \right). \tag{8.17}$$

It is seen that the maximum value of η is unity, obtained for $a = \tfrac{1}{2}$, corresponding to $u_{x,out} = 0$. The continuity equation (8.10) then implies an infinite area A_{out}, and it will clearly be difficult to defend the assumption of no induced radial motion.

In fact, for a free stream device of this type, the efficiency (8.17) is of little relevance since the input flux may not be independent of the details of the device. The input area A_{in}, from which streamlines would connect with a fixed converter area A, could conceivably be changed by altering the construction of the converter. It is therefore more appropriate to ask for the maximum power output for fixed A, as well as fixed input velocity $u_{x,in}$, this being equivalent to maximising the "power coefficient" defined by

$$C_p = E / (\tfrac{1}{2} \rho A (u_{x,in})^3) = 4a (1 - a)^2. \tag{8.18}$$

The maximum value is obtained for $a = 1/3$, yielding $C_p = 16/27$ and $u_{x,out} = u_{x,in}/3$. The areas are $A_{in} = (1 - a)A = 2/3\ A$ and $A_{out} = (1 - a)A/(1 - 2a) = 2A$, so in this case it is not unlikely that it may be a reasonable approximation to neglect the radial velocity component in the far wake.

The maximum found above for C_p is only a true upper limit with the assumptions made. By discarding the assumption of irrotational flow, it becomes possible for the converter to induce a velocity field, for which rot(u) is no longer zero. It has been shown that if the additional field is in the form of a vortex ring around the converter region, so that it does not contribute to the far wake, then it is possible to exceed the upper limit power coefficient 16/27 found above (cf. Chapter 10).

8.2 General elements of wind flow conversion

Conversion of wind energy into linear motion of a body has been utilised extensively, particularly for transportation across water surfaces. A large sail-ship of the type used in the 19th century would have converted wind energy at peak rates of a quarter of a megawatt or more.

The force on a sail or a wing (i.e. profiles of negligible or finite thickness) may be broken down into a component in the direction of the undisturbed wind (drag) and a component perpendicular to the undisturbed wind direction (lift). When referring to an undisturbed wind direction it is assumed that a uniform wind field is modified in a region around the sail or the wing, but that beyond a certain distance such modifications can be disregarded.

In order to determine the force components, Euler's equations may be used. It states that for any quantity A,

$$\frac{\partial}{\partial t}(\rho A) + \mathrm{div}(v\rho A) + \mathrm{div}(s_A) = S_A. \tag{8.19}$$

This equation states that the change in A within a given volume, per unit time, is due to gains from external sources, S_A, minus the macroscopic and molecular outflow (divergence terms). The macroscopic outflow is due to the velocity field v, and the microscopic transport is described by a vector s_A, the components of which give the transport by molecular processes in the directions of the co-ordinate axes. The density of the medium (here air) is denoted ρ and the time variable t.

If viscous and external forces are neglected, and the flow is assumed to be irrotational (so that Bernoulli's equation is valid) and steady (so that the time-derivative of the velocity potential vanishes), then the force on a segment of the airfoil (sail or wing) may be written

$$\frac{dF}{dz} = \oint_C P n \, ds = -\tfrac{1}{2}\rho \oint_C (v \cdot v) n \, ds.$$

Here dz is the segment length (cf. Fig. 8.2), the integration is along a closed contour, C, containing the airfoil profile, n is a unit vector normal to the contour [in the (x, y)-plane] and ds is the path-length increment, directed along the tangent to the contour, still in the (x, y)-plane. Taking advantage of the fact that the wind velocity v approaches a homogeneous field W (assumed to be along the x-axis) far from the airfoil, the contour integral may be reduced and evaluated (e.g. along a circular path),

$$dF/dz = \rho W \Gamma e_y, \tag{8.20}$$

$$\Gamma = \oint_C v \cdot ds \approx \pi c W \sin \alpha. \tag{8.21}$$

Here e_y is a unit vector along the y-axis, c is the airfoil chord length and α is the angle between the airfoil and W. In the evaluation of the circulation Γ, it has been assumed that the airfoil is thin and without curvature. In this case c and α are well defined, but in general the circulation depends on the details of the profile, although an expression similar to the right-hand side of (8.21) is still valid as a first approximation, for some average chord length and angle of attack. Equation (8.20) is known as the theorem of Kutta (1902) and Joukowski (1906).

The expressions (8.20) and (8.21) are valid in a co-ordinate system fixed relative to the airfoil (Fig. 8.2), and if the airfoil is moving with a velocity U, the velocities v and W are to be interpreted as relative ones, so that

$$W = u_{in} - U, \tag{8.22}$$

if the undisturbed wind velocity is u_{in}.

The assumption that viscous forces may be neglected is responsible for obtaining in (8.20) only a lift force, the drag force being zero. Primitive sail-ships, as well as primitive windmills, have been primarily aimed at utilising the drag force. It is possible, however, with suitably constructed airfoils, to make the lift force one or two orders of magnitude larger than the drag force and thereby effectively approach the limit where the viscous forces and hence the drag can be neglected. This usually requires careful "setting" of the airfoil, i.e. careful choice of the angle of attack, α, and in order to study operation at arbitrary conditions the drag component should be retained.

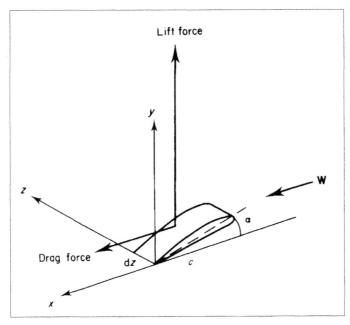

Figure 8.2. Forces on an airfoil segment.

It is customary to describe the drag and lift forces on an airfoil of given shape, as a function of α, in terms of two dimensionless constants, $C_D(\alpha)$ and $C_L(\alpha)$, defined by

$$dF_x/dz = \tfrac{1}{2}\,\rho\,C_D W^2 c,$$

$$dF_y/dz = \tfrac{1}{2}\,\rho\,C_L W^2 c.$$

(8.23)

The constants C_D and C_L are not quite independent of the size of the system, which is not unexpected since the viscous forces (friction) in air contribute most to turbulent motion on smaller scales. Introducing a quantity called the Reynolds number,

$$Re = Wc/\eta,$$

where η is the kinematic viscosity of air defined (Hinze, 1975) as a measure of the ratio between "inertial" and "viscous" forces acting between airfoil and air, the α-dependence of C_D and C_L for fixed Re, as well as the Re-dependence for the value of α which gives the highest lift-to-drag ratio, $L/D = C_L/C_D$, may look as shown in Figs. 8.3 and 8.4. The contours of these "high lift" profiles are indicated in Fig. 8.3.

Assuming that C_D, C_L and W are constant over the area $A = \int c\, dz$ of the airfoil, the work done by a uniform (except in the vicinity of the airfoil) wind field u_{in} on a device (e.g. a ship) moving with a velocity U can be derived from (8.23) and (8.22),

$$E = F \cdot U.$$

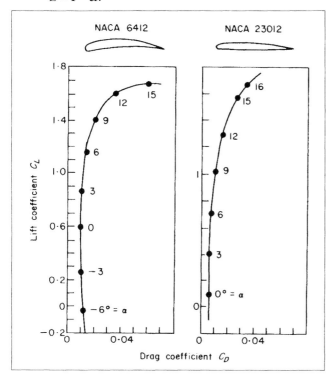

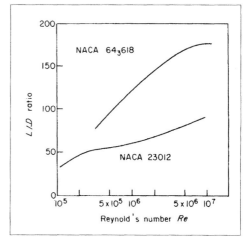

Figure 8.3. Lift and drag forces as a function of the angle of attack for two NACA airfoils (National Advisory Committee for Aeronautics; cf. e.g. Betz, 1959). The Reynolds number is $Re = 8 \times 10^6$.

Figure 8.4. Reynolds number dependence of the lift-to-drag ratio, defined as the maximum value (as a function of the angle of attack) of the ratio between the lift and drag coefficients C_L and C_D (based on Hütter, 1977).

The angle β between u_{in} and U (see Fig. 8.5) may be maintained by a rudder. The power coefficient (8.18) becomes

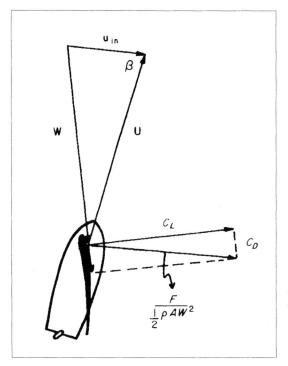

Figure 8.5. Velocity and force components for a sail-ship.

$$C_p = f (C_L \sin \beta - C_D (1 - \sin^2 \beta + f^2 - 2f \cos \beta)^{\frac{1}{2}}) (1 + f^2 - 2f \cos \beta)^{\frac{1}{2}},$$

with $f = U/u_{in}$. For $C_L = 0$ the maximum C_p is $4C_D/27$, obtained for $f = 1/3$ and $\beta = 0$, whereas the maximum C_p for high lift-to-drag ratios L/D is obtained for β close to $\frac{1}{2}\pi$ and f around $2C_L/(3C_D)$. In this case, the maximum C_p may exceed C_L by one to two orders of magnitude (Wilson and Lissaman, 1974).

It is quite difficult to maintain the high speeds U required for optimum performance in a linear motion of the airfoil, and it is natural to focus the attention on rotating devices, in case the desired energy form is shaft or electric power and not propulsion. Wind-driven propulsion in the past (mostly at sea) has been restricted to U/u_{in}-values far below the optimum region for high L/D airfoils (owing to friction against the water), and wind-driven propulsion on land or in the air has received little attention.

PROPELLER-TYPE CONVERTERS

CHAPTER 9

Propellers have been extensively used in aircraft to propel the air in a direction parallel to that of the propeller axis, thereby providing the necessary lift force on the aeroplane wings. Propeller-type rotors are similarly used for windmills, but here the motion of the air (i.e. the wind) makes the propeller, which should be placed with its axis parallel to the wind direction, rotate, thus providing the possibility of power extraction. The propeller consists of a number of blades that are evenly distributed around the axis (cf. Fig. 9.1), with each blade having a suitable aerodynamic profile usually designed to produce a high lift force, as discussed in Chapter 8. If there are two or more blades, the symmetrical mounting ensures a symmetrical mass distribution, but if only one blade is used it must be balanced by a counterweight.

9.1 Theory of non-interacting streamtubes

In order to describe the performance of a propeller-type rotor, the forces on each element of blade must be calculated, including the forces produced by the direct action of the wind field on each individual element as well as the forces arising as a result of interactions between different elements on the same or on different blades. Since the simple airfoil theory outlined in Chapter 8 deals only with the forces on a given blade segment, in the absence of the other ones and also without the inclusion of "edge effects" from "cutting out" this one segment, it is tempting as a first approximation to treat the different radial segments of each blade as independent. Owing to the symmetrical mounting of blades, the corresponding radial segments of different blades (if more than one) may for a uniform wind field be treated together, considering, as indicated in Fig. 9.1, an annulus-shaped *"stream-tube"* of flow, bordered by streamlines intersecting the rotor plane at radial distances r and $r+dr$. The approximation of independent contributions from each streamtube, implying that radially induced velocities (u_r in Fig. 8.1) are neglected, allows the total shaft power to be expressed as

$$E = \int_0^R \frac{dE}{dr}\, dr,$$

where dE/dr depends only on the conditions of the streamtube at r. Similar sums of independent contributions can in the same order of approximation be used to describe other overall quantities, such as the axial force component T and the torque Q (equal to the power E divided by the angular velocity Ω of the propeller).

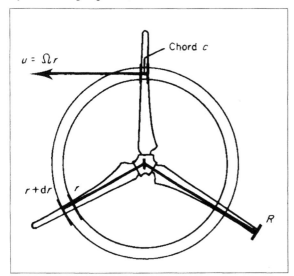

Figure 9.1. Definition of the streamtubes and blade integration variable for a propeller-type rotor.

In Fig. 9.2, a section of a wing profile (blade profile) is seen from a direction perpendicular to the cut. The distance of the blade segment from the axis of rotation is r, and its velocity $r\Omega$ is directed along the y-axis. This defines a coordinate system with a fixed x-axis along the rotor axis and moving y- and z-axes such that the blade segment is fixed relative to this coordinate system.

In order to utilise the method developed in Chapter 8, the apparent wind velocity W to be used in (8.23) must be determined. It is given by (8.22) only if the velocity induced by, and remaining in the wake of, the device, $u^{ind} = u_{out} - u_{in}$ (cf. Fig. 8.1), is negligible. Since the radial component u_r of u^{ind} has been neglected, u^{ind} has two components, one along the x-axis,

$$u^{ind}_x = u_{x,out} - u_{x,in} = -2au_{x,in}$$

[cf. (8.15)], and one in the tangential direction,

$u^{ind}_t = \omega^{ind}r = 2a'\Omega r,$

when expressed in a non-rotating coordinate system (the second equality is defining a quantity a', called the "tangential interference factor"). W is determined by the air velocity components in the rotor plane. From the momentum considerations underlying (8.15), the induced x-component u_0 in the rotor plane is seen to be

$$u_0 = \tfrac{1}{2}u^{ind}_x = -au_{x,in}. \tag{9.1}$$

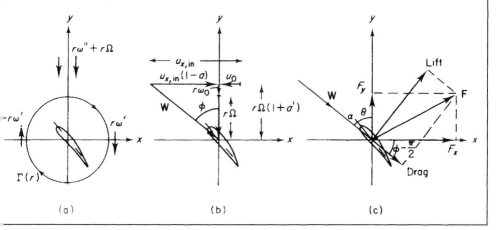

(a) (b) (c)

Figure 9.2. Velocity and force components of a rotating blade segment in a co-ordinate system following the rotation (which is in the direction of the y-axis): (a) determination of induced angular velocities; (b) determination of the direction of the apparent wind velocity; (c) determination of force components along x- and y-axes.

In order to determine $r\omega_0$, the induced velocity of rotation of the air in the rotor plane, one may use the following argument (Glauert, 1935). The induced rotation is partly due to the rotation of the wings and partly due to the non-zero circulation (8.21) around the blade segment profiles (see Fig. 9.2a). This circulation may be considered to imply an induced air velocity component of the same magnitude, $|r\omega'|$, but with opposite direction in front of and behind the rotor plane. If the magnitude of the component induced by the wing rotation is called $r\omega''$ in the fixed coordinate system, it will be $r\omega'' + r\Omega$ and directed along the negative y-axis in the coordinate system following the blade. It has the same sign in front of and behind the rotor. The total y-component of the induced air velocity in front of the rotor is then $-(r\omega'' + r\Omega + r\omega')$, and the tangential component of the induced air velocity in the fixed coordinate system is $-(r\omega'' - r\omega')$, still in front of the rotor. But here there should be no induced velocity at all. This follows, for example, from performing a closed line integral of the air velocity v along a circle with radius r and perpendicular to the x-axis. This integral should be zero because the air in front of the rotor has been assumed to be irrotational,

$$0 = \int_{circle\ area} \text{rot}\, v\, \mathrm{d}A = \oint v \cdot \mathrm{d}s = -2\pi r(r\omega'' - r\omega'),$$

implying $\omega'' = \omega'$. This identity also fixes the total induced tangential air velocity behind the rotor, $r\omega^{ind} = r\omega'' + r\omega' = 2r\omega''$, and in the rotor plane (note that the circulation component $r\omega'$ is here perpendicular to the y-axis),

$$r\omega_0 = r\omega'' = \tfrac{1}{2}r\omega^{ind} = \Omega\, ra'. \tag{9.2}$$

The apparent wind velocity in the co-ordinate system moving with the blade segment, W, is now determined as seen in Fig. 9.2b. Its x-component is u_x, given by (8.15), and its y-component is obtained by taking (9.2) to the rotating coordinate system,

$$W_x = u_x = u_{x,in}(1 - a),$$
$$W_y = -(r\omega_x + r\Omega) = -r\Omega\,(1 + a'). \tag{9.3}$$

The lift and drag forces (Fig. 9.2c) are now obtained from (9.2) (except that the two components are no longer directed along the coordinate axes), with values of C_D, C_L and c pertaining to the segments of wings at the streamtube intersecting the rotor plane at the distance r from the axis. As indicated in Figs. 9.2b and c, the angle of attack, α, is the difference between the angle ϕ, determined by

$$\tan \phi = -W_x / W_y, \tag{9.4}$$

and the pitch angle θ between blade and rotor plane,

$$\alpha = \phi - \theta. \tag{9.5}$$

Referred to the coordinate axes of Fig. 9.2, the force components for a single blade segment become

$$F_x = \tfrac{1}{2}\, \rho\, cW^2\, (C_D(\alpha)\, \sin \phi + C_L(\alpha)\, \cos \phi),$$
$$F_y = \tfrac{1}{2}\, \rho\, cW^2\, (-C_D(\alpha)\, \cos \phi + C_L(\alpha)\, \sin \phi), \tag{9.6}$$

and the axial force and torque contributions from a streamtube with B individual blades are given by

$$\mathrm{d}T/\mathrm{d}r = BF(r),$$
$$\mathrm{d}Q/\mathrm{d}r = BrF_y(r). \tag{9.7}$$

Combining (9.7) with equations expressing momentum and angular momentum conservation for each of the (assumed non-interacting) streamtubes, a closed set of equations is obtained. Equating the momentum change of the wind in the x-direction to the axial force on the rotor part intersected by an individual streamtube [i.e. (8.11) and (8.12) with $A = 2\pi r$], one obtains

$$dT/dr = -A\rho u_x u^{ind}_x = 2\pi r \rho u_{x,in} (1-a) 2au_{x,in}, \qquad (9.8)$$

and similarly equating the torque on the streamtube rotor part to the change in angular momentum of the wind (from zero to $r \times u^{ind}_t$), one gets

$$dQ/dr = A\rho u_x u^{ind}_t = 2\pi r^2 \rho u_{x,in} (1-a) 2a'\Omega r. \qquad (9.9)$$

Inserting $W = u_{x,in}(1-a)/\sin \phi$ or $W = r\Omega (1+a')/\cos \phi$ (cf. Fig. 9.2c) as necessary, one obtains a and a' expressed in terms of ϕ by combining (9.6) – (9.9). Since, on the other hand, ϕ depends on a and a' through (9.4) and (9.3), an iterative method of solution should be used. Once a consistent set of (a, a', ϕ)-values has been determined as a function of r (using a given blade profile implying known values of θ, c, $C_D(\alpha)$ and $C_L(\alpha)$ as a function of r), either (9.7) or (9.8) and (9.9) may be integrated over r to yield the total axial force, total torque or total shaft power $E = \Omega Q$.

One may also determine the contribution of a single streamtube to the power coefficient (8.18),

$$C_p(r) = \frac{\Omega \, dQ/dr}{\frac{1}{2}\rho u^3_{x,in} 2\pi r} = 4a'(1-a)\left(\frac{r\Omega}{u_{x,in}}\right)^2. \qquad (9.10)$$

The design of a rotor may utilise $C_p(r)$ for a given wind speed $u_{x,in}^{design}$ to optimise the choice of blade profile (C_D and C_L), pitch angle (θ) and solidity ($Bc/(\pi r)$) for given angular velocity of rotation (Ω). If the angular velocity Ω is not fixed (as it might be by use of a suitable asynchronous electrical generator, except at start and stop), a dynamic calculation involving $d\Omega/dt$ must be performed. Not all rotor parameters need to be fixed. For example, the pitch angles may be variable, by rotation of the entire wing around a radial axis, such that all pitch angles $\theta (r)$ may be modified by an additive constant θ_0. This type of regulation is useful in order to limit the C_p-drop when the wind speed moves away from the design value $u_{x,in}^{design}$. The expressions given above would then give the actual C_p for the actual setting of the pitch angles given by $\theta(r)_{design} + \theta_0$ and the actual wind speed $u_{x,in}$, with all other parameters left unchanged.

In introducing the streamtube expressions (9.8) and (9.9), it has been assumed that the induced velocities and thus a and a' are constant along the circle periphery of radius r in the rotor plane. As the model used to estimate the magnitude of the induced velocities (Fig. 9.2a) is based on being near to a rotor blade, it is expected that the average induced velocities in the rotor plane, u_0 and $r\omega_0$, are smaller than the values calculated by the above expressions, unless the solidity is very large. In practical applications, it is customary to compensate by multiplying a and a' by a common factor, F, less than unity and a function of B, r and ϕ (see e.g. Wilson and Lissaman, 1974).

Furthermore, edge effects associated with the finite length R of the rotor wings have been neglected, as have the "edge" effects at $r = 0$ due to the presence of the axis, transmission machinery, etc. These effects may be described in terms of "trailing vortices" shed from the blade tips and blade roots and moving in helical orbits away from the rotor in its wake. Vorticity is a name for the vector field rot$v$, and the vortices connected with the circulation around the blade profiles (Fig. 9.2a) are called "bound vorticity". This can "leave" the blade only at the tip or at the root. The removal of bound vorticity is equivalent to a loss of circulation I (8.21) and hence a reduction in the lift force. Its dominant effect is on the tangential interference factor a' in (9.9), and so it has the same form and may be treated on the same footing as the corrections due to finite blade number. (Both are often referred to as "tip losses", since the correction due to finite blade number is usually appreciable only near the blade tips, and they may be approximately described by the factor F introduced above.) Other losses may be associated with the "tower shadow", etc.

9.2 Model behaviour of power output and matching to load

A calculated overall C_p for a three-bladed, propeller-type wind energy converter is shown in Fig. 9.3, as a function of the tip-speed ratio,

$$\lambda = \Omega\, R\, /\, u_{x,in},$$

and for different settings of the overall pitch angle, θ_0. It is clear from (9.10) and the equations for determining a and a' that C_p depends on the angular velocity Ω and the wind speed $u_{x,in}$ only through the ratio λ. Each blade has been assumed to be an airfoil of the type NACA 23012 (cf. Fig. 8.3), with chord c and twist angle θ changing along the blade from root to tip, as indicated in Fig. 9.4. A tip-loss factor F has been included in the calculation of sets of corresponding values of a, a' and ϕ for each radial station (each "streamtube"), according to the Prandtl model described by Wilson and Lissaman (1974). The dashed regions in Fig. 9.3 correspond to values of the product aF (corresponding to a in the expression without tip loss) larger than 0.5. Since the axial air velocity in the wake is $u_{x,out} = u_{x,in}(1 - 2aF)$ [cf. (8.15) with $F = 1$], $aF > 0.5$ implies reversed or re-circulating flow behind the wind energy converter, a possibility that has not been included in the above description. The C_p-values in the dashed regions may thus be inaccurate and presumably overestimated.

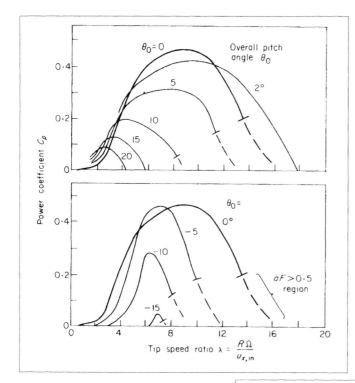

Figure 9.3. Dependence of power coefficient on tip-speed ratio for different settings of the overall pitch angle. The calculation is based on the wing geometry shown in Fig. 9.4 and the NACA 23012 profile of Fig. 8.3.

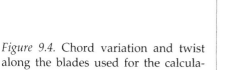

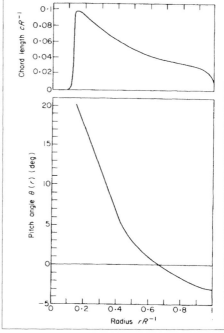

Figure 9.4. Chord variation and twist along the blades used for the calculations in this section.

Figure 9.3 shows that for negative settings of the overall pitch angle θ_0, the C_p distribution on λ-values is narrow, whereas it is broad for positive θ_0 of modest size. For large positive θ_0, the important region of C_p moves to smaller λ-values. The behaviour of $C_p(\lambda)$ in the limit of λ approaching zero is important for operating the wind energy converter at small rotor angular velocities and, in particular, for determining whether the rotor will be self-starting, as discussed in more detail below.

For a given angular speed Ω of the rotor blades, the power coefficient curve specifies the fraction of the power in the wind which is converted, as a function of the wind speed $u_{x,in}$. Multiplying C_p by $\frac{1}{2} \rho A u_{x,in}^3$, where the rotor area is $A = \pi R^2$ (if "coning" is disregarded, i.e. the blades are assumed to be in the plane of rotation), the power transferred to the shaft can be obtained as a function of wind speed, assuming the converter to be oriented ("yawed") such that the rotor plane is perpendicular to the direction of the incoming wind. Figures 9.5 and 9.6 show such plots of power E, as functions of wind speed and overall pitch angle, for two definite angular velocities (Ω = 4.185 and 2.222 rad s^{-1}, if the length of the wings is R = 27 m). For the range of wind speeds encountered in the planetary boundary layer of the atmosphere (e.g. at heights between 25 and 100 m), a maximum power level around 4000 W per average square metre swept by the rotor (10 MW total) is reached for the device with the higher rotational speed (tip speed $R\Omega$ = 113 ms^{-1}), whereas about 1000 W m^{-2} is reached by the device with the lower rotational velocity ($R\Omega$ = 60 m s^{-1}).

The rotor has been designed to yield a maximum C_p at about λ = 9 (Fig. 9.3), corresponding to $u_{x,in}$ = 12.6 and 6.6 m s^{-1} in the two cases. At wind speeds above these values the total power varies less and less strongly, and for negative pitch angles it even starts to decrease with increasing wind speeds in the range characterising "stormy weather". This phenomenon can be used to make the wind energy converter self-regulating, provided that the "flat" or "decreasing" power regions are suitably chosen and that the constancy of the angular velocity at operation is assured by, for example, coupling the shaft to a suitable asynchronous electricity generator (a generator allowing only minute changes in rotational velocity, with the generator angular speed Ω_g being fixed relative to the shaft angular speed Ω, e.g. by a constant exchange ratio $n = \Omega_g / \Omega$ of a gearbox). Self-regulating wind energy converters with fixed overall pitch angle θ_0 and constant angular velocity have long been used for AC electricity generation (Juul, 1961).

According to its definition, the torque $Q = E/\Omega$ can be written

$$Q = \frac{1}{2} \rho \, \pi R^3 \, u_{x,in}^2 \, C_p(\lambda) / \lambda, \tag{9.11}$$

and Fig. 9.7 shows C_p/λ as a function of λ for the same design as the one considered in Fig. 9.3. For pitch angles θ_0 less than about $-3°$, C_p/λ is negative for $\lambda = 0$, whereas the value is positive if θ_0 is above the critical value of about $-3°$. The corresponding values of the total torque (9.11), for a rotor with radius $R = 27$ m and overall pitch angle $\theta_0 = 0°$, are shown in Fig. 9.8, as a function of the angular velocity $\Omega = \lambda u_{x,in}/R$ and for different wind speeds. The small dip in Q for low rotational speeds disappears for higher pitch angles θ_0.

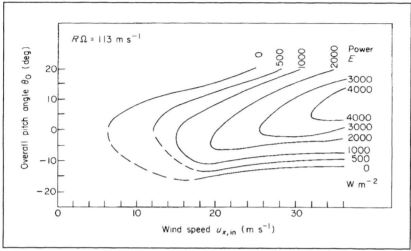

Figure 9.5. Power output of the wind energy converter described in Figs. 9.3 and 9.4, for a tip speed of 113 m s^{-1}, as a function of wind speed and overall pitch angle. The actual Reynolds number is in this case about 5×10^6 from blade root to tip, in reasonable consistency with the value used in the airfoil data of Fig. 8.3. Dashed lines indicate regions where a [defined in (9.3)] times the tip speed correction, F, exceeds 0.5 for some radial stations along the blades.

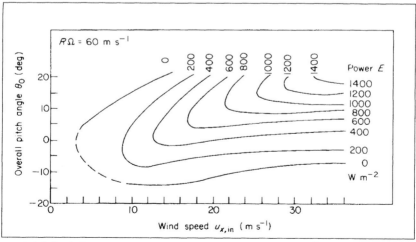

Figure 9.6 (below). Same as Fig. 9.5, but for a tip speed of 60 m s^{-1}.

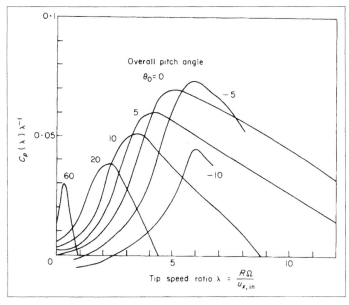

Figure 9.7. Variation of power coefficient over tip-speed ratio, as a function of tip-speed ratio, for the converter considered in the previous figures.

The dependence of the torque at $\Omega = 0$ on pitch angle is shown in Fig. 9.9 for a few wind speeds. Advantage can be taken of the substantial increase in starting torque with increasing pitch angle, in case the starting torque at the pitch angle desired at operating angular velocities is insufficient and provided the overall pitch angle can be changed. In that case a high overall pitch angle is chosen to start the rotor from $\Omega = 0$, where the internal resistance (friction in bearings, gearbox, etc.) is large. When an angular speed Ω of a few degrees per second is reached, the pitch angle is diminished to a value close to the optimum one (otherwise the torque will pass a maximum and soon start to decrease again, as seen from e.g. the $\theta_0 = 60°$ curve in Fig. 9.7). Usually, the internal resistance diminishes as soon as Ω is non-zero. The internal resistance may be represented by an internal torque, $Q_0(\Omega)$, so that the torque available for some load, "external" to the wind power to shaft power converter (e.g. an electric generator), may be written

$$Q^{available} = Q(\Omega) - Q_0(\Omega).$$

If the "load" is characterised by demanding a fixed rotational speed Ω (such as the asynchronous electric generator), it may be represented by a vertical line in the diagram shown in Fig. 9.8. As an example of the operation of a wind energy generator of this type, the dashed lines (with accompanying arrows) in Fig. 9.8 describe a situation with initial wind speed 8 m s^{-1}, assuming that this provides enough torque to start the rotor at the fixed overall pitch angle $\theta_0 = 0°$ (i.e. the torque produced by the wind at 8 m s^{-1} and $\Omega = 0$ is above the internal torque Q_0, here assumed to be constant). The excess torque makes the angular velocity of the rotor, Ω, increase along the specific curve for $u_{x,in}$ = 8 m s^{-1}, until the value Ω_0 characterising the electric generator is reached. At this point the load is connected and the wind energy converter begins to deliver power to the load area. If the wind speed later increases, the torque will increase along the vertical line at $\Omega = \Omega_0$ in Fig. 9.8, and if the wind speed decreases, the torque will decline along the same vertical line until it reaches the internal torque value Q_0. Then the angular velocity diminishes and the rotor is brought to a halt.

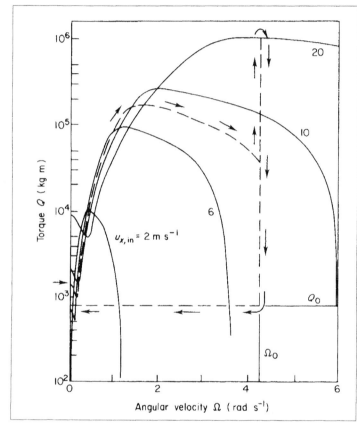

Figure 9.8. Torque variation with angular velocity for a wind energy converter of the type described in the previous figures, with radius R = 27 m and overall pitch angle $0°$. The dashed, vertical line represents operation at fixed angular velocity (e.g. with a generator). Q_0 illustrates the internal torque.

An alternative type of load may not require a constant angular velocity. Synchronous DC electric generators are of this type, providing increasing power output with increasing angular velocity (assuming a fixed exchange ratio between rotor angular velocity Ω and generator angular velocity Ω_g). Instead of staying on a vertical line in the torque–versus–Ω diagram, for varying wind speed, the torque now varies along some fixed, monotonically increasing curve characterising the generator. An optimum synchronous generator would be characterised by a $Q(\Omega)$ curve, which for each wind speed $u_{x,in}$ corresponds to the value of $\Omega = \lambda u_{x,in}/R$ which provides the maximum power coefficient C_p (Fig. 9.3). This situation is illustrated in Fig. 9.10, again indicating by a set of dashed curves the torque variation for an initial wind speed of 8 m s^{-1}, followed by an increasing and later again decreasing wind speed. In this case the $Q = Q_0$ limit is finally reached at a very low angular velocity, indicating that power is still delivered during the major part of the slowing-down process.

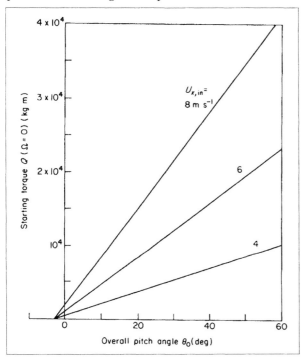

Figure 9.9. Starting torque as a function of overall pitch angle for the wind energy converter considered in the preceding figures.

9.3 Non-uniform wind velocity

The velocity field of the wind may be non-uniform in time as well as in spatial distribution. The influence of time variations in wind speed on power output of a propeller-type wind energy converter has been touched upon in the previous subsection, although a detailed investigation involving the actual time dependence of the angular velocity Ω was not included. In general, the direction of the wind velocity is also time dependent, and the conversion device should be able to successfully align its rotor axis with the long-range trends in wind direction, or suffer a power reduction that is not just the cosine to the angle β between the rotor axis and the wind direction ("yaw angle"), but involves calculating the performance of each blade segment for an apparent wind speed W and angle of attack α different from the ones previously used, and no longer axially symmetric. This means that the quantities W and α (or ϕ) are no longer the same inside a given annular stream-tube, but they depend on the directional position of the segment considered (e.g. characterised by a rotational angle ψ in a plane perpendicular to the rotor axis).

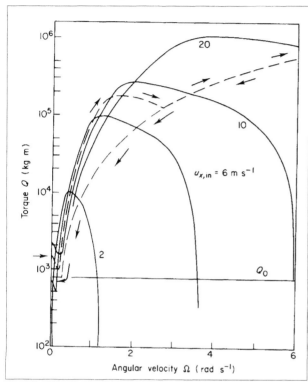

Figure 9.10. Same as Fig. 9.8, but here operation is illustrated for a load (e.g. asynchronous generator), which is optimised, i.e. corresponds to maximum C_p, at each wind speed.

Assuming that both the wind direction and the wind speed are functions of time and of height h (measured from ground level or from the height z_0, where the wind speed profile attains values near zero; cf. Sørensen, 2004), but that the direction remains horizontal, then the situation will be as depicted in Fig. 9.11. The coordinate system (x_0, y_0, z_0) is fixed and has its origin in hub height h_0, where the rotor blades are fastened. Consider now a blade element at a distance r from the origin, on the ith blade. The projection of this position on to the (y_0, z_0)-plane is turned the angle ψ_i from vertical, where

$$\psi_i = 2\pi/i + \Omega\, t, \tag{9.12}$$

at the time t. The coning angle δ is the angle between the blade and its projection onto the (y_0, z_0)-plane, and the height of the blade element above the ground is

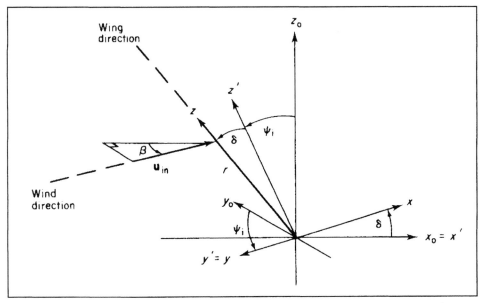

Figure 9.11. Definition of coordinate systems for a coning rotor in a non-uniform wind field.

$$h = h_0 + r \cos \delta \cos \psi_i, \qquad (9.13)$$

where h_0 is the hub height. The height h enters as a parameter in the wind speed $u_{in} = |\mathbf{u}_{in}(h, t)|$ and the yaw angle $\beta = \beta(h, t)$, in addition to time.

An attempt can now be made to copy the procedure used in the case of a uniform wind velocity along the rotor axis, i.e. to evaluate the force components for an individual streamtube both by the momentum consideration made in Section 8.1 and by the lift and drag approach of Section 8.2. The individual streamtubes can no longer be taken as annuli, but must be of an area A_s (perpendicular to the wind direction) small enough to permit the neglect of variations in wind velocity over the area at a given time. It will still be assumed that the flows inside different streamtubes do not interact, and also, for simplicity, the streamtubes will be treated as "straight lines", i.e. not expanding or bending as they pass the region of the conversion device (as before, this situation arises when induced radial velocities are left out).

Consider first the "momentum equation" (8.11), with the suffix s denoting "in the stream-wise direction" or "along the streamtube",

$$F_s = -J_m u_s^{ind}.$$

This force has components along the x-, y-, and z-directions of the local coordinate system of a blade turned the angle ψ from vertical (cf. Fig. 9.2). The

y-direction is tangential to the rotation of the blade element, and in addition to the component of F_s there may be an induced tangential velocity u_t^{ind} (and force) of the same type as the one considered in the absence of yaw (in which case F_s is perpendicular to the y-axis). The total force components in the local coordinate system are thus

$$F_x = - J_m u_s^{ind} (\cos \beta \cos \delta + \sin \beta \sin \delta \sin \psi),$$

$$F_y = - J_m u_s^{ind} \sin \beta \cos \psi + J_m u_t^{ind}. \qquad (9.14)$$

From the discussion of Fig. 8.1, (8.12) and (8.15),

$$J_m = A_s \rho u_s = A_s \rho u_{in}(1 - a),$$

$$u_s^{ind} = - 2a u_{in},$$

and in analogy with (9.2), a tangential interference factor a' may be defined by

$$u_t^{ind} = 2a' r \Omega \cos \delta.$$

However, the other relation contained in (9.2), from which the induced tangential velocity in the rotor plane is half the one in the wake, cannot be derived in the same way without the assumption of axial symmetry. Instead, the variation in the induced tangential velocity as a function of rotational angle ψ is bound to lead to crossing of the helical wake strains and it will be difficult to maintain the assumption of non-interacting streamtubes. Here, the induced tangential velocity in the rotor plane will still be taken as $\frac{1}{2}u_t^{ind}$, an assumption that at least gives reasonable results in the limiting case of a nearly uniform incident wind field and zero or very small yaw angle.

Second, the force components may be evaluated from (9.4)–(9.6) for each blade segment, defining the local coordinate system (x, y, z) as in Fig. 9.11 with the z-axis along the blade and the y-axis in the direction of the blade's rotational motion. The total force, averaged over a rotational period, is obtained by multiplying by the number of blades, B, and by the fraction of time each blade spends in the streamtube. Defining the streamtube dimensions by an increment dr in the z-direction and an increment $d\psi$ in the rotational angle ψ, each blade spends the time fraction $d\psi/(2\pi)$ in the streamtube, at constant angular velocity Ω. The force components are then

$$F_x = B (d\psi/2\pi) \tfrac{1}{2} \rho c W^2 (C_D(\alpha) \sin \phi + C_L(\alpha) \cos \phi) dr,$$

$$F_y = B (d\psi/2\pi) \tfrac{1}{2} \rho c W^2 (-C_D(\alpha) \cos \phi + C_L(\alpha) \sin \phi) dr, \qquad (9.15)$$

in the notation of (9.6). The angles ϕ and α are given by (9.4) and (9.5), but the apparent velocity W is the vector difference between the stream-wise velocity $u_{in} + \tfrac{1}{2} u_s^{ind}$ and the tangential velocity $r\Omega \cos \delta + \tfrac{1}{2} u_t^{ind}$, both taken in the rotor plane. From Fig. 9.11,

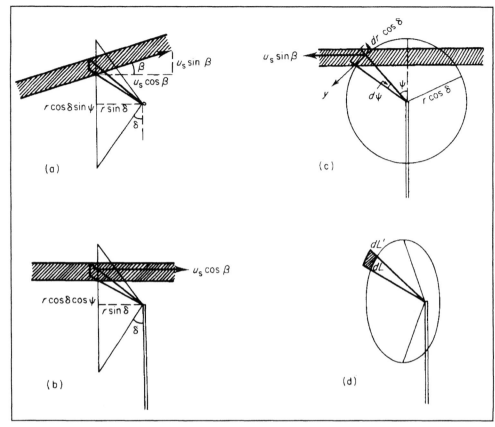

Figure 9.12. Streamtube definition (hatched areas) for a coning rotor in a non-uniform wind field: (a) view from top; (b) side view; (c) front view; (d) view along streamtube.

$$W_x = u_{in} (1 - a) (\cos \beta \cos \delta + \sin \beta \sin \delta \sin \psi),$$

$$W_y = u_{in} (1 - a) \sin \beta \cos \psi + r \Omega \cos \delta (1 + a'),$$

$$W_z = u_{in} (1 - a) (- \cos \beta \sin \delta + \sin \beta \cos \delta \sin \psi).$$

(9.16)

The appropriate W^2 to insert into (9.15) is $W_x^2 + W_y^2$. Finally, the relation between the streamtube area A_s and the increments dr and $d\psi$ must be established. As indicated in Fig. 9.12, the streamtube is approximately a rectangle with sides dL and dL', given by

$$dL = r \cos \delta (\cos 2\beta \cos 2\psi + \sin 2\psi)^{1/2} d\psi,$$

(9.17)

$$dL' = (\sin 2\delta (1 + \cos 2\beta) + \cos 2\delta (\cos 2\beta \sin 2\psi + \cos 2\psi))^{1/2} dr,$$

and

$$A_s = dL \, dL'.$$

Now, for each streamtube, a and a' are obtained by equating the x- and y-components of (9.14) and (9.15) and using the auxiliary equations for (W, ϕ) or (W_x, W_y). The total thrust and torque are obtained by integrating F_x and $r \cos \delta F_y$, over dr and $d\psi$ (i.e. over all streamtubes).

9.4 Restoration of wind profile in wake and implications for turbine arrays

For a wind energy converter placed in the planetary boundary layer (i.e. in the lowest part of the Earth's atmosphere), the reduced wake wind speed in the stream-wise direction, $u_{s,out}$, will not remain below the wind speed $u_{s,in}$ of the initial wind field, provided that this is not diminishing with time. The processes responsible for maintaining the general kinetic motion in the atmosphere by making up (by solar energy) for the kinetic energy lost by surface friction and other dissipative processes will also act in the direction of restoring the initial wind profile (speed as function of height) in the wake of a power extracting device by transferring energy from higher air layers (or from the "sides") to the partially depleted region (Sørensen, 1996). The large amounts of energy available at greater altitude (Sørensen, 2004) make such processes possible almost everywhere at the Earth's surface and not just at those locations where new kinetic energy is predominantly being created.

In the near wake, the wind field is non-laminar, owing to the induced tangential velocity component, u_t^{ind}, and owing to vorticity shedded from the wing tips and the hub region (as discussed in Section 9.1). It is then expected that these turbulent components gradually disappear further downstream in the wake, as a result of interactions with the random eddy motion of different scales present in the "unperturbed" wind field. "Disappear" here means "get distributed on a large number of individual degrees of freedom", so that no contributions to the time-averaged quantities considered remain. For typical operation of a propeller-type wind conversion device, such as the situations illustrated in Figs. 9.5 and 9.6, the tangential interference factor a' is small compared to the axial interference factor a, implying that the most visible effect of the passage of the wind field through the rotor region is the change in stream-wise wind speed. This is a function of r, the distance of the blade segment from the hub centre, as illustrated in Fig. 9.13, based on wind tunnel measurements. The induced r-dependence of the axial velocity is seen to gradually smear out, although it is clearly visible at a distance of two rotor radii in the set-up studied.

Figure 9.14 suggests that under average atmospheric conditions the axial velocity, $u_{s,in}$, will be restored to better than 90% at about 10 rotor diameters behind the rotor plane and better than 80% at a distance of 5–6 rotor diameters behind the rotor plane, but rather strongly dependent on the amount of turbulence in the "undisturbed" wind field.

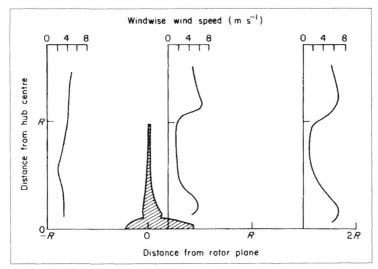

Figure 9.13. Wind tunnel results illustrating the distortion of the wind field along the radial direction (R is the rotor radius) at various distances from the rotor plane (based on Hütter, 1976).

A second wind energy converter may be placed behind the first one, in the wind direction, at a distance such that the wind profile and magnitude are reasonably well restored. According to the simplified investigations in wind tunnels (Fig. 9.14), supported by field measurements behind buildings, forests and fences, a suitable distance would seem to be 5–10 rotor diameters (increasing to over 20 rotor diameters if "complete restoration" is required). If there is a prevailing wind direction, the distance between conversion units perpendicular to this direction may be smaller (essentially determined by the induced radial velocities, which were neglected in the preceding subsections, but appear qualitatively in Fig. 8.1). If, on the other hand, several wind directions are important, and the converters are designed to be able to "yaw against the wind", then the distance required by wake considerations should be kept in all directions.

More severe limitations may possibly be encountered with a larger array of converters, say, distributed over an extended area with the average spacing X between units. Even if X is chosen so that the relative loss in streamwise wind speed is small from one unit to the next, the accumulated effect may be substantial, and the entire boundary layer circulation may become altered in such a way that the power extracted decreases more sharply than expected from the simple wake considerations. Thus, large-scale conversion of wind energy may even be capable of inducing local climatic changes.

A detailed investigation of the mutual effects of an extended array of wind energy converters and the general circulation on each other requires a combination of a model of the atmospheric motion with a suitable model of the disturbances induced in the wake of individual converters. In one of the first discussions of this problem, Templin (1976) considered the influence of an infinite two-dimensional array of wind energy converters with fixed average spacing on the boundary layer motion to be restricted to a change of the roughness length z_0 in the logarithmic expression (see Sørensen, 2004) for the wind profile, assumed to describe the wind approaching any converter in the array. The change in z_0 can be calculated from the stress τ^{ind} exerted by the converters on the wind, due to the axial force F_x in (9.6) or (9.8), which according to (8.14) can be written

$$\tau^{ind} = F_x/S = \tfrac{1}{2}\rho\,(u_{s,in})^2\,2(1-a)^2\,A/S,$$

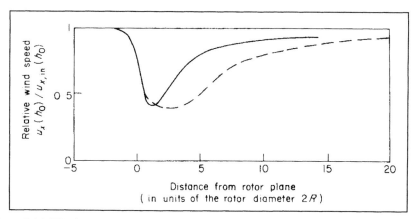

Figure 9.14. Wind tunnel results indicating the restoration at stream-wise wind speed behind a rotor (placed at distance 0, the rotor being simulated by a gauze disc). The approaching wind has a logarithmic profile (solid line) or is uniform with approximately laminar flow (dashed line) (based on Pelser, 1975).

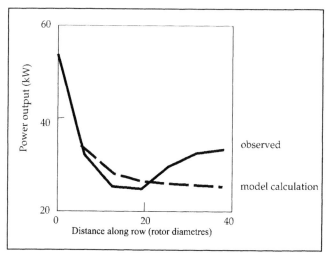

Figure 9.15. Calculated (see text) and measured power outputs from one row of turbines perpendicular to the wind direction, in a wind park located at Nørrekær Enge in Denmark. It consists in total of five rows with a turbine spacing of 6 rotor diameters (Taylor *et al.*, 1993).

with S being the average ground surface area available for each converter and A/S being the "density parameter", equal to the ratio of rotor-swept area to ground area. For a quadratic array with regular spacing, S may be taken as X^2. According to Sørensen (2004), $u_{x,in}$ taken at hub height h_0 can, in the case of a neutral atmosphere, be written

$$u_{x,in}(h_0) = \frac{1}{\kappa}\left(\frac{\tau^0 + \tau^{ind}}{\rho}\right)^{1/2} \log\left(\frac{h_0}{z'_0}\right),$$

where τ^0 is the stress in the absence of wind energy converters and z'_0 is the roughness length in the presence of the converters, which can now be determined from this equation and τ^{ind} from the previous one.

Figure 9.15 shows the results of a calculation for a finite array of converters (Taylor et al., 1993) using a simple model with fixed loss fractions (Jensen, 1994). It is seen that this model is incapable of reproducing the fast restoration of winds through the turbine array, presumably associated with the propagation of the enhanced wind regions created just outside the swept areas (as seen in Fig. 9.13). A three-dimensional fluid dynamics model is required for describing the details of array shadowing effects. Such calculations are in principle possible, but so far no convincing implementation has been presented. The problem is the very accurate description needed, of complex three-dimensional flows around the turbines and for volumes comprising the entire wind farm of maybe hundreds of turbines. This is an intermediate regime between the existing three-dimensional models for gross wind flow over complex terrain and the detailed models of flow around a single turbine used in calculations of aerodynamical stability.

9.5 Off-shore foundation issues

The possibility of placing wind turbines off-shore, typically in shallow waters of up to some 20 m depth, relies on use of low-cost foundation methods developed earlier for harbour and oil-well uses. The best design depends on the material constituting the local water floor, as well as local hydrological conditions, including strength of currents and icing problems. Breakers are currently used to prevent ice from damaging the structure. The most common structure in place today is a concrete caisson as shown in Fig. 9.16a, but in the interest of cost minimising, the steel solutions shown in Fig. 9.16b,c have received increasing attention and the monopile (Fig. 9.16c) was selected for the recent Samsø wind farm in the middle of Denmark (Birch and Gormsen, 1999; Offshore Windenergy Europe, 2003). Employing the suction effect (Fig. 9.16b) may help cope with short-term gusting forces, while the general stability must be ensured by the properties of the overall structure itself.

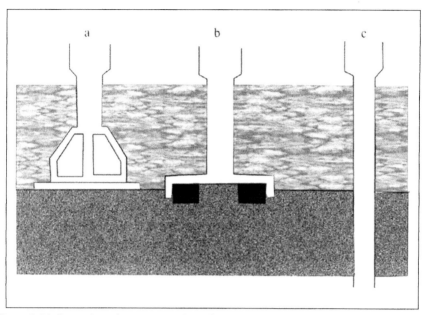

Figure 9.16. Some foundation types for off-shore wind turbine towers placed in shallow waters; a: sand-filled caisson standing on the bottom, b: suction buckets, c: steel monopile. These foundation types are in use at locations with clay-till water floor material.

9.6 Height scaling and frequency distribution of wind

In closing this chapter, some approximate relations often employed in simple wind power estimates will be mentioned. One describes the scaling of wind speeds with height, used e.g. to estimate speed u at a wind turbine hub height z from measured data at another height z_1 (Sørensen, 2004):

$$u(z) = u(z_1) \log((z - z_1)/z_0), \tag{9.18}$$

where the parameter z_0 describes the local roughness of the surface. The form (9.18) is valid only for stable atmospheres. Another often-used approximation is the Weibull distribution of wind speeds v over time at a given location (f is the fraction of time in a given speed interval; Sørensen, 1986):

$$f(v) = (k/c) \, (v/c)^{k-1} \, exp(-(v/c)^k), \tag{9.19}$$

where the parameter k is around 2 and c around 8, both with considerable variations from site to site.

CHAPTER 10

CROSS-WIND AND OTHER ALTERNATIVE CONVERTER CONCEPTS

Wind energy converters of the cross-wind type have the rotor axis perpendicular to the wind direction. The rotor axis may be horizontal as in wheel-type converters (in analogy to waterwheels) or vertical as in the panemones used in Iran and China. The blades (ranging from simple "paddles" to optimised airfoil sections) will be moving with and against the wind direction on alternative sides of the rotor axis, necessitating some way of emphasising the forces acting on the blades on one side. Possible ways are simply to shield half of the swept area, as in the Persian panemones (Wulff, 1966); to curve the "paddles" so that the (drag) forces are smaller on the convex than on the concave side, as in the Savonius rotor (Savonius, 1931); or to use aerodynamically shaped wing blades producing high lift forces for wind incident on the "front edge", but small or inadequate forces for wind incident on the "back edge", as in the Darrieus rotor (cf. Fig. 10.1) and related concepts. Another possibility is to allow for changes in the pitch angle of each blade, as, for example, achieved by hinged vertical blades in the Chinese panemone type (Li, 1951). In this case the blades on one side of a vertical axis have favourable pitch angles, while those on the other side have unfavourable settings. Apart from the shielded ones, vertical axis cross-wind converters are omnidirectional, that is, they accept any horizontal wind direction on equal footing.

10.1 Performance of a Darrieus-type converter

The performance of a cross-wind converter, such as the Darrieus rotor shown in Fig. 10.1, may be calculated in a way similar to that used in Chapter 9, by derivation of the unknown, induced velocities from expressing the forces in terms of both lift and drag forces on the blade segments and in terms of momentum changes between incident wind and wake flow. Also, the flow may be divided into a number of streamtubes (assumed to be non-interacting), according to assumptions on the symmetries of the flow field. Figure 10.1 gives an example of the streamtube definitions, for a two-bladed Darrieus rotor with angular velocity Ω and a rotational angle ψ describing the position of a given blade, according to (9.12). The blade chord, c, has been illustrated as constant, although it may actually be taken as varying to give the optimum performance for any blade segment at the distance r from the rotor axis. As in the propeller rotor case, it is not practical to extend the chord increase to the regions near the axis.

The bending of the blade, characterised by an angle $\delta(h)$ depending on the height h (the height of the rotor centre is denoted h_0), may be taken as a troposkien curve (Blackwell and Reis, 1974), characterised by the absence of bending forces on the blades when they are rotating freely. Since the blade profiles encounter wind directions at both positive and negative forward angles, the profiles are often taken as symmetrical (e.g. NACA 00XX profiles).

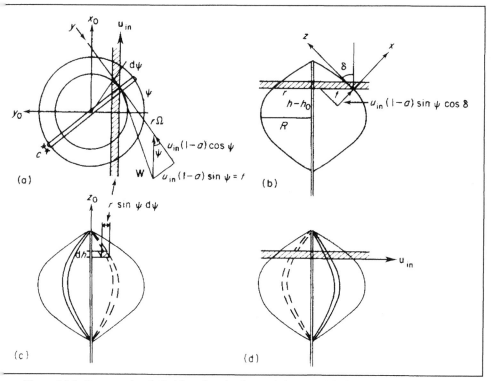

Figure 10.1. Streamtube definition (hatched areas) for two-bladed Darrieus rotor, and determination of apparent wind velocity, in the case of negligible cross-wind induced velocity ($u_{y0}^{ind} = 0$): (a) view from top; (b) view along tangent to blade motion; (c) view along streamtube; (4) view perpendicular to streamtube.

Assuming as in Section 9.1 that the induced velocities in the rotor region are half of those in the wake, the streamtube expressions for momentum and angular momentum conservation analogous to (9.8) and (9.9) are

$$F_{x0} = -J_m \, u_s^{ind} = 2\rho \, A_s \, (u_{in})^2 \, (1-a) \, a,$$
$$F_{y0} = J_m \, u_{c.w.}^{ind},$$

(10.1)

where the axial interference factor a is defined as in (9.1), and where the streamtube area A_s corresponding to height and angular increments dh and dψ is (cf. Fig. 10.1)

$$A_s = r \sin \psi \, d\psi dh.$$

The cross-wind induced velocity $u_{c.w.}^{ind}$ is not of the form (9.2), since the tangent to the blade rotational motion is not along the y_0-axis. The sign of $u_{c.w.}^{ind}$ will be fluctuating with time, and for low chordal ratio c/R (R being the maximum value of r) it may be permitted to put F_{y0} equal to zero (Lissaman, 1976; Strickland, 1975). This approximation will be made in the following. It will also be assumed that the streamtube area does not change by passage through the rotor region and that individual streamtubes do not interact (these assumptions being the same as those made for the propeller-type rotor).

The forces along the instantaneous x- and y-axes due to the passage of the rotor blades at a fixed streamtube location (h, ψ) can be expressed in analogy to (9.15), averaged over one rotational period,

$$F_x = B \, (d\psi/2\pi) \, \tfrac{1}{2} \rho \, c \, W^2 \, (C_D \sin \phi + C_L \cos \phi) \, dh/\cos \delta,$$
$$F_y = B \, (d\psi/2\pi) \, \tfrac{1}{2} \rho \, c \, W^2 \, (-C_D \cos \phi + C_L \sin \phi) \, dh/\cos \delta,$$

(10.2)

where

$$\tan \phi = - W_x / W_y; \qquad W^2 = W_x^2 + W_y^2$$

and (cf. Fig. 10.1)

$$W_x = u_{in} \, (1-a) \sin \psi \cos \delta,$$
$$W_y = - r \, \Omega - u_{in} \, (1-a) \cos \psi,$$
$$W_z = - u_{in} \, (1-a) \sin \psi \sin \delta.$$

(10.3)

The angle of attack is still given by (9.5), with the pitch angle θ being the angle between the y-axis and the blade centre chord line (cf. Fig. 9.2c). The force components (10.2) may be transformed to the fixed (x_0, y_0, z_0) coordinate system, yielding

$$F_{x0} = F_x \cos \delta \sin \psi + F_y \cos \psi,$$

(10.4)

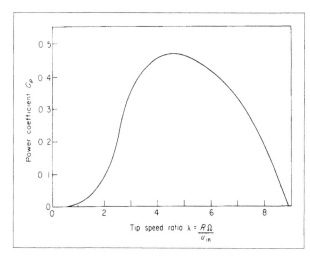

Figure 10.2. Power coefficient as a function of tip-speed ratio ("tip" means point furthest away from axis), for a two-bladed Darrieus rotor, with a chord ratio $c/R = 0.1$ and blade data corresponding to a symmetrical profile (NACA 0012 with Reynolds number Re = 3 × 10^6). A streamtube model of the type described in Fig. 10.1 has been used (based on Strickland, 1975).

with the other components F_{y0} and F_{z0} being neglected due to the assumptions made. Using the auxiliary relations given, a may be determined by equating the two expressions (10.1) and (10.4) for F_{x0}.

After integration over dh and dψ, the total torque Q and power coefficient C_p can be calculated. Figure 10.2 gives an example of the calculated C_p for a low-solidity, two-bladed Darrieus rotor with NACA 0012 blade profiles and a size corresponding to Reynolds number Re = 3 × 10^6. The curve is rather similar to the ones obtained for propeller-type rotors (e.g. Fig. 9.3), but the maximum C_p is slightly lower. The reasons why this type of cross-wind converter cannot reach the maximum C_p of 16/27 derived from (8.18) (the "Betz limit") are associated with the fact that the blade orientation cannot remain optimal for all rotational angles ψ_i, as discussed (in terms of a simplified solution to the model presented above) by Lissaman (1976).

Figure 10.3 gives the torque as a function of angular velocity Ω for a small three-bladed Darrieus rotor. When compared with the corresponding curves for propeller-type rotors shown in Figs. 9.8 and 9.10 (or generally Fig. 9.7), it is evident that the torque at $\Omega = 0$ is zero for the Darrieus rotor, implying that it is not self-starting. For application in connection with an electric grid or some other back-up system, this is no problem since the auxiliary power needed to start the Darrieus rotor at the appropriate times is very small compared with the wind converter output, on a yearly average basis. However, for application as an isolated source of power (e.g. in rural areas), it is a disadvantage, and it has been suggested that a small Savonius rotor should be placed on the main rotor axis in order to provide the starting torque (Banas and Sullivan, 1975).

Another feature of the Darrieus converter (as well as of some propeller-type converters), which is evident from Fig. 10.3, is that for application with a load of constant Ω (as in Fig. 9.8), there will be a self-regulating effect in that the torque will rise with increasing wind speed only up to a certain wind speed. If the wind speed increases further, the torque will begin to decrease. For variable-Ω types of load (as in Fig. 9.10), the behaviour of the curves in Fig. 10.3 implies that cases of irregular increase and decrease of torque, with increasing wind speed, can be expected.

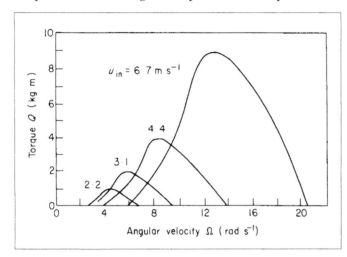

Figure 10.3. Torque as a function of angular velocity for a small three-bladed Darrieus rotor and at selected wind speeds. The experimental rotor had a radius $R = 2.25$ m, a chord ratio $c/R = 0.085$ and NACA 0012 blade profiles (based on Banas and Sullivan, 1975).

10.2 Augmenters

In the preceding sections, it was assumed that the induced velocities in the converter region were half of those in the wake. This is strictly true for situations where all cross-wind induced velocities (u_t and u_r) can be neglected, as shown in (8.13), but if suitable cross-wind velocities can be induced so that the total stream-wise velocity in the converter region, u_x, exceeds the value of $-\frac{1}{2}(u_{x,in} + u_{x,out})$ by a positive amount δu_x^{ind}, then the Betz limit on the power coefficient, $C_p = 16/27$, may be exceeded,

$$u_x = \frac{1}{2}(u_{x,in} + u_{x,out}) + \delta u_x^{ind}.$$

A condition for this to occur is that the extra induced stream-wise velocity δu_x^{ind} does not contribute to the induced velocity in the distant wake, u_x^{ind}, which is given implicitly by the above form of u_x, since

$$u_{x,out} = u_{x,in} + u_x^{ind} = u_{x,in}(1 - 2a).$$

The streamtube flow at the converter, (8.12), is then

$$J_m = \rho A_s u_x = \rho A_s u_{x,in}(1 - a + \tilde{a}) \tag{10.5}$$

with $\tilde{a} = \delta u_x^{ind}/u_{x,in}$, and the power (8.16) and power coefficient (8.18) are replaced by

$$E = \rho A_s \, (u_{x,in})^3 \, 2a \, (1-a) \, (1-a+\tilde{a}),$$
$$C_p = 4a \, (1-a) \, (1-a+\tilde{a}), \qquad\qquad\qquad (10.6)$$

where the streamtube area A_s equals the total converter area A, if the single-streamtube model is used.

10.3 Ducted rotor

In order to create a positive increment $\tilde{a}u_{x,in}$ of axial velocity in the converter region, one may try to take advantage of the possibility of inducing a particular type of cross-wind velocity, which causes the stream-tube area to contract in the converter region. If the streamtube cross section is circular, this may be achieved by an induced radial outward force acting on the air, which again can be caused by the lift force of a wing section placed at the periphery of the circular converter area, as illustrated in Fig. 10.4.

Figure 10.4 compares a free propeller-type rotor (top), for which the streamtube area is nearly constant (as was actually assumed in Sections 9.1, 9.2 and 10.1) or expanding as a result of radially induced velocities, with a propeller rotor of the same dimensions, shrouded by a duct-shaped wing-profile. In this case the radial inward lift force F_L on the shroud corresponds (by momentum conservation) to a radial outward force on the air, which causes the streamtube to expand on both sides of the shrouded propeller; in other words, it causes the streamlines passing through the duct to define a streamtube, which contracts from an initial cross section to reach a minimum area within the duct and which again expands in the wake of the converter. From (8.23), the magnitude of the lift force is

$$F_L^{duct} = \tfrac{1}{2} \, \rho \, W_{duct}^2 \, c_{duct} \, C_L^{duct},$$

with the duct chord c_{duct} defined in Fig. 10.4 and W_{duct} related to the incoming wind speed $u_{x,in}$ by an axial interference factor a_{duct} for the duct, in analogy with the corresponding one for the rotor itself, (8.15),

$$W_{duct} = u_{x,in} \, (1 - a_{duct}).$$

F_L^{duct} is associated with a circulation Γ around the shroud profile (shown in Fig. 10.4), given by (8.21). The induced velocity δu_x^{ind} inside the duct appears in the velocity being integrated over in the circulation integral (8.21), and it may to a first approximation be assumed that the average induced velocity is simply proportional to the circulation (which in itself does not depend on the actual choice of the closed path around the profile),

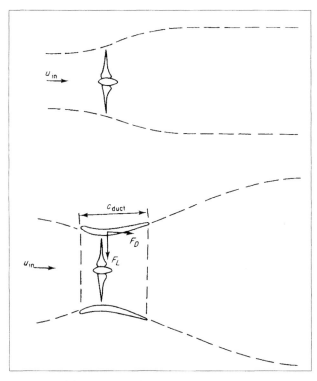

Figure 10.4. Peripheral streamlines for free stream (top) and ducted rotor (below).

$$\delta u_x^{\,ind} = k_{duct}\, \Gamma_{duct} / R, \qquad (10.7)$$

where the radius of the duct, R (assumed to be similar to that of the inside propeller-type rotor), has been introduced, because the path-length in the integral Γ_{duct} is proportional to R and the factor k_{duct} appearing in (10.7) is therefore reasonably independent of R. Writing $\Gamma_{duct} = (\delta\, W_{duct})^{-1}\, F_L^{\,duct}$ in analogy to (8.21), and introducing the relations found above,

$$\tilde{a}_{duct} = k_{duct}\, c_{duct}\, C_L^{\,duct}\, (1 - a_{duct}) / 2R. \qquad (10.8)$$

If the length of the duct, c_{duct}, is made comparable to or larger than R, and the other factors in (10.8) can be kept near to unity, it is seen from (10.6) that a power coefficient about unity or larger is possible. This advantage may, however, be outweighed by the much larger amounts of materials needed to build a ducted converter, relative to a free rotor.

Augmenters taking advantage of the lift forces on a suitably situated aerodynamic profile need not provide a fully surrounding duct around the simple rotor device. A vertical cylindrical tower structure (e.g. with a wing profile similar to that of the shroud in Fig. 10.4) may suffice to produce a reduction in the widths of the streamtubes relevant to an adjacently located rotor and thus may produce some enhancement of power extraction.

10.4 Rotor with tip-vanes

A formally appealing design, shown in Fig. 10.5, places tip-vanes of modest dimensions on the rotor blade tips (van Holten, 1976). The idea is that the tip-vanes act like a duct, without causing much increase in the amount of materials needed in construction. The smaller areas producing lift forces are compensated for by having much larger values of the apparent velocity W_{vane} seen by the vanes than W_{duct} in the shroud case. This possibility occurs because the tip-vanes rotate with the wings of the rotor (in contrast to the duct), and hence see air with an apparent cross-wind velocity given by

$$W_{vane} = R\Omega \, (1 - a_{vane}).$$

The magnitude of the ratio W_{vane}/W_{duct} is thus of the order of the tip-speed ratio) $\lambda = R\Omega/u_{x,in}$, which may be 10 or higher (cf. Section 9.2).

The lift force on a particular vane is given by $F_L^{vane} = \frac{1}{2} \rho \, W_{vane}^2 \, c_{vane} \, C_L^{vane}$, and the average inward force over the periphery is obtained by multiplying this expression by $(2\pi R)^{-1} \, Bb_{vane}$, where B is the number of blades (vanes), and where b_{vane} is the lengths of the vanes (see Fig. 10.5).

Using a linearised expression analogous to (10.7) for the axial velocity induced by the radial forces,

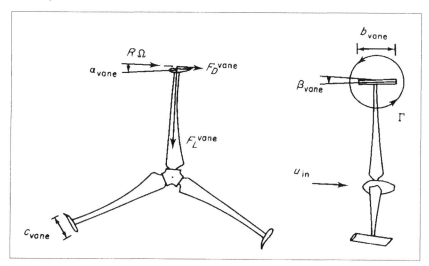

Figure 10.5. Geometry of propeller-type rotor with tip-vanes.

$$\delta u_x^{ind} = \tilde{a}_{vane} \, u_{x,in} = k_{vane} \, \Gamma / R, \tag{10.9}$$

the additional interference factor $\tilde{a}_{vane}$ to use in (10.6) may be calculated. Here Γ is the total circulation around the length of the tip-vanes (cf. Fig. 10.5), and not the circulation $\Gamma_{vane} = (\rho\, W_{vane})^{-1}\, F_L^{vane}$ in the plane of the vane lift and drag forces (left-hand side of Fig. 10.5). Therefore, Γ may be written

$$\Gamma = F_L^{vane}\, B\, b_{vane} / (2\pi R \rho\, w''),$$

i.e. equal to the average inward force divided by ρ times the average axial velocity at the tip-vane containing peripheral annulus,

$$w'' = u_{x,in}\,(1+a'').$$

Expressions for a'' have been derived by van Holten (1976). By inserting the above expressions into (10.9), $\tilde{a} = \tilde{a}_{vane}$ is obtained in the form

$$\tilde{a} = k_{vane}\,(1 - a_{vane})^2\, B\, c_{vane} b_{vane}\, \lambda^2\, C_L^{vane} / (4\pi R^2\,(1 + a'')). \tag{9.10}$$

The part $(Bc_{vane}b_{vane}/ \pi R^2)$ in the above expression is the ratio between the summed tip-vane area and the rotor-swept area. Taking this as 0.05 and [according to van Holten (1976) for $b_{vane}/R = 0.5$] $k_{vane}\,(1 - a_{vane})^2 / (1 + a'')$ as 0.7, the tip-speed ratio λ as 10 and $C_L^{vane} = 1.5$, the resulting $\tilde{a}$ is 1.31 and the power coefficient according to the condition that maximises (10.6) as function of a,

$$a = (2 + \tilde{a} - ((2 + \tilde{a})^2 - 3(1 + \tilde{a}))^{1/2}) = 0.43; \qquad C_p = 1.85.$$

The drag forces F_D^{vane} induced by the tip-vanes (cf. Fig. 10.5) represent a power loss from the converter, which has not been incorporated in the above treatment. According to van Holten (1976), this loss may in the numerical example studied above reduce C_p to about 1.2. This is still twice the Betz limit, for a tip-vane area that would equal the area of the propeller blades, for a three-bladed rotor with average chord ratio $c/R = 0.05$. It is conceivable that the use of such tip-vanes to increase the power output would in some cases be preferable to achieving the same power increase by increasing the rotor dimensions.

10.5 Other wind conversion concepts

A large number of alternative devices for utilisation of wind energy have been studied, in addition to those treated above and in the preceding sections. The lift-producing profile exposed to the wind may be hollow with holes through which inside air is driven out by the lift forces. Through a hollow tower structure, replacement air is then drawn to the wing profiles from sets of openings placed in such a way that the air must pass one or more turbine propellers in order to reach the wings. The turbine propellers provide the power output, but the overall efficiency based on the total dimensions of the device is low (Hewson, 1975). Improved efficiency of devices that try to concentrate the wind energy before letting it reach a modest-size propeller may be obtained by first converting the mainly axial flow of the wind into a flow with a large vorticity of a simple structure, such as a circular air motion. Such vortices are, for example, formed over the sharp edges of highly swept delta wings, as illustrated in Fig. 10.6a (Sforza, 1976). There are two positions along the baseline of the delta wing where propeller rotors can be placed in an environment with stream-wise velocities u_x/u_{in} of 2–3 and tangential velocities u_t of the same order of magnitude as u_{in} (Sforza, 1976).

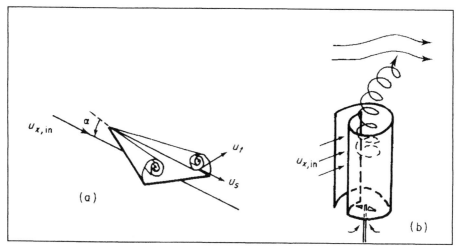

Figure 10.6. Augmenter concepts: (a) delta wing forming trailing vortices; (b) ducted vortex system (based on Sforza, 1976; Yen, 1976).

If the wake streamlines can be diffused over a large region, such that interaction with the wind field not contributing to those streamtubes passing through the converter can transfer energy to the slipstream motion and thereby increase the mass flow J_m through the converter, then further increase in power extraction can be expected. This is the case for the ducted system shown in Fig. 10.4 (lower part), but it may be combined with the vorticity concept described above to form a device of the general layout shown in Fig. 10.6b (Yen, 1976). Wind enters the vertical cylinder through a vertical slit and is forced to rotate by the inside cylinder walls. The vortex system created in this way (an "artificial tornado") is pushed up through the cylinder by pressure forces and leaves the open cylinder top. Owing to the strong vorticity, the rotating air may retain its identity high up in the atmosphere, where its diffusion extracts energy from the strong winds expected at that height. This energy is supposed to be transferred down the "tornado", strengthening its vorticity and thus providing more power for the turbines placed between the "tornado" bottom and a number of air inlets at the tower foot, which replace air that has left the top of the cylinder (cf. Fig. 10.6b). Neither of these two constructions has found practical applications.

Conventional (or unconventional) wind energy converters may be placed on floating structures at sea (where the wind speeds are often higher than on land) or may be mounted on balloons (e.g. a pair of counter-rotating propellers beside one other) in order to utilise the increasing wind speed usually found at elevations not accessible to ordinary building structures on the ground. In order to serve as a power source, the balloons must be guyed to a fixed point on the ground, with wires of sufficient strength.

10.6 Heat, electrical/mechanical power, and fuel generation

The wind energy converters described in the preceding sections have primarily been converting the power in the wind into rotating shaft power. The conversion system generally includes a further conversion step if the desired energy form is different from that of the rotating shaft.

Examples of this are the electric generators with fixed or variable rotational velocity, which were mentioned in connection with Figs 9.8 and 9.10. The other types of energy can, in most cases, be obtained by secondary conversion of electric power. In some such cases the "quality" of electricity need not be as high as that usually maintained by utility grid systems, in respect to voltage fluctuations and variations in frequency in the (most widespread) case of alternating current. For wind energy converters aimed at constant working angular velocity Ω, it is customary to use a gearbox and an induction-type generator. This maintains an a.c. frequency equal to that of the grid and constant to within about 1%. Alternatively, the correct frequency can be prescribed electronically. In both cases, reactive power is created (i.e. power which, like that of a condenser or coil, is phase shifted), which may be an advantage or a disadvantage, depending on the loads on the grid.

For variable-frequency wind energy converters, the electric output would be from a synchronous generator and in the form of variable-frequency alternating current. This would have to be subjected to a time-dependent frequency conversion, and for arrays of wind turbines, phase mismatch would have to be avoided. Several schemes exist for achieving this, for example, semiconductor rectifying devices (thyristors) which first convert the variable frequency AC to DC and then in a second step the DC (direct current) to AC (alternating current) of the required fixed frequency (cf. Chapter 26).

If the desired energy form is heat, "low-quality" electricity may first be produced, and the heat may then be generated by leading the current through a high ohmic resistance. Better efficiency can be achieved if the electricity can be used to drive the compressor of a heat pump (Chapter 6), taking the required heat from a reservoir of temperature lower than the desired one. It is also possible to convert the shaft power more directly into heat. For example, the shaft power may drive a pump, pumping a viscous fluid through a nozzle, such that the pressure energy is converted into heat. Alternatively, the shaft rotation may be used to drive a "paddle" through a fluid, in such a way that large drag forces arise and that the fluid is put into turbulent motion, gradually dissipating the kinetic energy into heat. If water is used as the fluid medium, this arrangement is called a "water-brake".

Windmill shaft power has traditionally been used to perform mechanical work of various kinds, including flour milling, threshing, lifting and pumping. Pumping of water, e.g. for irrigation purposes, with a pump connected to the rotating shaft, may be particularly suitable as an application of wind energy, since variable and intermittent power would, in most cases, be acceptable, as long as the average power supply in the form of lifted water over an extended period of time is sufficient.

In other cases an auxiliary source of power may be needed, so that the demand can be met at any time. This can be achieved for grid-based systems by trade of energy (Sørensen, 2004; 2005). Demand matching can also be ensured if an energy storage facility of sufficient capacity is attached to the wind energy conversion system. A number of such storage facilities will be mentioned in Part IX, and among them will be the storage of energy in the form of fuels, such as hydrogen. Hydrogen may be produced, along with oxygen, by electrolysis of water, using electricity from the wind energy converter. The detailed working of these mechanisms over time and space is simulated in the scenarios outlined in several studies (Sørensen, 2004; 2005; 2007).

The primary interest may also be oxygen, for example for dissolving into the water of lakes that are deficient in oxygen (say, as a result of pollution), or to be used in connection with "ocean farming", where oxygen may be a limiting factor in cases where nutrients are supplied in large quantities, e.g. by artificial upwelling. Such oxygen may be supplied by wind energy converters, while the co-produced hydrogen may be used to produce or move the nutrients. Again, in this type of application large power fluctuations may be acceptable.

CHAPTER 11

HYDRO AND TIDAL ENERGY CONVERSION

11.1 Conversion of water flows

Electricity generation from flowing or elevated water possessing potential, kinetic and/or pressure energy (8.2)–(8.4) can be achieved by means of a turbine, the general theory of which was outlined in Chapter 8. The particular design of turbine to be used depends on whether there is a flow J_m through the device, which must be kept constant for continuity reasons, or whether it is possible to obtain zero fluid velocity after the passage through the turbine.

The form of energy at the entrance of the turbine may be kinetic or pressure energy, causing the forces on the turbine blades to be a combination of "impulse" and "reaction" forces, which can be modified at ease. The potential energy of elevated water may be allowed to "fall", i.e. forming kinetic energy, or it may act as a pressure source through a water-filled tube connecting the elevated water source with the turbine placed below. Conversely, pressure energy may be transformed into kinetic energy by passage through a nozzle.

Typical classical turbine designs are illustrated in Fig. 11.1. For high specific energy differences $w_{in}-w_{out}$ (large "heads"), the Pelton turbine (Fig. 11.1a), which is a high-speed variant of the simple undershot waterwheel, may be used. It has the inflow through a nozzle, providing purely kinetic energy, and negligible w_{out} (if the reference point for potential energy is taken to correspond to the water level after passing the turbine). Also, the Francis turbine (Fig. 11.1b) is used with large water heads. Here the water is allowed to approach the entire rotor, guided to obtain optimum angles of attack, and the rotor moves owing to the reaction forces resulting from both the excess pressure at entrance and the suction at exit.

The third type of turbine, illustrated in Fig. 11.1c, can be used for low water heads. Here the rotor is a propeller, designed to obtain high angular speeds. Again, the angle of attack may be optimised by installation of guiding blades at the entrance to the rotor region. If the blade pitch angle is fixed, it is called a Nagler turbine. If it can be varied, it is called a Kaplan turbine.

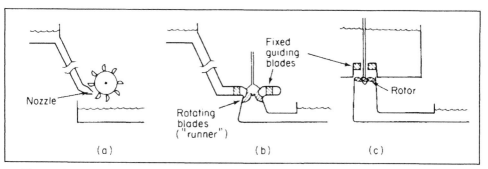

Figure 11.1. Pelton (a), Francis (b) and Kaplan (c) water turbines.

Figure 11.2 gives examples of actual efficiencies for the types of turbines described above (Fabritz, 1954) as functions of the power level. The design point for these turbines is about 90% of the rated power (which is set to 100% on the figure), so the power levels below this point correspond to situations in which the water head is insufficient to provide the design power level.

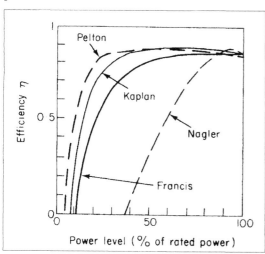

Figure 11.2. Efficiency of water turbines as a function of power level (based on Fabritz, 1954).

Pelton and Francis turbines have been used in connection with river flows with rapid descent, including waterfalls, and in many cases it has been possible by building dams to provide a steady energy source throughout most of the year. This implies storing the water at an elevated level in natural or artificial reservoirs and letting the water descend to the turbines only when needed. The operation of systems that include several sources of flow (springs, glacier or snow melt, rainy regions), several reservoirs of given capacity, a given number of turbine power stations and a load of electricity usage that also varies with time, has been studied and optimised by simulation techniques (see e.g. Jamshidi and Mohseni, 1976).

For many years, hydropower has been the most widely used renewable source of electricity and also – among all types of power plants including fossil and nuclear – the technology inviting the largest schemes of power plants rated at several gigawatts and often involving artificial water reservoirs of gigantic size. The development of such schemes in disregard of social and environmental problems has given hydropower a negative reputation. In developing countries thousands of people have been forcefully removed from their homes with no compensation to give way for flooded reservoirs, causing monumental environmental disruption (though this practice is not restricted to developing countries, as the examples of Tasmania or Norway show) and in some cases destroying priceless sites of archaeological value (e.g. in Turkey). In recent decades, it has become clear that in many cases there exist ways of minimising environmental damage, albeit at a higher cost. The environmental discussion below mentions Swiss efforts to use cascading systems to do away with any large reservoirs, accepting that the smaller reservoirs located along the flow of water and designed to minimise local impacts would provide somewhat less regulation latitude. Full consideration of these concerns is today in most societies a prerequisite for considering hydropower as a benign energy source.

Environmental issues related to hydro power

The environmental impact of non-regulated hydro-generation of power is mainly associated with preventing the migration of fish and other biota across the turbine area, but the building of dams in connection with large hydro facilities may have an even more profound influence on the ecology of the region, in addition to introducing accident risks. For large reservoirs, there has been serious destruction of natural landscapes and dislocation of populations living in areas to be flooded. There are ways to avoid some of the problems. Modular construction, where the water is cascaded through several smaller reservoirs, has been used, e.g. in Switzerland, with a substantial reduction in the area modified as a result. The reservoirs need not be constructed in direct connection with the generating plants, but can be separate installations placed in optimum locations, with a two-way turbine that uses excess electric production from other regions to pump water up into a high-lying reservoir. When other generating facilities cannot meet demand, the water is then led back through the turbines. This means that although the water cycle may be unchanged on an annual average basis, considerable seasonal modifications of the hydrological cycle may be involved. The influence of such modifications on the vegetation and climate of the region below the reservoir, which would otherwise receive a water flow at a different time, has to be studied in each individual case. The same may be true for the upper region, for example, owing to increased evaporation in the presence of a full reservoir.

Although these modifications are local, they can influence the ecosystems with serious consequences for humans. An example is provided by the building of the Aswan Dam in Egypt, which has allowed water snails to migrate from the Nile delta to the upstream areas. The water snails may carry parasitic worms causing schistosomiasis, and this disease has actually spread from the delta region to Upper Egypt since the building of the dam (Hayes, 1977).

It is unlikely that hydropower utilisation will ever be able to produce changes in the seasonal hydrological cycle, which could have global consequences, but no detailed investigation has yet been made.

Tidal energy flows

Kaplan (or Nagler) turbines are used in connection with low water heads (e.g. the local community power plants in China, cf. China Reconstructs, 1975) and tidal plants (André, 1976), and they may be used if ocean currents are to be exploited. The 240-MW tidal plant at *la Rance* in France has turbines placed in a dam structure across an inlet, which serves as a reservoir for two-way operation of the turbines (filling and emptying of the reservoir). According to André (1976), the turbine efficiency is over 90%, but the turbines are only generating electricity during part of the day, according to the scheme outlined in Fig. 11.3. The times at which generation is possible are determined by the tidal cycle, according to the simple scheme of operation, but modifications to suit the load variations better are possible, e.g. by using the turbines to pump water into the reservoir at times when this would not occur as a result of the tidal cycle itself. The installation has had several problems due to siltation, causing it to be operated most of the time only for water flows in one direction, despite its design to accept water inflow from both sides.

Environmental issues related to tidal power

Environmental impacts may arise from utilisation of tidal power. When the La Rance tidal plant in France was built in the 1960s, the upper estuary was drained for water during two years, a procedure that would hardly be considered environmentally acceptable today. Alternative building methods using caissons or diaphragms exist, but in all cases the construction times are long and careful measures have to be taken to protect the biosphere (e.g. when stirring up mud from the estuary seabed). Generally, the coastal environment is affected by the building and operation of tidal power plants, both during construction and to a lesser extent during operation, depending on the layout (fish bypasses etc., as known from hydropower schemes). Some fish species may be killed in the turbines, and the interference with the periodic motion of bottom sand may lead to permanent siltation problems (as it has at *la Rance*).

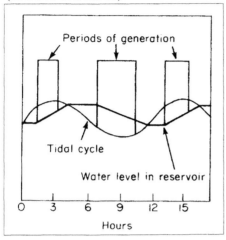

Figure 11.3. Operation of the tidal power plant at *la Rance* (based on André, 1976).

CHAPTER 12

MAGNETO HYDRO-DYNAMIC CONVERTERS

For the turbines considered in Chapters 8-11, it was explicitly assumed that no heat was added. Other flow-type converters are designed to receive heat during the process. An example of this is the (Brayton cycle) gas turbine, which was described in Chapter 3 from the point of view of thermodynamics. The gas turbine allows the conversion of heat into shaft power, but it may equally well be viewed as the successive conversion of heat into kinetic energy of flow and then of the kinetic energy of flow into shaft power.

The magnetohydrodynamic converter is another device converting heat into work, but delivering the work directly as electrical power without the intermediate steps of mechanical shaft power. The advantage is not in avoiding the shaft power to electrical power conversion, which can be done with small losses, but rather in avoiding a construction with moving parts, thereby permitting higher working temperatures and higher efficiency. The heat added is used to ionise a gas, and this conducting gas ("plasma") is allowed to move through an expanding duct, upon which an external magnetic field B is acting. The motion of the gas is sustained by a pressure drop between the chamber where heat is added and the open end of the expanding duct. The charged particles of velocity u in the plasma are subjected to a Lorentz force

$$F = \rho_{el}\, u \times B, \tag{12.1}$$

where the direction of this induced force is perpendicular to B and u, but opposite for the positive atoms and for the negatively charged electrons. Since the mass of an electron is much smaller than that of an atom, the net induced current will be in the direction given by a negative value of ρ_{el}. Assuming a linear relationship between the induced current J_{ind} and the induced electric field $E_{ind} = F/\rho_{el} = u \times B$, the induced current may be written

$$J_{ind} = \sigma u \times B, \tag{12.2}$$

where σ is the electrical conductivity of the plasma. This outlines the mechanism by which the magnetohydrodynamic (MHD) generator converts the kinetic energy of moving charges into electrical power associated with the induced current J_{ind} across the turbine. A more detailed treatment must take into account the contributions to the force (12.1) on the charges, which arise from the induced velocity component J_{ind} / ρ_{el}, as well as the effect of variations (if any) in the flow velocity u through the generator stage (see e.g. Angrist, 1976).

The generator part of the MHD generator has an efficiency determined by the net power output after subtraction of the power needed for maintaining the magnetic field B. Only the gross power output can be considered as given by (2.12). Material considerations require that the turbine be cooled, so in addition to power output there is a heat output in the form of a coolant flow, as well as the outgoing flow of cooled gas. The temperature of the outflowing gas is still high (otherwise recombination of ions would inhibit the functioning of the converter), and the MHD stage is envisaged as being followed by one or more conventional turbine stages. It is believed that the total power generation in all stages could be made to exceed that of a conversion system based entirely on turbines with moving parts, for the same heat input.

Very high temperatures are required for the ionisation to be accomplished thermally. The ionisation process can be enhanced in various ways. One is to "seed" the gas with suitable metal dust (sodium, potassium, caesium, etc.), for which case working MHD machines operating at temperatures around 2500 K have been demonstrated (Hammond et al., 1973). If the heat source is fossil fuel, and particularly if it is coal with high sulphur content, the seeding has the advantage of removing practically all the sulphur from the exhaust gases (the seeding metals are rather easily retrieved and must anyway be recycled for economic reasons).

CHAPTER

13

WAVE ENERGY CONVERTERS

The surface of a wavy ocean may be described by a function $\sigma = \sigma(x, y, t)$, giving the deviation of the vertical coordinate z from zero. Performing a spectral decomposition of a wave propagating horizontally along the x-axis yields

$$\sigma(x,t) = \int_{-\infty}^{\infty} \exp(-i(kx - \omega(k)t))S(k)dk, \tag{13.1}$$

where $S(k)$ represents the spectrum of σ. If ω is constant and positive, (13.1) describes a harmonic wave moving towards the right; if ω is constant and negative, the wave moves towards the left. By superimposing waves with positive and negative ω, but with the same spectral amplitude $S(k)$, standing waves may be constructed. If $S(k)$ in (13.1) is different from zero only in a narrow region of k-space, σ will describe the propagation of a disturbance (wave packet) along the ocean surface.

A lowest-order solution for σ under the influence of the local force of gravity g may be found using the infinitesimal wave approximation (Wehausen and Laitone, 1960),

$$\sigma = a \cos(k(x - Ut)) \exp(-2k^2 \eta t)), \tag{13.2}$$

where the connection between wave velocity (phase velocity) U, wave number k and wavelength λ is

$$U = \left(\frac{g}{k}\right)^{1/2}; \qquad k = \frac{2\pi}{\lambda} \tag{13.3}$$

No boundary condition has been imposed at the bottom, corresponding to a deep ocean assumption, $\lambda << h$, with h being the depth of the ocean. If the viscous forces are disregarded, the last exponential factor in (13.2) containing the viscosity η (= 1.8×10^{-6} m^2 s^{-1} for water) may be omitted.

The energy of wave motion is the sum of potential, kinetic and surface contributions (if surface tension is considered). For a vertical column of unit area and density ρ, one has (Sørensen, 2004)

$$W^{total} = \tfrac{1}{2}\, \rho\, g\, a^2\, \exp(-4k^2 \eta\, t). \tag{13.4}$$

This shows that a wave, which does not receive renewed energy input, will dissipate energy by molecular friction, with the rate of dissipation

$$D = -\frac{d\overline{W}^{total}}{dt}\bigg|_{t=0} = 2\rho g a^2 k^2 \eta. \tag{13.5}$$

Of course, this is not the only mechanism by which waves lose energy. Energy is also lost by the creation of turbulence on a scale above the molecular level. This may involve interaction with the air, possibly enhanced by the breaking of wave crests, or oceanic interactions due to the Reynold stresses (see Sørensen, 2004). Also, at the shore, surf formation and sand put into motion play a role in energy dissipation from the wave motion.

The power in the waves, i.e. the rate at which energy (from surface to ocean bottom) is crossing a unit of length at the surface, perpendicular to the direction of the wave motion, is from (13.4), with neglect of the viscosity factor,

$$P = \tfrac{1}{2}\, U\, W^{total} = \rho\, g\, U\, a^2/4 = \rho\, g^2\, T\, a^2/(8\pi), \tag{13.6}$$

where the last equality has introduced the wave period, T, which for harmonic waves equals the zero-crossing period, i.e. the time interval between successive crossings of zero height ($\sigma = 0$) in the same direction.

Once the wind has created a wave field, this may continue to exist for a while, even if the wind ceases. If only frictional dissipation of the type (13.5) is active, a wave of wavelength $\lambda = 10$ m will take 70 h to be reduced to half the original amplitude, while the time is 100 times smaller for $\lambda = 1$ m.

13.1 Wave energy conversion devices

A large number of devices for converting wave energy to shaft power or compression energy have been suggested, and some of them have been tested on a modest scale. Reviews of wave devices may be found, for example, in Leichman and Scobie (1975), Isaacs et al. (1976), Slotta (1976), Clarke (1981), Sørensen (1999a), Thorpe (2001), CRES (2002), and DEA Wave Program (2002). Below, the technical details will be given for two typical examples of actual devices: the oscillating water column device that has been in successful small-scale operation for many years for powering mid-sea buoys, and the Salter duck that theoretically has a very high efficiency, but has not been successful in actual prototyping experiments. However, first some general remarks will be offered.

Resource estimates (Sørensen, 2004) indicate that the most promising locations for a wave utilisation apparatus would be in the open ocean rather than in coastal or shallow water regions. Yet all three device types have been researched: (a) shore-fixated devices for concentrating waves into a modestly elevated reservoir, from which the water may drive conventional hydro-turbines; (b) near-shore devices making use of the oscillations of a water column or a float on top of the waves, relative to a structure standing at the sea floor; and (c) floating devices capturing energy by differential movements of different parts of the device.

As a first orientation towards the prospects of developing economically viable wave power devices, a comparison with off-shore wind power may be instructive. One may first look at the weight of the construction relative to its rated power. For an on-shore wind turbine, this number is around 0.1 kg/W_{rated}, while adding the extra foundation weight for off-shore turbines (except caisson in-fill) increases the number to just below 0.2 kg/W_{rated}. For 15 wave power devices studied by the DEA Wave Program (2002), the range of weight to rated power ratios is from 0.4 to 15 kg/W_{rated}. The two numbers below 1.0 are for device concepts not tested and for use on-shore or at low water depth, where the power resources are small anyway. So the conclusion is that the weight/power ratio is likely to be at least 2 but likely more than 5 times that of off-shore wind turbines, which to a first approximation is also a statement on the relative cost of the two concepts.

Using the same data, one may instead look at the ratio of actually produced power at a particular location and the weight of the device. For off-shore wind in the North Sea, this is around 20 $kWh\ y^{-1}\ kg^{-1}$. For the 15 wave devices, values of 0.1 to 10 are found, by using for all devices the same wave data estimated for a location some 150 km west of the city Esbjerg in Denmark. Clearly, it is not reasonable to use wave data for a location 150 km out into the ocean for wave power devices that must stand on the shore or at very shallow water. Omitting these cases, the resulting range reduces to 0.1-1.5 $kWh\ y^{-1}\ kg^{-1}$ or over 13 times less that for off-shore wind in the same region. Again, the simplistic translation from weight to cost indicates that wave energy is economically unattractive, because at the same location wind power can be extracted at much lower cost. Also, there are no particular reasons to expect the distribution of weight on less expensive materials (concrete or steel) and more expensive materials (special mechanical and electric equipment) to be substantially different. It is part of this argument, that where wave power would be feasible, off-shore wind power is also available, because it is the wind that creates the waves. Only at large water depths, say, over 30 m depth, where any kind of foundation would be problematic, might the wave devices floating on the surface be more attractive than wind. Yet the line of arguments presented above give no indications that the cost of such mid-ocean wave power extraction and cable transmission to land will be economically viable, unless all near-shore wind options have already been exploited.

Finally, there is the argument of the time-distribution of the power from wave devices. One might have expected the variations in wave energy with time to be smaller than for wind energy because of the accumulating nature of the resource. However, this is not true on a seasonal scale, where wave power exhibits very large variations with seasons (Sørensen, 2004). This is the case not only for mid-ocean sites, but also e.g. for the North Sea location mentioned above. Rambøll (1999) finds 6 times more average wave power in January than in June, where the corresponding factor for wind power is 2 (Sørensen, 2000a). Acceptance of wave power into grids serving electricity demands is thus going to be considerably more difficult than acceptance of wind, which roughly has the same seasonal variation as demand, at least on the Northern Hemisphere. Thus, in addition to an initial cost likely to be substantially higher than that of wind power, additional costs for energy storage or other supply-demand mismatch management must be considered.

In spite of these difficulties, much ingenuity has been exercised in constructing wave devices of all kinds: Air-compression systems, mechanical displacements converted into power, enclosed turbines like those used for hydropower or even the free-stream propeller turbines used for wind power, concentrators using ramps to the shore or using lens systems at sea. Few of the devices tested at full scale in actual sea have survived the most severe storms encountered, indicating the difficulty in acquiring enough structural strength without using uneconomically large amounts of materials. It is quite likely that structural failures could be avoided if materials usage and cost were not an issue.

13.2 Pneumatic converter

A wave energy conversion device that for some time has been in practical use, although on a fairly small scale, is the buoy of Masuda (1971), shown in schematic form in Fig. 13.1. Several similar wave power devices exist, based on an oscillating water column driving an air turbine, in some cases shore based and with only one chamber (see e.g. CRES, 2002). The buoy in Fig. 13.1 contains a tube extending downward, into which water can enter from the bottom, and a "double-action" air turbine, i.e. a turbine that turns in the same way under the influence of pressure and suction (as illustrated by the non-return valves).

The wave motion will cause the whole buoy to move up and down and thereby create an up-and-down motion of the water level in the centre tube, which in turn produces the pressure increases and decreases that make the turbine in the upper air chamber rotate. Special nozzles may be added in order to increase the speed of the air impinging on the turbine blades.

For a simple sinusoidal wave motion, as described by (13.2), but omitting the viscosity-dependent exponential factor, the additional variables σ_i and Z describing the water level in the centre tube and the vertical displacement of the entire buoy, respectively, may also be expected to vary sinusoidally, but with phase delays and different amplitudes,

$$\sigma = a \cos (kx - \omega t),$$
$$\sigma_i = \sigma_{i0} \cos (kx - \omega t - \delta_\sigma),$$
$$Z = Z_0 \cos (kx - \omega t - \delta_Z),$$

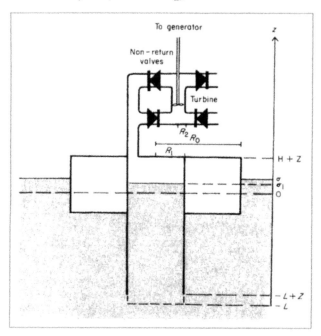

Figure 13.1. Masuda's pneumatic wave energy conversion device.

with $\omega = kU$ in terms of the wave number k and phase velocity U (13.3). The relative air displacement ρ_i in the upper part of the central tube (the area of which is $A_1 = \pi R_1^2$) is now (see Fig. 13.1)

$$\rho_i = \sigma_i - Z.$$

Assuming the airflow to be incompressible, continuity requires that the relative air velocity $d\rho_2/dt$ in the upper tubes forming the air chamber (of area $A_2 = \pi R_2^2$) be

$d\rho_2/dt = (A_1/A_2)\,d\rho_1/dt.$

It is now possible to set up the equations of motion for σ_1 and Z, equating the accelerations $d^2\sigma_1/dt^2$ and d^2Z/dt^2 multiplied by the appropriate masses (of the water column in the centre tube and of the entire device) to the sum of forces acting on the mass in question. These are buoyancy and pressure forces, as well as friction forces. The air displacement variable ρ_2 satisfies a similar equation involving the turbine blade reaction forces. The equations are coupled, but some of the mutual interaction can be incorporated as damping terms in each equation of motion. McCormick (1976) uses linear damping terms $-b(d\sigma_1/dt)$, etc., with empirically chosen damping constants, and drops other coupling terms, so that the determination of σ_1 and Z becomes independent. Yet the $d\rho_2/dt$ values determined from such solutions are in good agreement with measured ones, for small wave amplitudes a (or "significant heights" $H_S = 2 \times 2^{1/2}a$) and wave periods $T = 2\pi/\omega$ which are rather close to the resonant values of the entire buoy, T_0, or to that of the centre tube water column, T_1. Between these resonant periods the agreement is less good, probably indicating that couplings are stronger away from the resonances than near them, as might be expected.

The power transferred to the turbine shaft cannot be calculated from the expression (2.12) because the flow is not steady. The mass flows into and out of the converter vary with time, and at a given time they are not equal. However, owing to the assumption that the air is incompressible, there will be no build-up of a compressed air energy storage inside the air chamber, and therefore, the power can still be obtained as a product of a mass flux, i.e. that to the turbine,

$$J_m = \rho_a A_1 \, d\rho_1/dt$$

(ρ_a being the density of air), and a specific energy change Δw (neglecting heat dissipation). In addition to the terms (8.2)–(8.4) describing potential energy changes (unimportant for the air chamber), kinetic energy changes and pressure/enthalpy changes, Δw will contain a term depending on the time variation of the air velocity $d\rho_2/dt$ at the turbine entrance,

$$\Delta w \approx \tfrac{1}{2}\left(\left(\frac{d\rho_1}{dt}\right)^2 - \left(\frac{d\rho_2}{dt}\right)^2\right) + \frac{\Delta P}{\rho_a} - \int\left(\frac{d^2\rho_2}{dt^2}\right)d\rho,$$

where ΔP is the difference between the air pressure in the air chamber (and upper centre tube, since these are assumed equal) and in the free outside air, and ρ is the co-ordinate along a streamline through the air chamber (McCormick, 1976). Internal losses in the turbine have not been taken into account.

13. WAVE ENERGY CONVERTERS

Figure 13.2 shows an example of the simple efficiency of conversion, η, given by the ratio of $J_m\Delta w$ and the power in the incident waves, which for sinusoidal waves is given by (13.6). Since the expression (13.6) is the power per unit length (perpendicular to the direction of propagation, i.e. power integrated over depth z), it should be multiplied by a dimension of the device that characterises the width of wave acceptance. As the entire buoy is set into vertical motion by the action of the waves, it is natural to take the overall diameter $2R_0$ (cf. Fig. 13.1) as this parameter, such that

$$\eta = 8\pi J_m \Delta w/(2R_0\, \rho g^2 Ta^2) = 32\pi J_m\, \Delta w/(\rho g^2 TH_S^2 R_0). \tag{13.7}$$

The efficiency curves as functions of the period T, for selected significant wave heights H_S, which are shown in Fig. 13.2, are based on a calculation by McCormick (1976) for a 3650-kg buoy with overall diameter $2R_0 = 2.44$ m and centre tube diameter $2R_1 = 0.61$ m. For fixed H_S, η has two maxima, one just below the resonant period of the entire buoy in the absence of damping terms, $T_0 \approx 3$ s, and one near the air chamber resonant period, $T_1 \approx 5$ s. This implies that the device may have to be "tuned" to the periods in the wave spectra giving the largest power contributions, and that the efficiency will be poor in certain other intervals of T. According to the calculation based on solving the equations outlined above and with the assumptions made, the efficiency η becomes directly proportional to H_S. This can only be true for small amplitudes, and Fig. 13.2 shows that unit efficiency would be obtained for $H_S \approx 1.2$ m from this model, implying that non-linear terms have to be kept in the model if the power obtained for waves of such heights is to be correctly calculated. McCormick (1976) also refers to experiments, which suggest that the power actually obtained is smaller than that calculated.

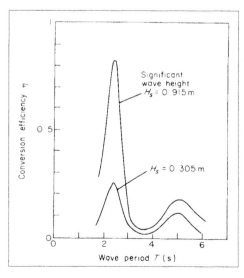

Figure 13.2. Efficiency of a pneumatic wave energy converter as function of wave period, calculated for two wave amplitudes (based on McCormick, 1976).

Masuda has proposed placing a ring of pneumatic converters (floating on water of modest depth) such that the diameter of the ring is larger than the wavelength of the waves of interest. He claims that the transmission of wave power through such a ring barrier is small, indicating that most of the wave energy will be trapped (e.g. multiply reflected on the interior barriers of the ring) and thus eventually be absorbed by the buoy and internal fluid motion in the central tubes. In other developments of this concept, the converters are moored to the sea floor or lean towards a cliff on the shore (Clarke, 1981; Vindeløv, 1994; Thorpe, 2001).

13.3 Oscillating cylinder or vane converter

Since water particles in a sinusoidal wave are moving in circular orbits, it may be expected that complete absorption of the wave energy is only possible with a device possessing two degrees of freedom, i.e. one allowing displacement in both vertical and horizontal directions. Indeed, Ogilvie (1963) has shown that an immersed cylinder of a suitable radius, moving in a circular orbit around a fixed axis (parallel to its own axis) with a certain frequency, is capable of completely absorbing the energy of a wave incident perpendicular to the cylinder axis, i.e. with zero transmission and no wave motion "behind" the cylinder.

Complete absorption is equivalent to energy conversion at 100% efficiency, and Evans (1976) has shown that this is also possible for a half-immersed cylinder performing forced oscillations in two dimensions, around an equilibrium position, whereas the maximum efficiency is 50% for a device that is only capable of performing oscillations in one dimension.

The system may be described in terms of coupled equations of motion for the displacement co-ordinates X_i ($i = 1, 2$),

$$m \, \mathrm{d}^2 X_i / \mathrm{d}t^2 = -d_i \, \mathrm{d}X_i / \mathrm{d}t - k_i X_i + \sum_j F_{ij},$$

where m is the mass of the device, d_i is the damping coefficient in the ith direction, k_i correspondingly is the restoring force or "spring constant" (the vertical component of which may include the buoyancy forces), and F_{ij} is a matrix representing the complete hydrodynamic force on the body. It depends on the incident wave field and on the time derivatives of the co-ordinates X_i, the non-diagonal terms representing couplings of the motion in vertical and horizontal directions.

The average power absorbed over a wave period T is

$$E = \sum_i T^{-1} \int_0^T \frac{\mathrm{d}X_i}{\mathrm{d}t} \sum_j F_{ij} \, \mathrm{d}t, \tag{13.8}$$

and the simple efficiency η is obtained by inserting each of the force components in (13.8) as forces per unit length of cylinder, and dividing by the power in the waves (13.6). One may regard the following as design parameters: the damping parameters d_i, which depend on the power extraction system; the spring constants k_i and the mass distribution inside the cylinder, which determine the moments and hence influence the forces F_{ij}. There is one value of each d_i for a given radius R of the cylinder and a given wave period, which must be chosen as a necessary condition for obtaining maximum efficiency. A second condition involves k_i, and if both can be fulfilled the device is said to be "tuned" and it will be able to reach the maximum efficiency (0.5 or 1, for oscillations in one or two dimensions). Generally, the second condition can be fulfilled only in a certain interval of the ratio R/λ between the radius of the device and the wavelength $\lambda = gT^2/(2\pi)$.

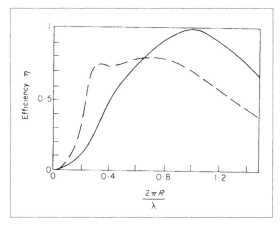

Figure 13.3. Efficiency of wave energy absorption by oscillating motion of a half-immersed cylinder of radius R. The calculation assumes parameters describing damping and coupling terms, such that the device becomes fully tuned at $2\pi R/\lambda = 1$ (solid line) or partially tuned at $2\pi R/\lambda = 0.3$ (dashed line) (based on Evans, 1976).

Figure 13.3 shows the efficiency as a function of $2\pi R/\lambda$ for a half-immersed cylinder tuned at $2\pi R/\lambda_0 = 1$ (full line) and for a partially tuned case, in which the choice of damping coefficient corresponds to tuning at $2\pi R/\lambda_0 = 0.3$, but where the second requirement is not fulfilled. It is seen that such "partial tuning" offers a possibility of widening the range of wave periods accepted by the device.

A system whose performance may approach the theoretical maximum for the cylinder with two degrees of freedom is the oscillating vane or cam of Salter (1974), illustrated in Fig. 13.4. Several structures of this cross section indicated are supposed to be mounted on a common backbone providing the fixed axis around which the vanes can oscillate, with different vanes not necessarily moving in phase. The backbone has to be long in order to provide an approximate inertial frame, relative to which the oscillations can be utilised for power extraction, for example, by having the rocking motion create pressure pulses in a suitable fluid (contained in compression chambers between the backbone and oscillating structure). The necessity of such a backbone is also the weakness of the system, owing to the large bending forces along the structure, which must be accepted during storms.

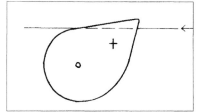

Figure 13.4. Salter's design of an oscillating vane wave energy converter. The cross section shows the backbone and axis of rotation as a small circle and the location of the centre of gravity as a cross. The waves are incident from the right-hand side.

Figure 13.5. Efficiency of Salter's device based on model experiments (solid line) and extrapolations (dashed line) (based on Mollison *et al.*, 1976).

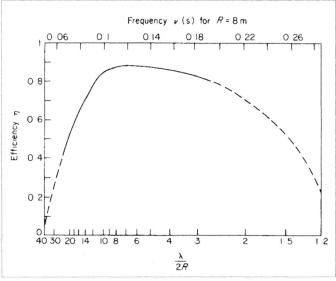

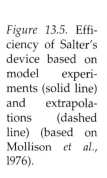

The efficiency of wave power absorption for a Salter "cam" or "duck" of the type shown in Fig. 13.4 is indicated in Fig. 13.5, based on measurements performed on a model of modest scale ($R = 0.05$ m). Actual devices for use in, say, a good wave energy location in the North Atlantic would need a radius of 8 m or more if tuned at the same value of $\lambda/(2\pi R)$ as the device used to construct Fig. 13.5 (Mollison *et al.*, 1976). Electricity generators of high efficiency are available for converting the oscillating vane motion (the angular amplitude of which may be of the order of half a radian) to electric power, either directly from the mechanical energy in the rocking motion or via the compression chambers mentioned above, from which the fluid may be let to an electricity-producing turbine. The pulsed compression is not incompatible with a continuous power generation (Korn, 1972). Experiments have shown that the structural strength needed for the backbone is a problem. For all wave devices contemplated placed in the favourable locations at mid-ocean, transmission to shore will be an additional problem for the economy of the scheme.

KYOCERA PV CAR foto BENT SØRENSEN

IV. Solar radiation
conversion processes

CHAPTER

14

PHOTOVOLTAIC CONVERSION

14.1 Primer on energy bands in semiconductors

This section offers a brief introduction to the quantum mechanics of energy bands in semiconductors. The subject is relevant for a detailed understanding of photovoltaic devices, but the following sections can be read without this background.

The electrons in a solid move in a potential, which for crystalline forms is periodic, corresponding to the lattice structure of the material. The forces on a given electron are electromagnetic, comprising the attractive interaction with the positively charged nuclei (the effective charge of which may be diminished by the screening effect of more tightly bound electrons) as well as repulsive interaction with other electrons. Owing to the small mass ratio between electrons and nuclei, the positions of the nuclei (R_i, i = 1,2,3...) may to a first approximation be regarded as fixed. This is the basis for saying that the electrons are moving in a "potential". When an electron is close to a definite nucleus, its wave function may be approximated by one of the atomic wave functions $\psi_{i,n(i)}$ ($r - R_i$) for this isolated nucleus. One may therefore attempt to describe an arbitrary electron wave function as a linear combination of such atomic wave functions ("orbitals"),

$$\psi(r) = \sum_{i=1}^{N} \sum_{n(i)} c_{i,n(i)} \, \psi_{i,n(i)}(r - R_i). \tag{14.1}$$

Here N is the total number of atomic nuclei, and r is the position vector of the electron.

A simple example is that of a diatomic molecule (N = 2) with only one important level in each atom,

$$\psi(r) = c_{1,n(1)} \, \psi_{1,n(1)}(r - R_1) + c_{2,n(2)} \, \psi_{2,n(2)}(r - R_2),$$

for which two normalised solutions exist, with opposite relative sign between the two terms. If the overlap

$$S = \int \psi_{1,n(1)}{}^{*}(r{-}R_1) \ \psi_{2,n(2)}(r{-}R_2) \, dr$$

(the asterisk denotes complex conjugation) is non-zero, the energies of the two superposition states will be different from those ($W_{1,n(1)}$ and $W_{2,n(2)}$) of the atomic wave functions $\int \psi_{1,n(1)}$ and $\psi_{2,n(2)}$. The most tightly bound solution, $\psi_b(r)$, will correspond to an energy lower than the lowest of the original ones $W_{1,n(1)}$ and $W_{2,n(2)}$, while the other solution, $\psi_a(r)$, will correspond to an energy higher than the highest of the original ones (see e.g. Ballhausen and Gray, 1965).

If the energies W_b and W_a of the bonding solution $\psi_b(r)$ and the anti-bonding solution $\psi_a(r)$ are calculated for various relative positions $R = |R_1{-}R_2|$, one may obtain a picture as shown in Fig. 14.1 (approximately describing an ionised hydrogen molecule $H_2{}^{+}$).

If the number of (say, identical) nuclei, N, is larger than two, but still placed in a regular geometrical relationship characterised by a single distance R (e.g. the distance between atomic nuclei in a cubic lattice or a diamond type lattice), the superposition type wave function (14.1) still gives an energy picture similar to Fig. 14.1, but now with N curves, the highest and lowest of which may look like the two curves for $N = 2$.

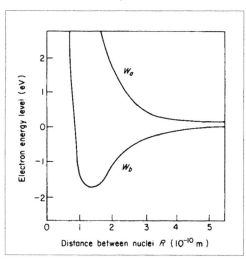

Figure 14.1. Electronic energy levels in a diatomic molecule as a function of inter-atomic distance (e.g. levels formed from 1s atomic orbitals in the $H_2{}^{+}$ molecule).

For a solid crystal, N may be of the order of 10^{24}, so the region between the lowest energy state and the highest one will be filled out with energy levels that for all practical purposes can be regarded as continuous. Such a region is called an *energy band*. For each pair of overlapping atomic orbits, $n(1)$ and $n(2)$, there will be a set of linear combinations of wave functions in the $N = 2$ case, with bonding and anti-bonding characteristics. In the large-N case, each combination of overlapping atomic orbits will lead to an energy band, such that the overall picture of energy levels as a function of lattice characteristics (in the simple case just R) will be one of a number of energy bands inside which all electron energies are "allowed", plus the spaces between energy bands in which no allowed energy is present. The energy band structure may, for example, resemble that shown in Fig. 14.2.

If the distance between the nuclei is large, an electron "feels" only one nucleus (plus a number of tightly bound "inner" electrons, if the nucleus has $Z > 2$) at a time, and the energy spectrum is identical to that of an individual atom (right-hand side of Fig. 14.2). As the assumed distance between nuclei becomes smaller, energy bands develop and become wider. When the energy bands begin to overlap and cross (left-hand side of Fig. 14.2), the basis for the linear assumption (14.1) breaks down, at least in the form discussed so far. If the restriction that the bands are formed from one definite atomic orbital [so that there is no sum over $n(i)$ in (14.1)], is relaxed, it may be possible still to use an expression of the form (14.1) to describe some features of the region of overlapping bands.

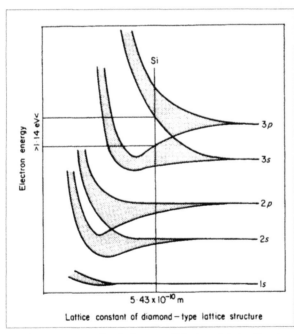

Figure 14.2. Energy band structure of a solid as a function of lattice constant (assumed to completely specify interatomic distances, as it does for crystals of diamond structure). Atomic physics labelling of orbits is indicated to the right. The band structure in the left-hand region of overlapping bands should be regarded as highly schematic. It may take different forms for different solids. The figure indicates how the overlapping 3s and 3p bands in silicon form the 1.14-eV energy gap further illustrated in Fig. 14.3.

As an example, it is possible to obtain the two separate bands in silicon (see Fig. 14.2) first by defining two suitable superpositions of 3s and 3p atomic orbits [corresponding to the summation over $n(i)$, this procedure being called hybridisation in molecular theory], and then by applying the band theory [corresponding to the summation over i in (14.1)]. It is evident that in this region near the band crossing point, it is possible to find energy gaps between adjacent bands, which are much smaller than in the non-overlapping region.

The electron potentials initially represent a situation with one electron per atom being considered. The simplest approximation of many-electron states in one atom is to regard the electrons as independent of each other, except that they have to obey the Pauli exclusion principle according to which there can be at most one electron in each quantum state. Because the electron has an intrinsic spin of $S = \frac{1}{2}$, there are two spin states ("spin-up" and "spin-down") for each spatial orbit. Thus, the periodic system would be derived by first placing two electrons in the 1s-orbit (cf. Fig. 14.2), then two in the 2s-orbit and then six in the 2p-orbit (comprising three degenerate spatial orbits, because the orbital angular momentum is $L = 1$), etc.

A more accurate solution of the Schrödinger equation of quantum mechanics is necessary in order to include the effects of electron–electron interactions, which make the entire energy spectrum depend on electron number in addition to the smooth dependence on the charge of the nucleus ("atomic number"). In a number of cases, these interactions even influence the succession of orbits being filled in the periodic system.

Similar statements can be made for the energy bands in many-atom lattices. The spectrum of energy bands can be used to predict the filling systematics, as a function of the number of electrons per atom, but again the spectrum is not independent of the degree of filling (for this reason no energy scale is provided in Fig. 14.2). Thus, a diagram such as Fig. 14.2 cannot be used directly to predict the atomic distance (e.g. represented by the lattice constant) that will give the lowest total energy and hence the ground state of the system. This is in contrast to the H_2^+ molecule (Fig. 14.1), for which the minimum of the W_b curve determines the equilibrium distance between the nuclei.

Knowing the lattice structure and the lattice constant (the dimension of an elementary cube in the lattice), a vertical cut in a diagram of the type in Fig. 14.2 will give the allowed and forbidden energy values, with the reservations made above. Filling the proper number of electrons per atom into the energy bands, with reference to the Pauli principle, the energy of the latest added electron may be determined, as well as the lowest available energy level into which an additional electron added to the system may go.

If the electrons available exactly fill a band, and if there is a substantial distance to the next higher level, the material is an electrical insulator (no electrons can change state, i.e. they cannot "move"). If a band is partially filled, the material is a good conductor (e.g. a metal). The continuum of levels in the partially filled band allows the electrons to move throughout the lattice. If the distance between the highest filled band and an empty band is small, the material is called a semiconductor. At zero absolute temperature a semiconductor is an insulator, but because of the thermal energy spread at higher temperatures, given by (4.1) because electrons are Fermi particles, some electrons will be excited into the higher band (the "conduction band"). The conductance of silicon increases by a factor 10^6 between 250 and 450 K.

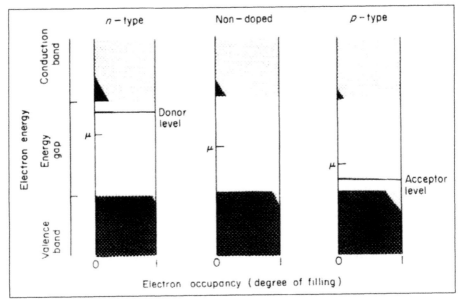

Figure 14.3. Energy band structure near the Fermi energy (μ) for a semiconductor material without impurities (middle column) or with n- or p-type doping (cf. text). The dark shading indicates the electron occupancy as a function of energy for a finite temperature (occupancy equal to unity corresponds to the maximum number of electrons at a given energy or energy interval, which is consistent with the exclusion principle for Fermi-type particles).

The bands corresponding to 1s, 2s and 2p atomic orbits are completely filled in Si, but then four electrons remain per atom. According to Fig. 14.2, the 3s and 3p bands are overlapping and allow two mixed bands to be constructed, with an intermediate gap of about 1 eV (1.6×10^{-5} J). The lower band (the "valence band") can hold four electrons per atom, so that this band will be completely full and the other one empty at zero temperature.

If a few of the atoms in a silicon lattice are replaced by an atom with higher Z (atomic number), e.g. phosphorus, the additional electrons associated with the impurity cannot be accommodated in the valence band, but will occupy a discrete level (named the "donor level") just below the conduction band (the energy depression being due to the larger attractive force from the atom of higher Z). A semiconductor material with this type of impurity is called n-type. The electrons in the donor-level are very easily excited into the conduction band. Adding n-type impurities makes the Fermi level [μ_i in (4.1)] move upwards from the initial position approximately half way between the valence and conduction bands.

Impurities with lower Z (e.g. Al in a Si-lattice) lead to electron vacancies or "holes", which are associated with an energy slightly higher than the top of the valence band, again due to the Z-dependence of the attractive force. These "acceptor levels" easily transfer holes to the valence band, a process that may, of course, alternatively be described as the excitation of electrons from the valence band into the acceptor level. Semiconductor material with this type of impurity is called p-type. The holes can formally be treated as particles like the electrons, but with positive charges and provided that the state of the pure semiconductor at zero temperature is taken as reference ("vacuum state" in quantum theory).

The energy diagrams of "doped" semiconductors of n- and p-types are sketched in Fig. 14.3. For a more complete account of basic semiconductor theory, see, for example, Shockley (1950).

14.2 Photovoltaic conversion

Conversion of radiant energy (light quanta) into electrical energy can be achieved with the use of semiconductor materials, for which the electron excitation caused by impinging light quanta has a strongly enhancing effect on the conductivity.

It is not sufficient, however, that electrons are excited and are able to move more freely, if there is no force to make them move. Such a force would arise from the presence of a gradient of electrical potential, such as the one found in a p–n junction of doped semiconductor materials (a p–n junction is a junction of a p-type and an n-type semiconductor, as further described below). A p–n junction provides an electrical field that will cause the electrons excited by radiation (such as solar) to move in the direction from the p-type to the n-type material and cause the vacancies (holes) left by the excited electrons to move in the opposite direction. If the electrons and holes reach the respective edges of the semiconductor material, the device is capable of delivering electrical power to an external circuit. The motion of electrons or holes receives competition from recombination processes (electrons being recaptured into vacancies), giving importance to such factors as overall dimensions and electron mobility in the material used.

The p–n junction

An essential constituent of photovoltaic cells is the p–n junction. A refresher on the semiconductor physics needed for understanding the p–n junction is given in Section 4.A at the end of this chapter. When a p-type and an n-type semiconductor are joined so that they acquire a common surface, they are said to form a p–n junction. This will initially cause electrons to flow in the n to p direction because, as seen in Fig. 14.3 of Section 14.1, the electron density in the conduction band is higher in n-type than in p-type material and because the hole density in the valence band is higher in the p-type than in the n-type material (the electron flow in the valence band can also be described as a flow of positive holes in the direction p to n).

This electron flow builds up a surplus of positive charge in the n-type material and a surplus of negative charge in the p-type material, in the neighbourhood of the junction (mainly restricted to distances from the junction of the order of the mean travelling distance before recombination of an electron or a hole in the respective materials). These surplus charges form a dipole layer, associated with which is an electrostatic potential difference, which will tend to hinder any further unidirectional electron flow. Finally, an equilibrium is reached in which the potential difference is such that no net transfer of electrons takes place.

Another way of stating the equilibrium condition is in terms of the Fermi energy [cf. Fig. 14.3 and (4.1)]. Originally, the Fermi energies of the p- and n-type materials, μ_p and μ_n, are different, but at equilibrium $\mu_p = \mu_n$. This is illustrated in Fig. 14.4, and it is seen that the change in the relative positions of the conduction (or valence) bands in the two types of material must equal the electron charge, $- e$, times the equilibrium electrostatic potential.

The number of electrons in the conduction band may be written

$$n_c = \int_{E_c}^{E'_c} n'(E) f(E)\, dE, \tag{14.2}$$

where E_c and E_c' are the lower and upper energy limit of the conduction band, $n'(E)$ is the number of quantum states per unit energy interval (and, for example, per unit volume of material, if the electron number per unit volume is desired), and finally, $f(E)$ is the Fermi–Dirac distribution (4.1). If the electrons in the conduction band are regarded as free, elementary quantum mechanics gives (see e.g. Shockley, 1950)

$$n'(E) = 4\pi h^{-3} (2m)^{3/2} E^{1/2}, \tag{14.3}$$

where h is Planck's constant and m is the electron mass. The corrections for electrons moving in condensed matter, rather than being free, may to a first approximation be included by replacing the electron mass by an "effective" value.

If the Fermi energy is not close to the conduction band,

$$E_c - \mu >> kT,$$

the Fermi–Dirac distribution (4.1) may be replaced by the Boltzmann distribution,

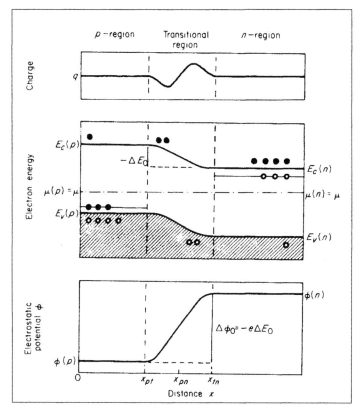

Figure 14.4. Schematic picture of the properties of a p–n junction in an equilibrium condition. The x-direction is perpendicular to the junction (all properties are assumed to be homogeneous in the y- and z-directions). The charge (top) is the sum of electron charges in the conduction band and positive hole charges in the valence band, plus charge excess or defect associated with the acceptor and donor levels. In the electron energy diagram (middle), the abundance of minority charge carriers (closed circles for electrons, open circles for holes) is schematically illustrated. The properties are further discussed in the text.

$$f_B(E) = \exp(-(E - \mu)/kT).\tag{14.4}$$

Evaluating the integral, (14.2) then gives an expression of the form

$$n_c = N_c \exp(-(E_c - \mu)/kT).\tag{14.5}$$

The number of holes in the valence band is found in an analogous way,

$$n_v = N_v \exp(-(\mu - E_v)/kT),\tag{14.6}$$

where E_v is the upper limit energy of the valence band.

The equilibrium currents in a p–n junction such as the one illustrated in Fig. 14.4 can now be calculated. Considering first the electron currents in the conduction band, the electrons thermally excited into the conduction band in the p-region can freely flow into the n-type materials. The corresponding current, $I_0^-(p)$, may be considered proportional to the number of electrons in the conduction band in the p-region, $n_c(p)$, given by (14.5),

$$I_0^-(p) = \alpha N_c \exp\left(- (E_c(p) - \mu(p)) / kT\right), \tag{14.7}$$

where the constant α depends on electron mobility in the material and on the electrostatic potential gradient, grad ϕ. The electrons excited into the conduction band in the n-type region will have to climb the potential barrier in order to move into the p-region. The fraction of electrons capable of doing this is given by a Boltzmann factor of the form (14.5), but with the additional energy barrier $\Delta E_0 = - \Delta \phi_0 / e$ ($-e$ being the electron charge),

$$n_c(n) = N_c \exp(- (E_c(n) - \mu(n) - \Delta E_0) / kT).$$

Using $-\Delta E_0 = E_c(p) - E_c(n)$ (cf. Fig. 14.4) and considering the current $I_0^-(n)$ as being proportional to $n_c(n)$, the corresponding current may be written

$$I_0^-(n) = \alpha' N_c \exp\left(- (E_c(n) - \mu(n)) / kT\right), \tag{14.8}$$

where α' depends on the diffusion parameter and on the relative change in electron density, $n_c^{-1}\mathrm{grad}(n_c)$, considering the electron motion against the electrostatic potential as a diffusion process. The statistical mechanical condition for thermal equilibrium demands that $\alpha = -\alpha'$ (Einstein, 1905), so (14.7) and (14.8) show that the net electron current,

$$I_0^- = I_0^-(p) + I_0^-(n),$$

becomes zero precisely when

$$\mu(p) = \mu(n),$$

which is then the condition for thermal equilibrium. The same is true for the hole current,

$$I_0^+ = I_0^+(p) + I_0^+(n).$$

If an external voltage source is applied to the p–n junction in such a way that the n-type terminal receives an additional electrostatic potential $\Delta\phi_{ext}$ relative to the p-type terminal, then the junction is no longer in thermal equilibrium, and the Fermi energy in the p-region is no longer equal to that of the n-region, but satisfies

$$\mu(p) - \mu(n) = e^{-1}\Delta\phi_{ext} = \Delta E_{ext} \tag{14.9}$$

if the Boltzmann distributions of electrons and of holes are to maintain their shapes in both p- and n-regions. Similarly $E_c(p) - E_c(n) = - (\Delta E_0 + \Delta E_{ext})$, and assuming that the proportionality factors in (14.7) and (14.8) still bear the relationship $\alpha = -\alpha'$ in the presence of the external potential, the currents are connected by the expression

$$I^-(n) = -I^-(p) \exp(\Delta E_{ext} / kT).$$

The net electron current in the conduction band then becomes

$$I^- = I^-(n) + I^-(p) = -I^-(p) (\exp(\Delta E_{ext} / kT) - 1). \qquad (14.10)$$

For a positive $\Delta\phi_{ext}$, the potential barrier that electrons in the n-region conduction band (see Fig. 14.4) have to climb increases and the current $I^-(n)$ decreases exponentially (ΔE_{ext} negative, "reverse bias"). In this case, the net current I^- approaches a saturation value equal to $I^-(p)$, according to (14.10).

For negative $\Delta\phi_{ext}$, (positive ΔE_{ext}, "forward bias"), the current $I^-(n)$ increases exponentially with the external potential. In both cases $I^-(p)$ is assumed to remain practically unchanged, when the external potential of one or the other sign is applied, considering that $I^-(p)$ is primarily limited by the number of electrons excited into the conduction band in the p-type material, a number that is assumed to be small in comparison with the conduction band electrons in the n-type material (cf. Figs. 14.4 and 14.3).

The contributions to the hole current, I^+, behave similarly to those of the electron current, and the total current I across a p–n junction with an external potential $\Delta\phi_{ext} = -e\,\Delta E_{ext}$ may be written

$$I = I^- + I^+ = -I(p) (\exp(\Delta E_{ext} / kT) - 1). \qquad (14.11)$$

The relationship between current and potential is called the "characteristic" of the device, and the relation (14.11) for the p–n junction is illustrated in Fig. 14.5 by the curve labelled "no light". The constant saturation current $I(p)$ is sometimes referred to as the "dark current".

Solar cells

A p–n junction may be utilised to convert solar radiation energy into electric power. A solar cell is formed by shaping the junction in such a way that, for example, the p-type material can be reached by incident solar radiation, e.g. by placing a thin layer of p-type material on top of a piece of n-type semiconductor. In the dark and with no external voltage, the net current across the junction is zero, as was shown in the previous subsection, i.e. the intrinsic potential difference $\Delta\phi_0$ is unable to perform external work.

However, when irradiated with light quanta of an energy $E_{light} = h\nu = hc/\lambda$ (h is Planck's constant, c is the velocity of light and ν and λ are the frequency and wavelength of radiation), which is larger than the energy difference between the conduction and valence band for the p-type material,

$$E_{light} \geq E_c(p) - E_v(p),$$

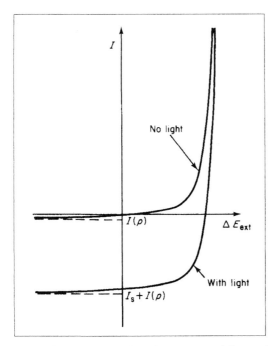

Figure 14.5. Characteristics (i.e. current as a function of external voltage) of a p–n junction, in the dark and with applied light. The magnitude of the short-circuit current, I_s, is a function of light intensity and spectral distribution.

then electrons may be photo-excited from the valence band into the conduction band. The absorption of light quanta produces as many holes in the valence band of the p-type material as electrons in the conduction band. Since in the dark there are many fewer electrons in the p-type conduction band than holes in the valence band, a dramatic increase in the number of conduction band electrons can take place without significantly altering the number of holes in the valence band. If the excess electrons are sufficiently close to the junction to be able to reach it by diffusion before recombining with a hole, then the current in this direction exceeds $I_0'(p)$ of (14.8) by an amount I_s, which is the net current through the junction in case of a short-circuited external connection from the n-type to the p-type material. The photo-induced current is not altered if there is a finite potential drop in the external circuit, since the relation between the current (14.11) and the external potential drop $e\,\Delta E_{ext}$ was derived with reference only to the changes in the n-region.

An alternative n–p type of solar cell may consist of a thin n-type layer exposed to solar radiation on top of a p-type base. In this case, the excess holes in the n-type valence band produce the photo-induced current I_s.

The total current in the case of light being absorbed in the p-type material and with an external potential drop is then

$$I = I_s - I(p)\,(\exp(-\Delta\,\phi_{ext}/kT) - 1). \tag{14.12}$$

The short-circuit current I_s depends on the amount of incident light with frequencies sufficient to excite electrons into the conduction band, on the fraction of this light actually being absorbed, and on the conditions for transporting the excess electrons created in the conduction band, in competition with electron–hole recombination processes. I_s may be written as the sum of a conduction and a diffusion-type current, both related to the number of excess electrons in the conduction band, n_c^{ind}, induced by the absorption of light,

$$I_s = e(m_c E_e n_c^{ind} + k_c \, dn_c^{ind}/dx), \tag{14.13}$$

where e is the numerical value of the electron charge (1.6×10^{-19} C), m_c is the mobility of conduction band electrons [e.g. 0.12 $m^2 V^{-1} s^{-1}$ for silicon (Loferski, 1956), the dependence on the degree of doping being displayed in Fig. 14.6], E_e is the local electrical field, k_c is the diffusion constant, for which a typical value is $k_c = 10^{-3}$ $m^2 s^{-1}$ (Loferski, 1956)], and x is the depth below the solar cell surface, assumed to be the only significant co-ordinate (as in Fig. 14.4).

The excess electron number density, $n_c^{ind}(x)$, at a depth x, changes when additional electrons are photo-excited, when electrons are carried away from x by the current I_s, and when electrons recombine with holes,

$$\frac{\partial n_c^{ind}(x)}{\partial t} = \int \sigma(\nu) n_{ph}(\nu) \exp(-\sigma(\nu)x) \, d\nu + \frac{I}{e} \frac{\partial I_s}{\partial x} - n_c^{ind}(x) \frac{1}{\tau_c}. \tag{14.14}$$

Here $\sigma(\nu)$ is the cross section for absorption of light quanta ("photons") in the p-type material, and $n_{ph}(\nu)$ is the number of photons at the cell surface ($x = 0$) per unit time and unit interval of frequency ν. The absorption cross section is zero for photon energies below the semiconductor energy gap, $h\nu < E_c(p) - E_v(p)$, i.e. the material is transparent to such light. The most energetic light quanta in visible light could theoretically excite more than one electron per photon (e.g. 2–3 in Si with an energy gap slightly over 1 eV), but the probability for exciting just one electron to a higher energy is higher, and such a process is usually followed by a transfer of energy to other degrees of freedom (e.g. lattice vibrations and ultimately heat), as the excited electron approaches the lower part of the conduction band, or as the hole left by the electron de-excites from a deep level to the upper valence band. Thus, in practice the quantum efficiency (number of electron–hole pairs per photon) hardly exceeds one.

The last parameter introduced in (14.14), τ_c, is the average lifetime of an electron excited into the conduction band, before recombination [τ_c may lie in the interval 10^{-11} to 10^{-7}, with 10^{-9} being a typical value (Wolf, 1963)]. The lifetime τ_c is connected to the cross section for recombination, σ_c, and to the mean free path l_c of electrons in the conduction band by

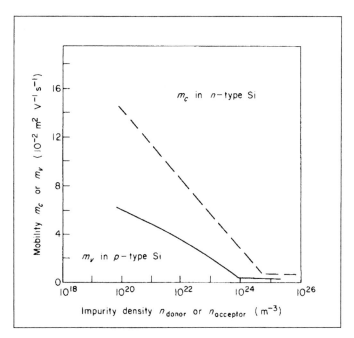

Figure 14.6. Mobility of minority carriers in Si at room temperature (about 300 K), extrapolated from measurements (Wolf, 1963). The mobility plotted is the "conduction mobility", equal to the conductivity divided by the number of minority carriers and by the electron charge. The attenuation of the flow of actual carriers by recombination effects (trapping) is not considered.

$$l_c = \sigma_c^{-1} = v_c \tau_c N_a,$$

where v_c is the average thermal velocity of the electrons, $v_c = (2kT/m)^{1/2}$ (*m* being the electron mass, *k* being Boltzmann's constant and *T* being the absolute temperature) and N_a is the number of recombination centres ("acceptor impurities"; cf. Fig. 14.3).

The boundary conditions for solving (14.14) may be taken as the absence of excess minority carriers (electrons or holes) at the junction $x = x_{pn}$,

$$n_c^{ind}(x_{pn}) = 0,$$

and a prescribed (material-dependent) excess electron gradient at the surface $x = 0$. This gradient, $(dn_c^{ind}/dx)\,|_{x=0}$, is often expressed in terms of a surface recombination velocity, s_c, through (14.13) by writing the left-hand side

$$I_s = s_c n_c^{ind}(0).$$

Typical values of s_c are of the order of 10^3 m s^{-1} (Wolf, 1963, 1971).

For n–p type solar cells, expressions analogous to the above can be used.

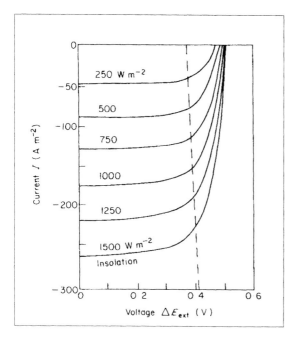

Figure 14.7. Characteristics of Cu$_2$S–CdS solar cell at 300 K for different intensities of incident radiation with typical clear sky solar frequency distribution. The points of intersection with the dashed line give the voltage and current leading to maximum power output for given solar radiation (based on Shirland, 1966).

Once $n_c^{ind}(x)$ has been found, I_s can be calculated. Figure 14.5 shows an example of the total current through a p–n junction, as a function of applied voltage but for a fixed rate of incoming solar radiation on the p-type surface. The short-circuit current I_s increases linearly with intensity of light, if the spectral composition is kept constant, for the entire interval of intensities relevant for applications of solar cells at or near Earth. This is illustrated in Fig. 14.7 for a solar cell based on a p–n heterojunction, with the p-type material being Cu$_2$S and the n-type material CdS.

For an open-ended solar cell (i.e. no external circuit), the difference in electrical potential between the terminals, $V_{oc} = \Delta\phi_{ext}(I=0)$, is obtained by putting I equal to zero in (14.12),

$$V_{oc} = kTe^{-1} \left(\log(I_s/I(p)) + 1\right). \tag{14.15}$$

The amount of electrical power, E, delivered by the irradiated cell to the external circuit is obtained by multiplication of (14.12) by the external voltage,

$$E = (\Delta\phi_{ext})I = \Delta\phi_{ext} \left(I_s - I(p)(\exp(-e\Delta\phi_{ext}/kT) - 1)\right). \tag{14.16}$$

From $\partial E/\partial (\Delta\phi_{ext}) = 0$, the external voltage V_{opt} may be found, which leads to the maximum value of power, E_{max}. In the situations of interest, V_{opt} is a slowly varying function of the amount of incident radiation, as illustrated by Fig. 14.7. The corresponding current may be denoted I_{opt}.

The efficiency of solar cell radiant-to-electrical energy conversion is the ratio of the power E delivered and the incident energy, denoted E_+^{sw} (eventually for a tilted orientation of the solar cell), $\eta = E/E_+^{sw}$. In terms of the flux of photons of given frequency incident on the solar cell [introduced in (14.14)], the non-reflected energy flux at the surface may be written [a is the albedo (fraction of incident radiation reflected) of the cell surface]

$$E_+^{sw}(1-a) = \int_0^\infty h\nu n_{ph}(\nu)\,d\nu, \tag{14.17}$$

where h is Planck's constant. For a given semiconductor material, the maximum fraction of the energy (14.17), which can be absorbed, is

$$\int_{h\nu=E_c(p)-E_v(p)}^\infty h\nu n_{ph}(\nu)\,d\nu.$$

The part of the integral from zero up to the energy gap (i.e. the part above a certain wavelength of light) constitutes a fundamental loss. The same can be said of the energy of each light quantum in excess of the semiconductor energy gap $E_c(p) - E_v(p)$, assuming a quantum efficiency of at most one, i.e. that all such quanta are indeed absorbed (which may not be true if their energy is, say, between the upper limit of the conduction band and the lower limit of the following band) and that all excess energy is spent in exciting lattice degrees of freedom (vibrational phonons) that do not contribute to the photovoltaic process. In that case the energy flux available for photoconversion is only

$$(E_c(p) - E_v(p)) \int_{h\nu=E_c(p)-E_v(p)}^\infty n_{ph}(\nu)\,d\nu = E^{avail}. \tag{14.18}$$

Further losses in addition to reflection and insufficient or excess photon energy may be associated with imperfections in the junction materials or in the current extraction system, causing heat formation or light re-emission rather than electrical power creation. Both heat creation (in the lattice) and re-radiation may take place in connection with the recombination of photo-excited electrons and holes. Since many of these processes are highly temperature dependent, the maximum efficiency that can be obtained in practice is also temperature dependent. Examples of maximum theoretical efficiencies, as well as those obtained in practice, will be given below.

Rather than being p- and n-doped materials of the same elemental semiconductor, the solar cell junction may be based on different materials ("heterojunction") or on a metal and a semiconductor ("Schottky junction").

The discussion of individual types of solar cells will be taken up below, after presenting a few more general energy conversion methods.

Design of photovoltaic converters

A photovoltaic converter consists of a number of solar cells connected in a suitable way, plus eventually some auxiliary equipment such as focusing devices in front of the cells and tracking systems. The maximum efficiency of a solar cell is given by the ratio of the maximum power output (14.16) and the incident radiation flux,

$$\max(\eta) = \max(E) / E_{s,\gamma}^{sw}. \tag{14.19}$$

It is smaller than unity for a number of reasons. First, as discussed in the beginning of the chapter, radiation of frequency below the semiconductor band gap is not absorbed. Second, according to (14.18), the excess energy of radiation with frequencies above the semiconductor band gap is not available to the photovoltaic conversion process. This loss would be small if the solar spectrum was peaked across the band gap, but most semiconductor gaps only correspond to a limited part of the broad solar spectrum.

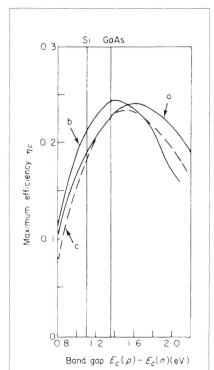

Figure 14.8. Early calculation of maximum efficiency for simple p-on-n solar cell: (a) outside the Earth's atmosphere (E^{sw} = 1350 W m^{-2}); (b) at the Earth's surface under standard conditions (E^{sw} = 890 W m^{-2}, air mass one, water content 0.02 m^3 m^{-2} and major absorption bands included in the calculation); (c) overcast condition (E^{sw} = 120 W m^{-2}) (based on Loferski, 1956).

Third, as seen e.g. in Fig. 14.7, the maximum power output is less than the maximum current times the maximum voltage. This reduction in voltage, necessary in order to get a finite current, is analogous to the necessity of a finite dissipation term in a thermodynamic engine in order to get energy out in a finite time (Chapter 2). The expression (14.16) does not fit measured power values in detail, and it has been suggested that a second exponential term be added in the current–voltage relation (14.12), of similar form but with $\Delta\phi_{ext}$ replaced by $\frac{1}{2}\Delta\phi_{ext}$ (Sah *et al.*, 1957). The origin of such a term is thermal generation and recombination of carriers in the vicinity of the junction. Fourth, the external potential (times the electron charge) has a maximum value (14.15), which is smaller than the semiconductor gap $E_c(p)$ – $E_c(n)$, since it equals the difference in Fermi level between the *p*- and *n*-regions [cf. (14.9)]. This loss may be diminished by increasing the number of impurities (and hence the number of minority carriers) in both the *p*- and the *n*-regions. However, if these are increased above a certain level, the increased recombination probability offsets the voltage gain.

Figure 14.8 shows an early example of calculated maximum efficiency as a function of the semiconductor band gap, including the above-mentioned losses, for radiation conditions corresponding to the top of the Earth's atmosphere (a), for a clear sky day at the Earth's surface with average atmospheric absorption and scattering conditions (b), and for a situation with cloud-covered sky (c). It is shown that for common solar cell materials such as Si or GaAs, the maximum efficiency is larger for the spectral composition of clear-day solar radiation at ground level than for the solar spectrum not disturbed by the Earth's atmosphere. The performance for scattered solar radiation (overcast case) is not substantially impaired.

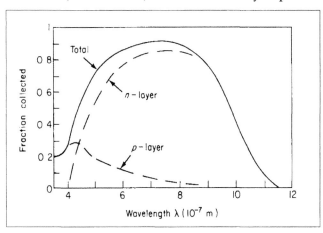

Figure 14.9. Spectral collection efficiency for a simple *p*-on-*n* solar cell. The curve labelled "total" is based on measurements, and the individual contributions from the *p*-layer (thickness 5×10^{-7} m) and the *n*-layer have been calculated. The total thickness is about 4.5×10^{-4} m (based on Wolf, 1963).

The wavelength dependence of the collection efficiency (i.e. the efficiency including the first two loss terms discussed above, but not the last two), is shown in Fig. 14.9 for a simple silicon cell consisting of a thin (5×10^{-7} m) *p*-layer on top of an *n*-layer base (of thickness 4.5×10^{-4} m). The total efficiency curve is based on measurements, whereas the separate contributions from the *p*- and *n*-layers have been calculated (Wolf, 1963). Short wavelengths are absorbed in the p-layer and give rise to a current of electrons directed towards the junction, but the bulk of the solar wavelengths are not absorbed until the photons reach the n-layer base. They give rise to a hole current towards the junction. The efficiency is not independent of cell temperature, as indicated in Fig. 14.10. The currently dominating silicon material exhibits an absolute loss in efficiency of 0.4–0.5% for each °C of temperature rise. Figure 14.11 adds more recent data on the temperature dependence of different types of solar cells, based upon measurements.

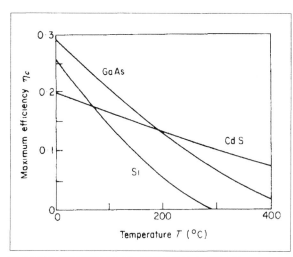

Figure 14.10 (left). Calculated temperature dependence of solar cell efficiency (based on Wysocki and Rappaport, 1960).

Figure 14.11 (below). Solar cell efficiency as function of operating temperature, normalised to a typical 25°C efficiency for each cell type. Based on Sørensen (2000b), based on Ricaud (1999), Dutta *et al.* (1992), Wysocky and Rappaport (1960), Yamamoto *et al.* (1999) and Rijnberg *et al.* (1998).

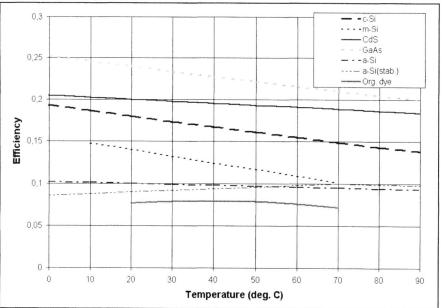

The data shown in Fig. 14.11 has, for each type of solar cell, been normalised to a typical absolute efficiency for current commercial or near-commercial versions of the type of device in question. The early theoretical calculations (Wysocki and Rappaport, 1960) are largely confirmed by current measurements, and the mechanisms are thus well understood, at least for conventional photovoltaic devices. The temperature dependence is chiefly due to band-gap effects, which explains why the slope of the crystalline silicon (c-Si) and multicrystalline silicon (m-Si) are identical (Yamamoto *et al.*, 1999). In other words, the grain boundaries do not give rise to additional temperature effects. Cd-S cells have a lower but still significant temperature gradient, whereas the temperature effect for amorphous silicon cells and organic dye-sensitised TiO_2 cells is very small.

The temperature effect is negative with increasing working temperature for all devices except two: the organic cells show a maximum near 40°C (Rijnberg *et al.*, 1998) and the amorphous silicon-hydrogen cells (a-Si) show a reversal of temperature trends after annealing (Dutta *et al.*, 1992). This positive temperature coefficient only persists until the un-degraded efficiency is reached, and it requires annealing as opposed to light soaking treatment, which causes the development of a stronger negative temperature coefficient. The modest temperature dependence is conveniently modelled by a power expansion of the efficiency,

$$\eta = \eta(298K) + a\,(T - 298K) + b\,(T - 298K)^2. \tag{14.20}$$

The operating temperature dependence of the solar energy to electricity conversion efficiency suggests that cooling the cell by extracting heat may improve the electric performance of the cell and thereby pay for some of the extra expense of the heat extraction equipment. Typical operating temperatures for un-cooled cells are about 50°C. Figure 14.11 shows that improvement can be obtained for crystalline or multicrystalline silicon photovoltaic (PV) cells, but not notably for dye-sensitised cells or amorphous PV cells. On the other hand, in order to make use of the heat it should preferably be collected at higher temperatures, which would indicate that the best solutions are those with little operating temperature effect of the electricity yields.

In cells currently produced, each loss factor is carefully minimised, and the resulting efficiencies have increased over time, for all types of solar cells, as discussed below.

Monocrystalline silicon cells

The photovoltaic cell principles described above form the basis for monocrystalline cells, which are cells constructed from single crystals, usually in the form of ingots sliced into a number of cells.

A number of improvements have brought the cell efficiency of state-of-the-art monocrystalline silicon cells up to about 25%. The light capture is improved through trapping structures that minimise reflection in directions not benefiting the collection area and by backside designs reflecting light rays back into the active areas (see e.g. Fig. 14.12). The doping degree is altered near electrodes (n^+ and p^+ areas), and a thin oxide layer further helps to prevent electrons reaching the surface rather than the electrode (this process being termed "passivation"). Further, top electrodes may be buried in order not to produce shadowing effects for the incoming light (Green, 1992).

Figure 14.13 shows the measured characteristics, i.e. current as a function of voltage, for a cell of the type shown in Fig. 14.12.

Simulation of the light trapping and electron transport processes in one, two or three dimensions has helped in selecting the best geometry and degree of doping (Basore, 1991; Müller *et al.*, 1992). Figure 14.14 gives an example of one-dimensional simulation of the variation in cell performance as a function of doping degree for a cell of the kind shown in Fig. 14.12.

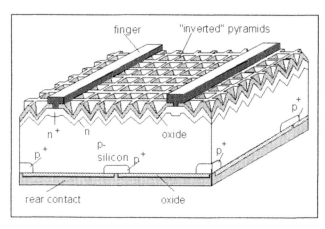

Figure 14.12. Structure of a monocrystalline silicon cell with passivated emitter and a locally diffused rear structure ("PERL"), used to obtain a 23% module efficiency (from Green *et al.*, 1998, used with permission).

Figure 14.13. Current-voltage curve for PERL cell similar to that of Fig. 14.12 (but with cell efficiency 22%), as measured at the Sandia Laboratories (USA) at 1006 W m^{-2} airmass 1.5 simulated radiation, for a 4 cm^2 cell. Key findings are V_{oc} = 696 mV (14-16), I_s = 160 mA (4.64) and a fill factor of 0.79 (Wenham *et al.*, 1995). The fill factor is the ratio between the area under the curve and $I_s V_{oc}$.

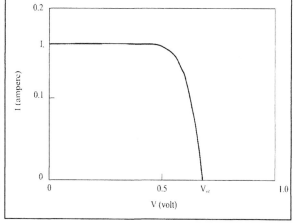

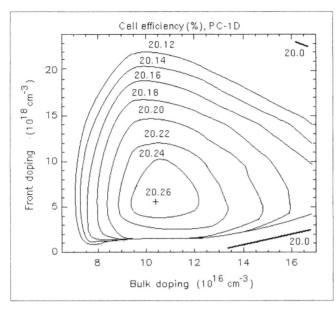

Figure 14.14. Calculated efficiency as a function of doping parameters for a simplified silicon cell of the type depicted in Fig. 14.12. The one-dimensional finite–element model used is described in the text (Sørensen, 1994).

The model takes into account the 10 μm facet depth and uses a curved one-dimensional path through the cell. The two most important doping parameters (impurities per unit volume) are the uniform bulk doping of the p-material and the n-doping at the front, assumed to fall off as an error function, thereby simulating the average behaviour of near-electrode sites and sites away from electrodes. Transport and recombination are calculated in a finite-element model (Basore, 1991). The backside doping is kept at 2×10^{19} cm^{-3}.

Multicrystalline cells

Another technology for producing solar cells uses multicrystalline (sometimes referred to as "polycrystalline") materials, instead of the single-crystal materials. Multicrystalline materials consist of small domains or grains of crystalline material, randomly oriented relative to each other. The crystal grains in multicrystalline materials sustain conductivity in the same way as single crystals do, but the transport of electrons across grain boundaries induces losses, reduces conductivity and thus makes the cells less efficient. On the other hand, they can be produced by simpler methods than those needed for monocrystals, e.g. by evaporating suitable coatings onto a substrate. This field is in rapid development, as it is becoming possible to deposit only a few atomic layers onto a substrate. With suitable techniques (such as using magnetic fields to align grains) it may soon be possible in this way to form near-monocrystalline layers without having to grow crystals.

It was initially believed that the additional losses at grain boundaries would necessarily make the efficiency of multicrystalline cells substantially lower than what could be obtained by crystalline materials. Actually, the difference has narrowed as a result of better understanding of the options for optimising performance of complex cell structures. One problem has been the damage inflicted upon multicrystalline cells by attempting to copy to them some of the efficiency-improving techniques that have worked well for monocrystalline cells (surface texturing, rear passivation by oxide layers). Yet, etching of inverted pyramids on the surface of multicrystalline cells has improved efficiency considerably (Stock *et al.*, 1996), and recently, less damaging honeycomb texture patterns have brought the efficiency up to 20% (Zhao *et al.*, 1998). This is a trend likely to induce the long-predicted change from expensive ingot-grown monocrystalline cell materials to deposition techniques for multicrystalline materials on suitable backings, much more suited for mass production and price reduction efforts. However, the development away from single-crystalline solar cell materials is slower than anticipated because of the higher maturity of the crystalline industry processes. Figure 14.15 shows the structure of the 20% efficient multicrystalline cell, obtaining over 90% absorption of incoming radiation. The advantages of thin-film multicrystalline solar cells over monocrystalline ones would on basic principles seem to more than compensate for the remaining 5% efficiency difference. This does not exclude that crystalline and multicrystalline technologies will continue to co-exist in the marketplace for a while.

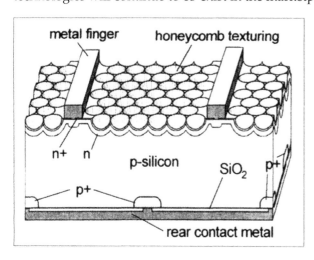

Figure 14.15. Texturing used to raise multicrystalline cell efficiency to 20% in a PERL-type cell (from Zhao *et al.*, 1998, used with permission).

Stacked cells

Instead of basing a solar cell on just a single p–n junction, it is possible to stack several identical or different cells on top of each other. The use of different cell materials aims at capturing a wider range of frequencies than is possible with a single junction. In this case materials of different band gaps will be stacked (Yazawa *et al.*, 1996; Takamoto *et al.*, 1997). However, it is not clear if the extra cost can be justified by the often modest efficiency improvement obtained. Stacking identical cells has been tried, with the aim of being able to use lower quality material (e.g. thinly sprayed crystalline cells in contrast to ingot-grown ones) and still get an acceptable overall efficiency by stacking several layers of low individual efficiency (Green, 1994). The concept has not worked out satisfactorily, e.g. due to difficulties in handling losses at the interfaces of the low-quality material sheets.

Amorphous cells

Amorphous semiconductor materials exhibit properties of interest for solar cell applications. While elemental amorphous silicon has a fairly uniform energy distribution of electron levels, composite materials have been constructed that exhibit a pronounced energy gap, i.e. an interval of energy essentially without any states, as in a crystal. Spear and Le Comber (1975) first produced such an amorphous material, which was later proved to be a silicon–hydrogen alloy, with the distribution of energy states shown in Fig. 14.16 as curve a. Ovshinsky (1978) produced a silicon–fluorine–hydrogen alloy with further depression of gap states (Fig. 14.16, curve b). The gap is about 1.6 eV wide and thus should be more favourable with respect to the solar energy spectrum than the 1.1-eV gap of crystalline silicon. Furthermore, doping (introduction of boron or phosphorus atoms) has proved possible, so that *p*- and *n*-type amorphous semiconductors can readily be made, and a certain amount of "engineering" of materials with exactly the desired properties with regard to gap structure, doping efficiency, conductivity, temperature sensitivity and structural stability (lifetime) can be performed.

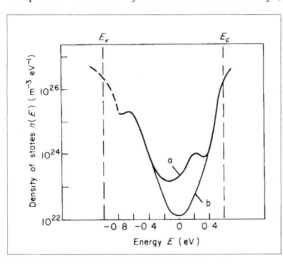

Figure 14.16. Density of electron states as a function of excitation energy for silicon-based amorphous materials: (a) silicon–hydrogen alloy; (b) silicium–fluorine–hydrogen alloy (based on Spear and Le Comber, 1975; Ovshinsky, 1978).

14. PHOTOVOLTAIC CONVERSION

A theoretical description of the gap occurrence and electron conduction in amorphous materials was first presented by Street and Mott (1975) and followed up by Fritsche (1977) and Pfister and Scher (1977). The basis is the occurrence in amorphous material of defects of particular affinity to attract or reject electrons (e.g. lone pair sites), and the transport of electrons is thought of as a quasi-random "hopping" between sites, some of which are capable of "trapping" an electron for a shorter or longer period of time. Abundance of broken or "dangling" bonds may give rise to serious trapping problems (low conductivity), and the success obtained by incorporating hydrogen seems due to its occupying and neutralising such sites.

Not very long after the theoretical description of amorphous solar cells, Japanese scientists succeeded in creating designs of such cells that were suited for industrial production (Hamakawa *et al.*, 1981) and soon found a market in powering calculators and similar small-scale devices, where the cell cost was relatively unimportant. A typical structure of a commercial amorphous cell is illustrated in Fig. 14.17. Band gaps in the range from 1.0 to 3.6 eV can be engineered with different silicon alloys (SiGe, Si, SiC), and such cells may be stacked to obtain a broader frequency acceptance (Ichikawa, 1993; Hamakawa, 1998).

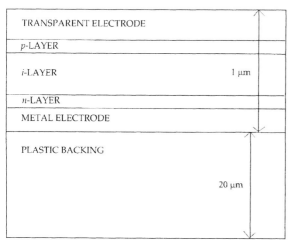

Figure 14.17. Structure of the most common a-Si cell. Several *pin*-layers may be stacked.

However, the simplest version has just one type of material: an intrinsic layer of an a-Si:H compound is the main area of light absorption, and adjacent *p*- and *n*-type layers ensure the transport to the electrodes, of which the front one is made of a transparent material. The whole structure is less than 1 μm thick and is deposited onto a plastic backing material. Maximum efficiencies of around 13% have been demonstrated (Fig. 14.18), but one problem has persisted: because the structure of the material is without order, bombardment with light quanta may push atoms around, and the material degrades with time. Current practice is to degrade commercial cells before they leave the factory, thereby decreasing the efficiency by some 20%, but in return obtaining reasonable stability over a 10-year period under average solar radiation conditions (Sakai, 1993). Several layers of p-, i- and n-layers may be stacked, and the highest efficiency is obtained by replacing the amorphous n-layers by a multicrystalline pure Si or silicon compound layer (Ichikawa, 1993; Ma et al., 1995).

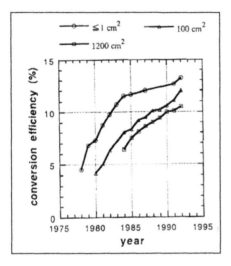

Figure 14.18. Development in a-Si solar cell efficiency for different cell area (before degradation; Sakai, 1993).

Other materials and other thin-film cells

Use of materials from the chemical groups III and V, such as GaAs, CdS and CdTe, instead of silicon allows better engineering of band gaps in crystalline solar cells to suit particular purposes and brings forward new properties suitable for certain tasks (notably space applications and use in concentrating collectors). Important considerations in selecting materials include temperature dependence where crystalline silicon cell efficiency drops rather fast with the increasing temperature likely to prevail (despite possible active cooling) for cells operating at high levels of solar radiation (see Figs. 14.10 and 14.11).

The GaAs band gap of 1.43 eV is well suited for the solar spectrum, and with a tandem cell of GaAs plus $GaInP_2$ an efficiency of over 30% has been reached (Bertness et al., 1994; Deb, 1998). At the moment these cells are expensive and are mainly used in space. However, thin-film versions may be developed as in the Si case, as they already have for CIS cells (copper-indium-diselenide). The highest efficiency obtained so far is about 17% for a Cu (In,Ga) Se_2 structure (Tuttle et al., 1996).

Among a range of non-traditional designs of solar cell is the use of spherical droplets. This idea grew out of an idea of reusing scrap material from the microelectronics industry, but it also may increase acceptance of light from light from different directions and reduce reflection that otherwise would have to be dealt with by surface texturing (Maag, 1993; Drewes, 2003; ATS, 2003).

Module construction

Individual cells based on monocrystalline materials typically have areas of 100 cm^2 (limited by techniques for growing ingots of monocrystalline material). Multicrystalline and photoelectrochemical cells, where semiconductor material is deposited on a backing template (usually glass), may have larger cell size, and for amorphous cells, there are essentially no limitations. Amorphous cells have been produced on rolls of flexible plastic backing materials, with widths of 1–2 m and rolls of any length. The same may also become possible for other thin-film types, such as spray-deposited multicrystalline materials.

It is customary to assemble cells into modules by a mixture of parallel and series connections, so that the resulting voltages become suitable for standard electricity handling equipment, such as inverters transforming the DC currents into AC currents of grid specifications and quality. Alternatively, microprocessor inverters may be integrated into each module or even into each cell in order to minimise transport losses. In recent years, specific inverters optimised for solar cell applications have been produced, with an inverter efficiency increase from 90% to some 98% (IEA, 1999).

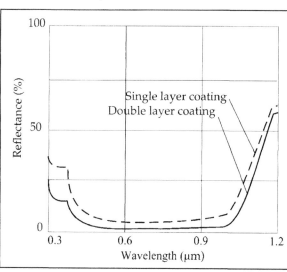

Figure 14.19. Front reflection for cell of the type shown in Fig. 14.12 (single-layer anti-reflection coating) and the same with double-layer coating (Zhao et al., 1995).

The solar cell technology is then characterised by two main solar radiation conversion efficiencies: the efficiency of each cell and the efficiency of the entire module sold to the customer. The latter is currently about 5% lower than the former, notably because of phase mismatch between the individual cell current components, but this does not need to be so, and the difference between the two efficiencies is expected to diminish in the future.

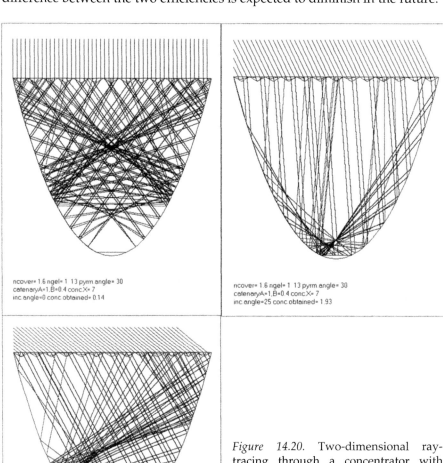

ncover= 1.6 ngel= 1.13 pyrm angle= 30
catenaryA=1.B=0.4 concX= 7
inc angle=0 conc obtained= 0.14

ncover= 1.6 ngel= 1.13 pyrm angle= 30
catenaryA=1.B=0.4 concX= 7
inc angle=25 conc obtained= 1.93

ncover= 1.6 ngel= 1.13 pyrm angle= 30
catenaryA=1.B=0.4 concX= 7
inc angle=50 conc obtained= 0.59

Figure 14.20. Two-dimensional ray-tracing through a concentrator with glass cover textured with 30° inverted pyramids, for incident angles 0°, 25° and 50°. Only for incident angles around 25° does the absorber at the bottom receive more light than would have hit it if there had been no concentrator at all.

Optical subsystem and concentrators

As indicated by the devices shown in Figs. 14.12 and 14.15, optical manipulation of incoming solar radiation is in use for non-concentrating solar cells. They serve the purpose of reducing the reflection on the surface to below a few percent over the wavelength interval of interest for direct and scattered solar radiation, as seen from Fig. 14.19. The collection efficiency is far better than the one shown in Fig. 14.19, exceeding 90% for wavelengths between 0.4 and 1.05 µm, and exceeding 95% in the interval 0.65–1.03 µm.

For large-factor concentration of light onto a photovoltaic cell much smaller than the aperture of the concentrator, the principles valid for thermal systems (Chapters 16 and 17) apply unchanged. Most of these concentrators are designed to focus the light onto a very small area. Thus tracking the Sun (in two directions) is essential, with the implications that scattered radiation largely cannot be used and that the expense and maintenance of non-stationary components have to be accepted.

One may think that abandoning very high concentration would allow the construction of concentrators capable of accepting light from most directions (including scattered light). However, this is not so easy. Devices that accept all angles have a concentration factor of unity (no concentration), and even if the acceptance angular interval is diminished to, say, 0–60°, which would be suitable because light from directions with a larger angle is anyway reduced by the incident cosine factor, only a very small concentration can be obtained. (Examples such as the design by Trombe are discussed in Meinel and Meinel, 1976).

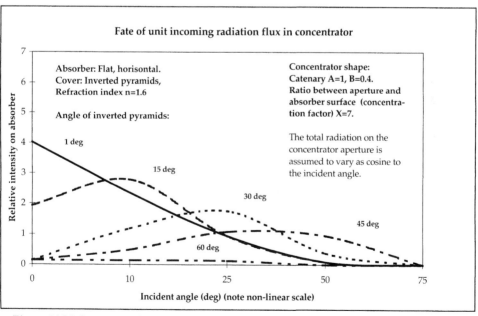

Figure 14.21. Intensity of rays on the absorber surface relative to that experienced in the absence of the concentrator (times cosine of incident angle, in order to reflect the different intensity of rays from different directions) for the $X = 7$ concentrator with a refracting, inverted pyramid cover as illustrated in Fig. 14.20.

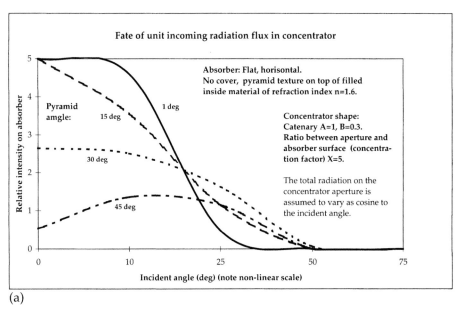

(a)

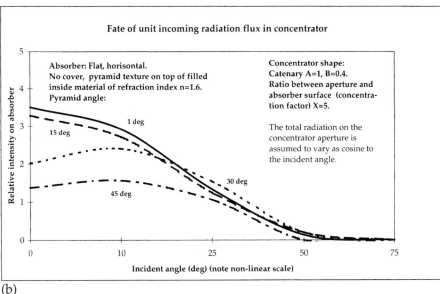

(b)

Figure 14.22a-c. Intensity of rays on the absorber surface relative to that experienced in the absence of the concentrator (times cosine of incident angle, in order to reflect the different intensity of rays from different directions) for different depths of the X = 5 concentrator without cover, but with a refracting inside material having pyramid texture on top, as illustrated in Fig. 14.24 (*a* and *b* above, *c* on next page).

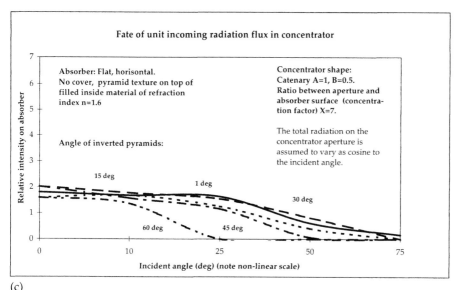

(c)

Figure 14.22c. See preceding page for details.

The difficulty may be illustrated by a simple two-dimensional ray-tracing model of an (arbitrarily shaped) absorber, i.e. the photovoltaic cell, sitting inside an arbitrarily shaped concentrator (trough) with reflecting inner sides, and possibly with a cover glass at the top, that could accommodate some texturing or lens structure. The one component that unfortunately is not available is a semi-transparent cover that fully transmits solar radiation from one side and fully reflects the same wavelengths of radiation hitting the other side. With such a cover, nearly 100% of the radiation coming from any direction could reach the absorber, the only exception being rays that are cyclically reflected into paths never reaching the absorber.

Figure 14.20 shows some ray-tracing experiments for a catenary-shaped concentrator holding a flat absorber at the bottom, with a concentration factor $X = 7$ (ratio of absorber and aperture areas, here lines). The cover has an inverted pyramid shape. This is routinely used for flat-plate devices (see Fig. 14.12), with the purpose of extending the path of the light rays inside the semiconductor material so that absorption is more likely. For the concentrator, the purpose is only to make it more likely that some of the rays reach the absorber, by changing their direction from one that would not have led to impact onto the absorber. However, at the same time, some rays that otherwise would have reached the absorber may now become diverted. Thus, not all incident angles benefit from the concentrator, and it has to be decided whether the desired acceptance interval is broad (say, 0–60°) or can be narrowed, e.g. to 0–30°. As illustrated, the concentration of rays towards the centre absorber is paid for by up to half of the incident rays being reflected upward after suffering total reflection on the lower inside of the cover glass.

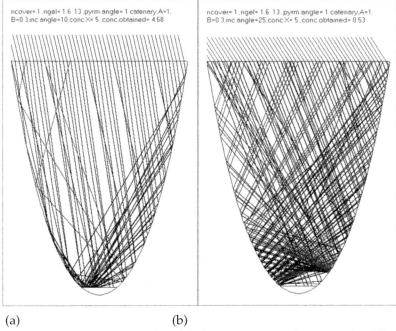

ncover= 1 ,ngel= 1.6 13 ,pyrm.angle= 1 .catenary,A=1.
B=0 3,inc.angle=10,conc.X= 5 ,conc.obtained= 4.68

ncover= 1 ,ngel= 1.6 13 ,pyrm.angle= 1 .catenary,A=1.
B=0 3,inc.angle=25,conc.X= 5 ,conc.obtained= 0.53

    (a)   (b)

Figure 14.23a,b. Two-dimensional model of concentrator without cover but filled with a refracting material (refraction index n_{gel} = 1.6), with slight pyramid texture at top. Incident angles are 10° and 25°.

The result of varying some of the parameters is shown in Fig. 14.21. An inverted pyramid angle of 0° corresponds to a flat-plate glass cover, and the figure shows that in this situation the concentration is limited to incident angles below 20°, going from a factor of four to unity. With increasing pyramid angle, the maximum concentration is moved towards higher incident angles, but its absolute value declines. The average concentration factor over the ranges of incident angles occurring in practice for a fixed collector is not even above unity, meaning that a photovoltaic cell without the concentrating device would have performed better.

Another idea is to direct the cover pyramids upwards, but to avoid "stretching out" the rays again at the lower glass boundary by having the entire trough filled with the refractive substance (which might be a gel or plastic material, textured at the upper surface). This is illustrated in Figs. 14.22a–c for an X = 5 concentrator. For a deep catenary-shaped collector, all rays up to 10° may be led to the absorber, but in this case (Figs. 14.22a and 14.23), incident angles above 25° are not accepted. To get reasonable acceptance, pyramid angles around 30° would again have to be chosen, and Figs. 14.22a–c show that the concentration penalty in trying to extend the acceptance interval to angles above 35° (by decreasing the depth of the concentrator, see Fig. 14.24) is severe. The best choice is still the deep catenary shape, with maximum concentration between 2 and 3 and average concentration over the interesting interval of incident angles (say, to 60°) no more than about 1.5. This is better than the inverted pyramid structure in Fig. 14.20, but it is still doubtful whether the expense of the concentrator is warranted. The collector efficiency as measured relative to total aperture area is the ratio of absorber intensity and X, i.e. considerably less than the cell efficiency.

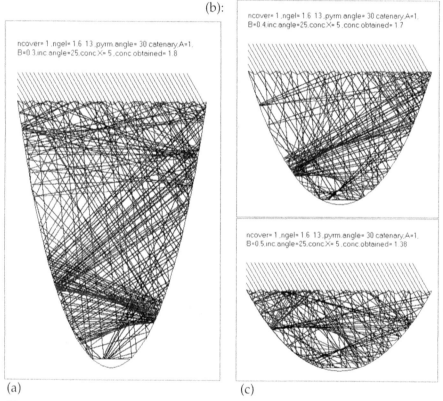

(a)   (c)

Figure 14.24a–c. Two-dimensional ray-tracing through a concentrator without cover, but filled with a refracting material (refraction index $n_{gel} = 1.6$), having its top pyramid-textured, for an incident angle of 25° and three catenary depths.

Generation 24-h a day at constant radiation intensity could be achieved by placing the photovoltaic device perpendicular to the Sun at the top of the Earth's atmosphere. If it were placed in a geosynchronous orbit, it has been suggested that the power generated might be transmitted to a fixed location on the surface of the Earth after conversion to microwave frequency (see e.g. Glaser, 1977).

Use of residual energy

Given that practical efficiencies of photovoltaic devices are in the range of 10–30%, it is natural to think of putting the remaining solar energy to work. This has first been achieved for amorphous cells, which have been integrated into windowpanes, such that at least a part of the energy not absorbed is passed through to the room behind the pane. Of course, conductors and other non-transparent features reduce the transmittance somewhat. The same should be possible for other thin-film photovoltaic materials, including the emerging multicrystalline silicon cells.

Another possibility, particularly for photovoltaic panels not serving as windows, is to convert the solar energy not giving rise to electricity into heat (as some of it in actuality already is). One may think of cooling the modules of cells by running pipes filled with water or another suitable substance along their bottom side, carrying the heat to a thermal store, or alternatively by an air flow above the collector (but below a transparent cover layer). The energy available for this purpose is the incoming radiation energy minus the reflected and the converted part. Reflection from the back side of the semiconductor material, aimed at increasing the path-length of the photons in the material, could be chosen to optimise the value of the combined heat and power production, rather than only the power production. Examples of such combined cycle solar panels (photovoltaic-thermal or PVT-panels) have been modelled extensively (Sørensen, 2000b, 2002a and 2004).

For concentrator cells, active cooling is needed in any case, because of the temperature dependence of the characteristics of the photovoltaic process (cf. Fig. 14.10).

The maximum electrical efficiency of a single junction photovoltaic device implied by semiconductor physics is around 40% (see the discussion earlier in this chapter). If reflections can be minimised and many different light absorption materials are stacked, the theoretical Carnot efficiency (2.4) for the temperature of the Sun relative to a reference temperature at the surface of the Earth, i.e. about 95%, may be approached. Still, considerations of power flow optimisation will diminish this efficiency in practical applications. Honsberg et al. (2002) and Green (2002) find a thermodynamical value of some 87% for an infinite stack of cells, to be used as a starting point for discussing the losses deriving from semiconductor physical arguments.

For a present generation solar cell with electric efficiency below 40%, the possible associated heat gain is of the order of magnitude another 40%, and thus the system offers interesting possibilities for decentralised applications in climate regimes with solar radiation without too strong seasonality and a heat demand for space heating or even just hot water.

PHOTO-ELECTROCHEMICAL CONVERSION

CHAPTER 15

Photoelectrochemistry is an area of confluence between solar cell technology, discussed here, and battery or fuel cell technology, discussed in Part V. Organic solar cells are a special kind of photoelectrochemical (PEC) devices that try to take advantage of inexpensive organic materials as opposed to the more expensive metals and doped semiconducting materials used in the photovoltaic devices discussed above. Suitable organic materials can by photosynthesis trap sunlight and convert radiation into other forms of energy. It has been attempted to copy this process to man-made devices in various ways. Calvin (1974) considered a double membrane that would separate the ionised reactants of a photo-excitation process,

$$A + h\nu \to A^* ; \qquad A^* + B \to A^+ + B^- .$$

In addition, a transport system is needed to get the ions to an electrode. No practical version of this idea has been produced. The same is the case for a concept aimed at both removing CO_2 from the atmosphere and at the same time producing methanol (Jensen and Sørensen, 1984). The absorption of solar radiation is used to fix atmospheric carbon dioxide to a ruthenium complex [Ru], which is then heated with water steam,

$$[Ru]CO_2 + 2H_2O + 713 \text{ kJ mol}^{-1} \to [Ru] + CH_3OH + \tfrac{1}{2}O_2.$$

One scheme that has been realised is the attachment of a ruthenium complex as a dye to TiO_2, which may then transport electrons formed by photo-excitation in the Ru-complex to an electrode. The process is similar to the dye-excitation processes used in conventional photographic prints and was first proposed by Moser (1887). He called the dye substance enhancing the absorption of solar radiation above what can be achieved by the TiO_2 a "sensitiser". The Ru-complex is restored by a redox process in an electrolyte joining the other electrode, with use of a platinum catalyst as indicated in Fig. 15.1. Because a mono-layer of even a very efficiently absorbing dye will absorb less than 1% of incoming solar radiation, the dye absorption layer is made three dimensional by adhering the dye to a nanostructured TiO_2 network of nodules, as first demonstrated by Tsubomura et al. (1976). They also introduced the liquid electrolyte, capable of penetrating into the cavities in the nanostructured, sensitised titanium dioxide and providing the necessary contact for transfer of the replacement electron to the dye. The effective absorption surface may be increased by three orders of magnitude and provides an overall cell efficiency to about 10%. It should still be possible for the TiO_2 nodules to transfer the absorbed electron to the back electrode through a series of transport processes.

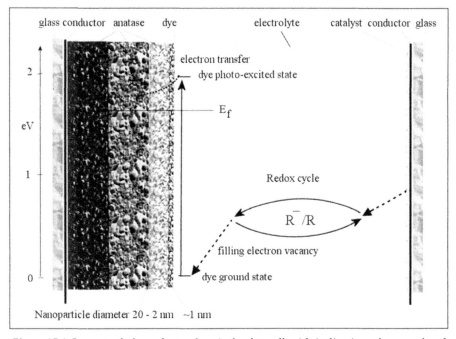

Figure 15.1. Layout of photoelectrochemical solar cell with indication of energy levels (E_f is the Fermi level of the semiconductor material). Solar radiation is absorbed in a dye layer, creating an excited electron state, from which an electron is transferred to the semiconductor at left and replenished from the counter electrode through a redox cycle in an electrolyte to the right (Sørensen, 2003a).

The material presently favoured for the anode nanoparticles is TiO_2 in the form of anatase (Fig. 15.2). Compared to the other forms of titanium dioxide (rutile and brookite), anatase better accommodates the dye molecules and forms nodules rather than flakes. The large-side dimension of the unit cell shown in Fig. 15.2 is about 0.2 nm. Several other semiconductor materials have been investigated, but so far anatase has shown the best overall properties and is used fairly universally. Electron transport through the anatase layers follows conventional solid-state physics, except for the issue of nodule coherence. A simple modelling effort has successfully described the transport as random walk, rather than governed by hopping models (Nelson *et al.*, 2001).

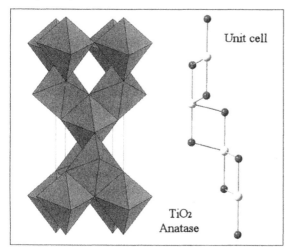

Unit cell

TiO₂
Anatase

Figure 15.2. Anatase structure (first determined by Horn *et al.*, 1970) and unit cell (Sørensen, 2003a).

On the cathode side, a redox couple is used to supply the electron to replace the one being excited in the dye and transferred to the anatase before it decays back to the dye ground state. The electrolyte is typically acetonitrile (C_2H_3N), and the redox couple is iodine/tri-iodine (I^-/I_3^-), which has been used rather exclusively since the work of Tsubomura *et al.* (1976). This does not seem ideal, as the difference between the anatase Fermi level and the I^-/I_3^- chemical potential, which determines the cell open-circuit voltage, is only about 0.9 eV, as compared with typical dye excitation energies of 1.8 eV. Many efforts have been directed at finding more appropriate redox shuttle systems, but so far none have shown overall properties making them preferable to the I^-/I_3^- couple (Wolfbauer, 1999). Electrolyte and redox couple integrity and lifetimes are of concern. A comparison to batteries is appropriate, and battery lifetimes are rarely as long as desired for solar cells that may be incorporated directly into building components and structures.

A further energy loss takes place at the cathode, where application of a catalyst is required in order to obtain the desired rate of electron transfer from electrode to electrolyte. As in batteries and fuel cells, traditionally preferred catalysts are based on platinum, but alternatives are under investigation. Generally speaking, the use of liquid electrolytes and catalysts is undesirable, and the much slower electron transfer through the electrolyte and its redox couple (as compared with the semiconductor transport) is likely to be the overall limiting factor for current in the device. However, the reason for this choice is also obvious. The cell is produced by deposition of anatase layers on the anode and subsequent annealing, processes requiring temperatures well over 100°C. The dye is then applied, either by a soaking or by a flushing process, creating the huge intrinsic surface for solar collection. Typical dye melting points are 80-100°C, so applying a second semiconductor material (if one with appropriate properties could be found) from the other side at appropriate temperatures would destroy the cell. More gentle application not requiring high temperatures is not likely to allow the surface contact area between dye and semiconductor to be equally large on both sides.

An alternative would be to find another material not requiring high temperatures for penetrating into the cavity structure of the initial semiconductor plus dye layers. Possible candidates would be conducting polymers or the ion-carrying polymers used in fuel cells. Actual achievements of 2-3% energy conversion efficiency have been obtained with two types of polymer systems. One uses a gel network polymer as electrolyte (Ren *et al.*, 2001). The other is a type of plastic solar cell, where the already known ability of ^{60}C-molecules to absorb solar radiation (Sariciftci *et al.*, 1992) is used to create a fairly large absorption area of ^{60}C sensitiser embedded in a suitable polymer (Shaheen *et al.*, 2001; Yu *et al.*, 1995).

The choice of sensitiser is ideally based upon fulfilment of requirements including at least the following:

• high absorption capability over the range of spectral frequencies characteristic of sunlight

• energetically suitable excited states

• good attachment to semiconductor nanoparticles, which ensures rapid electron transfer (in competition with de-excitation and back-transfer from semiconductor surface to dye sensitiser)

• easily accepting replacement electron from electrolyte

• dye lifetime consistent with stipulated device life

The search for optimised sensitisers has usually focused on a particular family of molecules. For example, one group (O'Regan and Grätzel, 1991; Nazeeruddin *et al.*, 1993, 2001; Shklover *et al.*, 1998) has looked at metal complexes based on ruthenium polypyridines, meticulously synthesising one variant after the other, adding rings, thiocyanate ligands and carboxylate groups in different combinations. The size of the molecule, in combination with its excitation spectrum, determines the frequencies of solar radiation that can be absorbed and the associated cross sections. The "black dye" (1 ruthenium atom, 3 pyridine rings, 3 thiocyanate ligands and 3 carboxylate groups) has led to the currently highest overall conversion efficiency of 10% for laboratory cells (area about 10^{-4} m^2). For comparison, an efficiency of 5% was claimed for a large cell (of the order of 1 m^2) based on sensitisers patented by Grätzel and developed for industrial production (STI, 2002). Subsequent failure of the company's first large-scale installation in Newcastle (Australia) has diminished expectations for this type of cell. An earlier favourite was the "N3 dye" (Ru, 2 bipyridine rings, 2 thiocyanate ligands and 4 carboxylate groups). It is particularly willing to transfer an excited electron to an anatase surface, a fact that has been attributed to its attachment to the anatase surface by two carboxylate binding sites at approximately the same spacing as the "indents" in one anatase surface. However, the light absorption stops below 800 nm, implying smaller efficiency for many potential real-life collector sites. Figure 15.3 compares the spectral sensitivities of the two dyes mentioned above, plus the coumarin-derivative organic dye considered in the following.

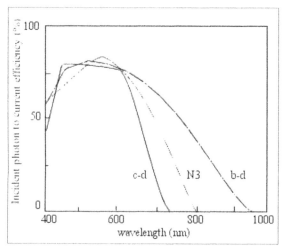

Figure 15.3 (left). Spectral sensitivity [c-d: coumarin derivative, N3 and b-d (black dye) ruthenium complexes] (based upon Hara *et al.*, 2001; Nazeeruddin *et al.*, 2001).

Figure 15.4 (below). The structure of N3 ($RuS_2O_8N_6C_{26}H_{24}$, below left) and b-d ($RuS_3O_6N_6C_{21}H_{22}$, below right) ruthenium sensitisers synthesised by Nazeeruddin *et al.* (1993; 2001). A modest structure optimisation has been performed (Sørensen, 2004b).

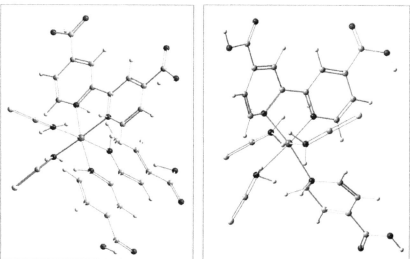

Figure 15.4 gives the molecular structure of the purple N3 and the black ruthenium dye (Nazeeruddin *et al.* 1993, 2001). Several variants have been stud-

ied (see e.g. Shklover *et al.*, 1998; Zakeeruddin *et al.*, 1997). The structure of modified sensitiser molecules is often roughly given by that of known components plus some general rules of thumb. Measurements of spectra such as the nuclear magnetic resonance (NMR) spectra will help determine the positions of specific atoms (e.g. hydrogen atoms), but not always in a unique way. The addition of new features to a dye molecule in order to enhance solar absorption or help electron transfer out of the molecule may give rise to structures not amenable to simple guesses, e.g. due to isomerism or for other reasons such as energy surfaces in configuration space with more than one minimum.

Quantum-mechanical modelling of the molecular structure* will lead to theoretical predictions of likely structures, as outcomes of optimisation studies. These involve following a path of steepest descent in the potential energy surface, eventually using second-order derivatives in order to stabilise the sizes of jumps made for each of the iterations (Sørensen, 2003a).

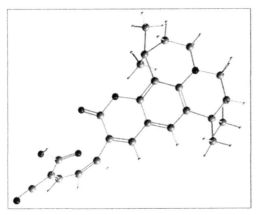

Figure 15.5. Optimised structure of coumarin-derivative organic dye (gross formula $H_{26}C_{25}N_2O_4$) synthesised by Hara *et al.* (2001) and yielding overall efficiencies of around 6% when used in PEC solar cells (Sørensen, 2003a).

Figure 15.5 shows the result of a theoretical optimisation (Sørensen, 2003a) for a modification of a coumarin dye, being part of a family of molecules studied by Hara *et al.* (2001) as potentially inexpensive, purely organic candidates for dyes to use with TiO_2 semiconductor nanoparticles in solar cells.

Figure 15.6 shows the molecular orbits near the Fermi level for the coumarin-derivative dye, obtained from a quantum-mechanical Hartree-Fock

---

* So-called *ab initio* calculations in quantum chemistry involve solving the Schrödinger equation under several simplifying assumptions: nuclear motion is neglected, the basis functions used are linear combinations of Gaussian functions for each atom, and far fewer than needed for completeness. Interactions are first treated by the Hartree-Fock method, implying that each electron is assumed to move in the mean field from all the other particles. Higher electron correlations are added in a perturbative way or by the phenomenological density functional theory.

self-consistent field calculation of the best ground state configuration (using software of Frisch *et al.*, 1998 and including a basis of 641 molecular states or about 3 for each physical orbit). It is seen that moving an electron from the highest occupied orbit (HOMO) to the lowest unoccupied one (LUMO) involves reducing the electron density in the coumarin part of the molecule and increasing it along and particularly at the end of the "arm" attached to the molecule. However, at this level of approximation, the energy difference (LUMO minus HOMO) is still nearly 8 eV, as opposed to an experimental value of 1.8 eV (Hara *et al.*, 2001). The second unoccupied orbit (Fig. 15.6 bottom) is quite different in nature.

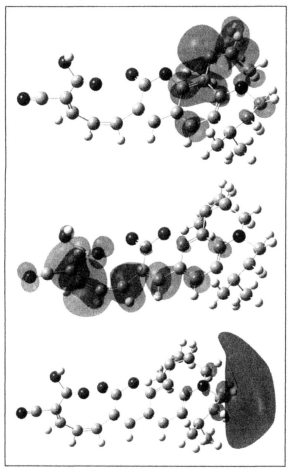

Figure 15.6. Electron density of HOMO (top), LUMO (middle) and second unoccupied molecular orbital (bottom) for coumarin-derivative dye, based on self-consistent field (SCF) calculation (Sørensen, 2003a).

In order to estimate more realistically the energy of the first excited state of the coumarin-derivative dye molecule, a number of calculations have been performed (Sørensen, 2003a). Interactions not included in the self-consistent ground state calculation may be added in a time-dependent Hartree-Fock calculation for one or more excited states (TDHF; Casida *et al.*, 1998), and further improvement is obtained by performing the TDHF calculation on top of ground state calculations including further interaction, such as including exchange forces in a density functional method (Kohn and Sham, 1965; Becke, 1993). Figure 15.7 shows the excitation energy of the lowest spin-0 and spin-1 excited states using successive improvements in the sophistication of the interactions included and in the number of basis states used for the calculations (Sørensen, 2003b).

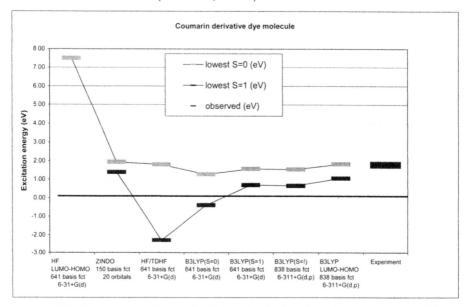

Figure 15.7. Measured (right-hand side) and calculated excitation energies of first excited states of spin 0 or 1 for the coumarin-derivative molecule shown in Fig. 15.5. The abscissa indicates acronyms for the type of calculation made, with complexity increasing from left to right (except the last, which by comparison with the first column shows the correlations built into the molecular orbits calculated with inclusion of exchange forces). It is seen that two of the calculations give unphysical states of negative energy, indicating the fragile nature of the approximations that effectively use other electron interactions for the excited state calculation than for the ground state calculation (because at a realistic level of approximation, the same approach will not yield both). The problem is solved in the subsequent columns by using basis states more appropriate for the spin-1 calculations (Sørensen, 2003b).

The fact that the first excited state comes down from an initial 8 eV molecular orbital energy difference to the observed 1.8 eV indicates that it must contain a substantial amount of correlations, or in other words that this excitation must comprise collective involvement of a large number of electrons.

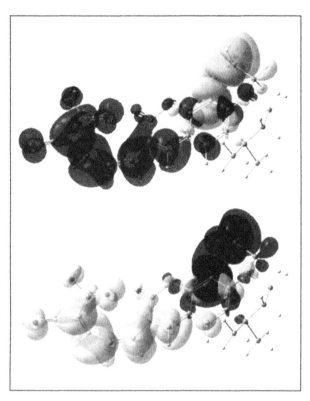

Figure 15.8. Calculated electron density difference between the first excited singlet and ground state of coumarin-derivative dye (top: positive values enhanced; below: negative values enhanced; Sørensen, 2003a).

This is borne out in Fig. 15.8, showing the electron density difference between the first singlet excited state of the coumarin-derivative, and the ground state, using the large basis TDHF calculation. The excited state is made up by a dozen significant molecular orbital (MO) excitation-pairs. The density difference shows the expected effect, already surmised from the HF ground state MO's, that the excitation moves electron density from the coumarin core to the peripheral arm added to the molecule. It would then be natural to assume that this is from where the transfer to the anatase surface, to which the dye adheres, takes place. This interpretation is supported by the large dipole moment found in the calculations (13.7 debyes).

The precise attachment of the dye to the anatase surface might be investigated by a combined optimisation of the dye plus a chunk of the semiconductor, with the distance and rotation angles of the dye relative to the surface as parameters (cf. Fig. 15.9). A recent study of a similar material has revealed the nature of surface distortion and surface Ti and O molecules (Erdman *et al.*, 2002). The distortion only penetrates one layer down, in contrast to the over 10-nm-thick space charge regions in surfaces of solids. The conclusion from the present calculation is that both dye and surface are modified, with the dye "arm" being bent to better attach to the surface, and the surface atoms to accommodate the dye particle (but even in the absence of the dye, the surface of the lattice structure would be different from the regular interior used as starting point for the optimisation).

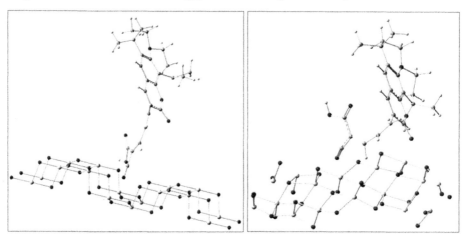

Figure 15.9. Attachment of coumarin-derivative dye to and modification of anatase surface. Left: initial guess, right: optimised (the depth of TiO_2 surface layer distortion cannot be determined by modelling just the top layer; Sørensen, 2003a, 2004b).

A number of further "handles" on the quantum chemical calculations is the comparison of measured spectra (IR, NMR, etc.) with those predicted by the calculations. The understanding of complex molecules has progressed rapidly as a result of the combination of spectral studies in the laboratory with quantum model calculations.

The photoelectrochemical dye and nanostructure technique has several applications beyond the possibility of forming solar cells. Among these are smart windows (Bechinger *et al.*, 1996; Granqvist *et al.*, 1998; STI, 2002), energy storage (Hauch *et al.*, 2001; Krasovec *et al.*, 2001), environmental monitors (Kamat and Vinodgopal, 1998; Yartym *et al.*, 2001), hydrogen production (Khaselev and Turner, 1998; Luzzi, 1999; Mathew *et al.*, 2001), computer and television screens also suitable for outdoor use (using the dye as light emitter rather than absorber; Rajeswaran *et al.*, 2000; Tang, 2001; Müller *et al.*, 2003), and three-dimensional data storage (Dwayne-Miller *et al.*, 2001). Weaknesses of the dye cells include durability (especially if organic materials are used), while weaknesses of the polymer cells include the extremely low efficiency, which is difficult to offset by low production costs, due to the large physical surface having to be covered (a feature that itself excludes many applications such as building-integrated systems serving a decent percentage of the building's loads).

SOLAR THERMAL CONVERSION

16.1 Heat generation

Conversion of solar energy to heat requires a light-absorbing material, a *collector*, which is able to distribute the absorbed radiant energy over internal degrees of freedom associated with kinetic energy of motion at the molecular level (e.g. lattice vibrations in case of a solid). Absorption of solar energy will raise the temperature of the collector or transfer energy to a reservoir, if the collector is connected to one. The collector will also emit radiation, and it may lose heat energy by conduction and convection processes. The frequency spectrum of the emitted radiation will correspond to the Planck spectrum,

$$
\frac{\mathrm{d}F}{\mathrm{d}\nu} = \frac{2h\nu^3}{c^2}\frac{1}{e^{h\nu/kT}-1}.
\tag{16.1}
$$

Here F is the power radiated per unit area and into a unit of solid angle, ν is the frequency of radiation, $h = 6.6 \times 10^{-34}$ J s is Planck's constant, $c = 3 \times 10^8$ m s^{-1} is the velocity of electromagnetic radiation (light) in vacuum and $k = 1.38 \times 10^{-23}$ JK^{-1} is Boltzmann's constant. This distribution may be used to describe the heat re-radiation from the collector surface with T equal to the collector temperature T_c, provided the collector is in a state that allows the definition of a thermodynamic temperature.

Collectors may be designed so that they achieve a large net absorption by minimising reflection and transmission and try to avoid losses, e.g. by operating the collector at temperatures not much above ambient air temperatures or, if higher load temperatures are required, by reducing the heat loss rates, e.g. by suitable transparent covers and insulation.

One may distinguish between "passive" and "active" systems, according to whether energy is specifically added (from pumps, etc.) in order to bring the collector heat gain to the load areas or not. A passive system need not be characterised by the absence of definite heat flow paths between collectors and load areas, but such flows should be "natural", i.e. they should not depend on other energy inputs provided by man. There may be borderline cases in which the term "natural circulation" would be difficult to define.

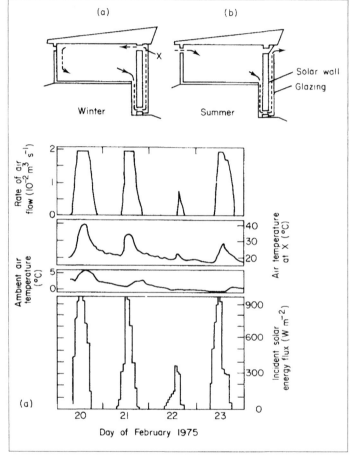

Figure 16.1. Solar wall type of passive heating and cooling system. Top cross sections show air flows during (a) winter and (b) summer. Curves below show, for a selected period of a few days operation, the air flow-rate and temperature [at the location X in (a)], ambient temperature outside the house, and solar radiation. The inside wall temperature remains above 20ºC during the period considered. The house considered is situated at 43°N latitude (based on Trombe, 1973; Stambolis, 1976).

Another kind of passive solar heat system uses the heat capacity of walls facing the sun during the daytime. The walls absorb radiation and accumulate it (the heat capacity of building materials increases roughly in proportion to mass), and at night they lose heat to their colder surroundings, including the inside area, which is thus heated. The wall's heat capacity also serves to cool the building during at least the first part of the daytime, if the wall temperature after the night's cooling off is lower than the next day's ambient temperature. More elaborate versions of such solar wall systems, directing the natural convection according to conditions (night/day, summer/winter), are shown in Figs. 16.1 and 16.2. These systems require a little "active" help in opening and closing vents, shutters or covers. Being dependent on a daily cycle, these systems are of most interest in climatic regions where the daily solar input is substantial also during the cold season.

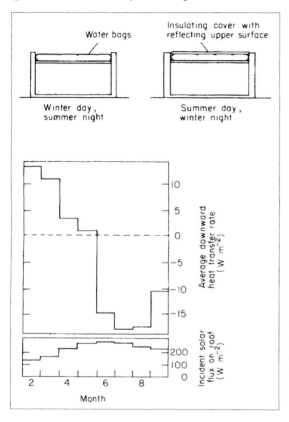

Figure 16.2. Solar roof-type passive heating and cooling system. The top cross sections show operation of cover panel at day and night, during summer and winter. Below are heat transfer rates through the roof (positive downward) and solar radiation, both averaged month by month during an 8-month period. The house is situated at 35°N latitude (based on Hay and Yellot, 1972; Stambolis, 1976).

Figure 16.3. Schematic cross section through a solar pond (cf. Tabor, 1967). The temperature profile may have a secondary peak within the top hundredths of a metre, due to absorption of ultraviolet radiation in this layer, and depending on the re-radiation conditions (effective sky temperature relative to temperature of pond surface layer).

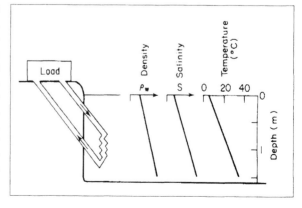

Greenhouses are also passive solar-collecting systems, as are water heaters based on water bags (e.g. placed on roofs) or on flat-plate collectors (see below) delivering heat to water, which is transferred by natural circulation to a storage tank lying higher than the collector. A saline pond can play the role of the collector, with a heat exchanger at the bottom of the pond transferring the collected energy to a working fluid, which can be circulated to the load area by natural or forced circulation. The solar pond itself (see Fig. 16.3) contains water with a high content of dissolved salts, causing the formation of a salinity and density gradient, preventing the mixing of surface and bottom water. The water absorbs some solar radiation, but if the pond is shallow most of the absorption takes place at the bottom, thereby creating a stable temperature gradient increasing towards the bottom, because heat is transferred upward only by slow processes.

Flat-plate collectors

The term "flat-plate collector" is used for absorbers with a generally flat appearance, although the collecting surface need not be flat in detail (it might have V-shaped carvings or even focusing substructure). The side of the collector facing the sun may have a cover system (e.g. one or more layers of glass), and a mass flow J_m^c of some fluid (e.g. water or air) passes the absorber and is supposed to carry the heat, which is transferred from the absorber plate to the mass flow, to the load area (i.e. the place of actual usage) or to some temporary energy storage. The general layout is shown in Fig. 16.4.

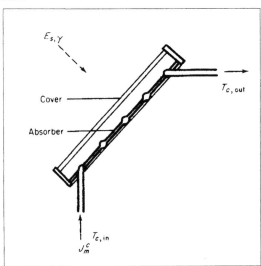

Figure 16.4. Example of flat-plate solar collector. The performance may not be given simply by the net incident radiation flux $E_{s,\gamma}$ (s is tilt angle and γ is azimuth angle), because the transmission–absorption product of the cover system may be different for different components of the incident radiation.

The absorber is characterised by an absorptance, $\alpha_\lambda(\Omega)$, which may depend on wavelength and on the direction of incident light. In many situations, it may be assumed in a first approximation that the absorptance is independent of direction and of wavelength ("grey surface"). For a black painted surface, this α may be around 0.95, but, if the surface is structure-less, the assumption that α is independent of direction breaks down when the angle between the normal to the surface and the direction of incidence exceeds about 60°. Towards 90°, $\alpha_\lambda(\Omega)$ actually approaches zero (Duffie and Beckman, 1974). From basic physics (Sørensen, 2004), the emittance ε equals the absorptance (both are properties of the surface in the grey-surface ap-proximation), and high absorptance thus implies high emittance for all wavelengths, including those characterising the thermal emission from the absorber of temperature T_c. In order to reduce this loss, use is made of sur-face treatments causing the emittance to assume two distinct values – a high one for the short wavelengths of the solar spectrum (implying a high solar absorptance α^{sw}) and a low one (α^{lw}) for the longer wavelengths characteris-ing the emission from typical absorber temperatures, assumed to have a spectrum approximately equal to the black-body spectrum for that tempera-ture. Such surfaces are called *selective surfaces,* and their wavelength-dependent absorptance/emittance may resemble that shown in Fig. 16.5, exhibiting the regions that can be characterised by $\alpha^{sw} \approx 0.95$ and $\alpha^{lw} \approx 0.1$. A further discussion of selective surface technologies may be found in Meinel and Meinel (1976).

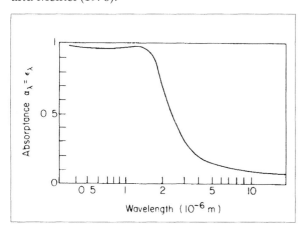

Figure 16.5. Spectral absorp-tance (or emittance) of a commercial "black-chrome" type of selective surface (based on Masterson and Seraphin, 1975).

As indicated in Fig. 16.4, the absorber may be covered by material that reduces heat losses and at the same time transmits most of the incoming radiation. The cover may consist, for example, of one or more layers of glass. The transmittance of glass depends on the type of glass and in particular on minority constituents such as Fe_2O_3. Figure 16.6 gives an example of the wavelength dependence of the transmittance, τ_λ, through ordinary window glass, defined as the fraction of incident light that is neither reflected nor absorbed,

$$\tau_\lambda = 1 - \rho_\lambda - \alpha_\lambda. \qquad (16.2)$$

For a collector of the type considered, multiple reflections may take place between the absorber and the cover system as well as within the cover system layers, if there are more than one. The reflections are mostly specular (angle of exit equal to angle of incidence) at glass surfaces, but mostly diffuse (hemispherically isotropic) at the absorber surface. Thus, the amount of radiation absorbed by the absorber plate, in general, cannot be calculated from knowledge of the total incoming flux but must be calculated for each incident direction from which direct, scattered or reflected light is received, as well as for each wavelength. The total absorbed flux is then obtained by integration.

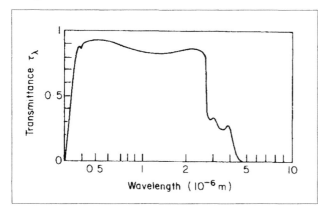

Figure 16.6. Spectral transmittance, $(1 - \rho_\lambda - \alpha_\lambda)$, of 2.8×10^{-3} m thick glass layer with a Fe_2O_3 content of 0.1 (based on Dietz, 1954).

In many cases a sufficiently good approximation can be obtained by considering only the two wavelength intervals, denoted "short" and "long" wavelengths above (the dividing wavelength being about 3×10^{-6} m), in view of the near-constancy of the absorptance and transmittance of the relevant materials (cf. Figs. 16.5 and 16.6). The net short-wavelength energy gain by the collector may then be written

$$E_c^{sw} = A \int E_{c+}^{sw}(\Omega) \, P^{t.a.}(\Omega) \, d\Omega, \qquad (16.3)$$

in terms of the incoming flux from a given direction Ω, $E_{c+}^{sw}(\Omega)$, and the "transmission–absorption product", $P^{t.a.}(\Omega)$, describing the fraction of incident short-wavelength radiation from the direction Ω, which gets transmitted through the cover system and gets absorbed by the absorber plate, the area of which is denoted A. It has been assumed that A serves as a simple proportionality constant, implying that incident radiation as well as transmission–absorption properties are uniform over the entire collector area. For a cover system consisting of N layers, e.g. of glass, characterised by a refraction index relative to air, n (~1.5 for glass), and an extinction coefficient, x, such that the reduction in intensity of radiation from traversing a distance d

through the material is $\exp(-xd)$, the total thickness of each cover layer being L, then $P^{t.a.}(\Omega)$ may be approximately written (see e.g. Duffie and Beckman, 1974)

$$P^{t.a.}(\Omega) = \frac{1-\rho}{1+(2N-1)\rho} e^{-xNL\cos\theta} \frac{\alpha^{sw}}{1-(1-\alpha^{sw})\rho_d(N)}, \qquad (16.4)$$

where θ is the polar angle between the incident direction Ω and the normal to the collector surface, and ρ is the reflectance of one cover layer, given by the Fresnel formula in case of unpolarised light,

$$\rho = \frac{1}{2}\left(\frac{\sin^2(\theta'-\theta)}{\sin^2(\theta'+\theta)} + \frac{\tan^2(\theta'-\theta)}{\tan^2(\theta'+\theta)} \right),$$

with the polar angle of refraction, θ', given by

$$\theta' = \mathrm{Arcsin}\left(\frac{\sin\theta}{n} \right).$$

The factorisation of (16.4) into a factor describing the multiple reflection through the N-layer cover system in the absence of absorption, times a factor describing the attenuation of the intensity from passage through glass, and a final factor describing the absorption in the absorber plate, after multiple reflections back and forth between the plate and cover system, is a valid approximation for most practical applications, owing to the smallness of the extinction product xL. The final factor in (16.4) may be rewritten as

$$\frac{\alpha^{sw}}{1-(1-\alpha^{sw})\rho_d} = \alpha^{sw} \sum_{i=1}^{\infty} ((1-\alpha^{sw})\rho_d)^i,$$

revealing that it is the sum of terms corresponding to i reflections back and forth between the plate and cover (considering only the lower cover surface), and assuming the non-absorbed fraction, $(1-\alpha^{sw})$, from each impact on the absorber plate to be diffusely reflected from the cover inside surface with a reflectance ρ_d different from that of specular reflection, ρ. In order to make up for processes in which light reflected back from the absorber gets transmitted to and reflected from cover layers other than the lowest one, and then eventually reaches back to become absorbed by the absorber plate, effective values of ρ_d depending on N may be used, as indicated in (16.4). Duffie and Beckman (1974) suggest that the diffuse reflectance ρ_d may be approximated by the specular reflectance for an incident polar angle of $60°$, using values of 0.16, 0.24, 0.29 and 0.32 for $N = 1, 2, 3$ and 4.

For the direct (un-scattered) part of the incident radiation flux, only one direction Ω is permitted, and the integration in (16.3) should be left out. For specular reflection on to the collector, the same is true in simple cases (cf.

Seitel, 1975), but for complicated geometry of the specularly reflecting objects in the surroundings, more than one direction may have to be taken into account. In general cases of reflection on to the collector, as well as for scattered radiation from the atmosphere, the integration in (16.3) would generally have to be kept, but approximations such as multiplying the total scattered *or* reflected flux at the outside collector surface by the transmission–absorption product for an "average" angle of incidence, such as θ = 60°, are sometimes used, permitting the total short-wavelength gain by the collector plate to be written

$$E_c^{sw} = A(D_{s,\gamma}P^{t.a.}(\theta) + (d_{s,\gamma}+R_{s,\gamma})<P^{t.a.}>), \tag{16.5}$$

e.g. with $<P^{t.a.}> = P^{t.a.}(\theta = 60°)$.

The net long-wavelength energy gain by the collector is the difference between the long-wavelength radiation received from the surroundings and the long-wavelength radiation emitted by the collector system. Considering the temperature T_c of the absorber plate as constant, the thermal radiation from the plate is proportional to T_c^4 (Sørensen, 2004), and the net gain of an absorber without cover would be

$$E_c^{lw}(N{=}0) = A\varepsilon^{lw}\sigma(T_e^4 - T_c^4), \tag{16.6}$$

where the temperature T_e of the atmospheric environment, in general, is lower than the ambient air temperature, T_a.

Finally, sensible heat (that associated with a heat capacity) may be exchanged between the collector and its surroundings through conduction and convection processes. The back and edge heat transfers are usually expressed in the same way as heat losses from buildings,

$$E_{back}^{sens} = -A\,U_{back}\,(T_c - T_b), \tag{16.7}$$

where U_{back} is a constant depending on the insulating properties of the materials separating the back side of the absorber plate from the environment of temperature T_b, which may be equal to the ambient temperature, T_a, if the collector is mounted freely or equal to the indoor temperature, T_L, if the collector forms an integral part of a building wall or roof. U_{back} may be assumed to include the "edge" heat losses from the sides of the collector, mainly of conduction type.

The net exchange of sensible energy at the front of the collector depends on the number of cover layers and on the wind speed, V, since the convection of heat is largely limited by the rate at which warm air at the top of the front cover is removed. In the absence of cover, the front (or top) exchange of heat may be written

$$E_{back}^{sens}(N{=}0) = -A f_1 (T_c - T_a), \tag{16.8}$$

where f_1 is a polynomial in V with empirical coefficients, as suggested by Duffie and Beckman (1991),

$$f_1 = max \ (5.0; \ 3.95 \ V[\text{m s}^{-1}]^{\ 0.6}) \ [\text{W K}^{-1}\text{m}^{-2}]. \tag{16.9}$$

With glass covers, the expressions (16.6) and (16.8) must be modified, for example, by explicitly considering the heat transfer equations connecting each layer and solving for steady-state solutions with constant (but different) temperatures of each layer. Klein (1975) has parametrised these solutions in an explicit form, given by

$$E_c^{lw} = \frac{A\sigma(T_e^4 - T_c^4)}{(\varepsilon_p^{lw} + c_1 N(1 - \varepsilon_p^{lw}))^{-1} + N(2 + (f_2 - 1)/N) - N)/\varepsilon_g^{lw}}, \tag{16.10}$$

where ε_p^{lw} and ε_g^{lw} are the long-wavelength emittances of the absorber plate and of the cover glasses, $c_1 = 0.05$ and

$$f_2 = (1 + 0.089 f_1 - 0.1166 f_1 \ \varepsilon_p^{lw}) \ (1 + 0.07866N), \tag{16.11}$$

with f_1 inserted from (16.7) the units indicated (Duffie and Beckman, 1991). The relation replacing (16.6) is

$$E_{top}^{sens} = -A(T_c - T_a) \cdot$$

$$\left\{ \left(\frac{NT_c}{f_3} f_4^{-c_2} + f_1^{-1} \right)^{-1} + \sigma(T_c + T_e)(T_c^2 + T_e^2) \left((\varepsilon_p^{lw} + 0.00591Nf_1)^{-1} + \frac{2N + f_2 - 1 + 0.133\varepsilon_p^{lw}}{\varepsilon_g^{lw}} - N \right)^{-1} \right\}. \tag{16.12}$$

Here $c_2 = 0.43(1 - 100/T_c)$ and

$$f_3 = 520 \ [\text{K}] \ (1 - 0.000051 s^2), \tag{16.13}$$

with s being the tilt angle of the collector or $70°$, whichever is smaller. Further,

$$f_4 = (T_c - T_a)/(N + f_2), \tag{16.14}$$

unless this expression becomes smaller than 1, in which case f_4 should rather be equal to 1. All temperatures in (16.10) and (16.12) should be inserted in K. The parametrisations are clearly phenomenological, and substantial changes between editions of the Duffie and Beckman book, notably affecting winter results, emphasise the *ad hoc* nature of the procedure.

The total net energy gain of the absorber plate of the collector is the sum of the contributions specified above,

$$E_c^{gain} = E_c^{sw} + E_c^{lw} + E_{top}^{sens} + E_{back}^{sens}. \tag{16.15}$$

Stalled and operating collector

If the mass flow $J_m{}^c = dm/dt$ through the solar collector is zero, the entire energy gain (16.15) will be used to raise the temperature of the collector. This situation of a "stalled collector" is of considerable interest for determining the maximum temperature that the materials used in constructing the collector must be able to withstand, for example, in a situation of pump failure or complete loss of circulating fluid. Denoting the heat capacity of the "dry" collector C', taken per unit area of collector, the equation determining the time development of the stalled-collector plate temperature becomes

$$AC' \, dT_c/dt = E_c^{sens}(T_c). \tag{16.16}$$

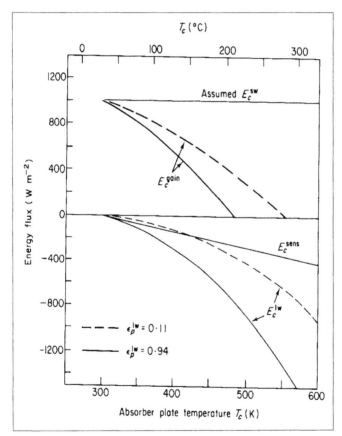

Figure 16.7. Energy flux components for a stalled collector, as a function of plate temperature T_c for a selective and a non-selective absorber surface (dashed and solid lines). There are two cover glasses of emittance $\varepsilon_g^{lw} = 0.88$, there is no wind (the curves are sensitive to wind speed changes), and all environmental temperatures have been taken as 300 K. The plate absorptance is $\alpha^{sw} = 0.94$, and the collector is placed horizontally. The curve E_c^{sens} represents the sum of fluxes reaching the top and back side of the collector (both positive towards the collector).

As T_c rises, the negative terms in (16.15) increase, and if the incident energy can be regarded as constant, an equilibrium situation will result in which the heat losses exactly balance the gains and the temperature thus remains constant,

$$E_c^{gain}(T_{c,max}) = 0. \tag{16.17}$$

The determination of $T_{c,max}$ and the magnitude of the individual terms in (16.15), as a function of T_c, are illustrated in Fig. 16.7 for a high incoming flux E_c^{sw}, environmental temperatures $T_e = T_a = T_L = 300$ K, and no wind, based on the explicit expressions (16.7)–(16.15). Two collectors are considered, one with selective surface ($\varepsilon_p^{lw} = 0.11$) and one with non-selective black surface ($\varepsilon_p^{lw} = \alpha^{sw} = 0.94$). Both have two layers of cover glass. The corresponding equilibrium temperatures are approximately $T_{c,max} = 550$ K and $T_{c,max} = 480$ K. If wind were present, the convective heat loss (16.12) would be much higher and the maximum temperature correspondingly lower.

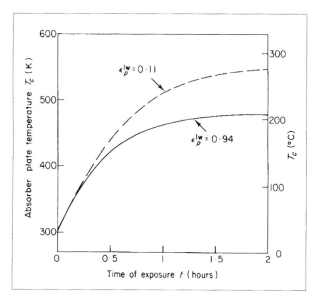

Figure 16.8. Plate temperature for a stalled, dry collector as a function of time. The incident flux is constant $E_c^{sw} = 1000$ W m^{-2}, and the heat capacity C' in (16.16) has been taken as 10^4 J m^{-2} K^{-1}, corresponding to a fairly light-weight construction. Other parameters are as in Fig. 16.7. About 7 min are required to reach operating temperatures (some 50 K above ambient). The time scale is proportional to C', so results for other values of C' are easily inferred.

The time required to reach a temperature T_c close to $T_{c,max}$ can now be calculated from (16.16) by integration, as illustrated in Fig. 16.8. While the maximum temperature is independent of the heat capacity of the collector, the time scales linearly with C'. The value $C' = 10^4$ J m^{-2} K^{-1} used in Fig. 16.8 corresponds to a fairly light collector (e.g. absorber plate of extruded aluminium), in which case the asymptotic temperature region is reached after about 1 h of constant high incoming radiation. The shortness of this period makes the assumption of a constant E_c^{sw} acceptable. The amount of time needed to reach operating temperatures of about 50°C is roughly 7 min. It would double if the ambient temperature were at the freezing point of water and would further increase if the radiation were lower than the 1000 W m^{-2} assumed in the example used in Figs. 16.7 and 16.8. A short response time of the collector is also of relevance in situations of variable radiation (e.g. caused by frequent cloud passage).

In case of an operating collector, $J_m{}^c$ is no longer zero and a fluid carries energy away from the absorber (cf. Fig. 16.4). The temperature of the outgoing fluid, $T_{c,out}$, will in this case be higher than that of the incoming fluid, $T_{c,in}$, and it is no longer consistent to assume that the plate temperature T_c is uniform over the plate. The change in plate temperature must be at least as large as the change in fluid temperature, between the points where the fluid is receiving its first and last heat transfer contribution from the absorber plate. Because of the non-linear relation between plate temperature and heat loss, it is not correct to evaluate the losses at the mean plate temperature, T_c. Still, this is often done in approximate calculations, choosing

$$\overline{T}_C = \tfrac{1}{2}(T_{c,in} + T_{c,out}),\tag{16.18}$$

and assuming further that the average fluid temperature is also the average plate temperature. Incomplete heat transfer between absorber plate and fluid channels, as well as the non-linearity of loss functions, may be simulated by adding a constant term to (16.18), or by using a factor larger than ½. The transfer of heat to the fluid may be effectuated by passing the fluid along the absorber plate (used e.g. in most collectors with air as a working fluid) or by the fluid's passage through pipes in highly heat-conducting contact with the absorber plate (most often used when the working fluid is water, oil, etc.). In the latter case, the non-uniform heat extraction constitutes a further cause of non-uniform plate temperature (the temperature gradients towards the pipe locations being, of course, the reason for obtaining a heat flow from any plate position to the relatively small area of contact with the fluid-carrying pipes).

If the fluid of fixed inlet temperature $T_{c,in}$ is allowed to pass only once through the collector (e.g. for hot water production), and if the simplifying assumptions mentioned above are made, then the equation replacing (16.16) for an operating collector will be

$$E_c{}^{gain}(\overline{T}_c) = AC'\, d\overline{T}_c/dt + J_m{}^c\, C_p{}^c\,(T_{c,out} - T_{c,in}),\tag{16.19}$$

or in a steady-state situation just

$$E_c{}^{gain}(\overline{T}_c) = J_m{}^c\, C_p{}^c\,(T_{c,out} - T_{c,in}).$$

$C_p{}^c$ is the heat capacity of the fluid flowing through the collector. This equation determines $T_{c,out}$, but the left-hand side depends on $T_{c,out}$, through (16.18) or some generalisation of this relation between the flow inlet and outlet temperatures and the effective average plate temperature. Therefore, it is conveniently solved by iteration,

$$T_{c,out}{}^{(i+1)} = T_{c,in} + E_c{}^{gain}\,(\tfrac{1}{2}\,(T_{c,in} + T_{c,out}{}^{(i)}))\,/\,(J_m{}^c C_p{}^c).\tag{16.20}$$

A better approximation for the transfer of heat from the absorber plate to the fluid channels may be obtained along the lines described below for the general problem of heat exchange encountered in several places with application of the solar collectors in heat supply systems, for example, involving a heat storage that is not part of the collector flow circuit.

Flat-plate collector with heat storage

The general layout of a flat-plate solar collector with heat storage is shown in Fig. 16.9. There are two heat exchangers on the collector side and two on the load side. If the actual number is less, for example, if the storage is in terms of a fluid that can be circulated directly through the collector, then the corresponding heat transfer coefficient [h in (6.3)] should be taken as infinitely large. While the relationship between temperatures at every heat exchanger may be taken from (6.3), the net energy transfer to the storage is associated with a change in the storage temperature (the average value of which is $\overline{T}_s$), which may be obtained from the equation accounting for the relevant energy fluxes, assuming the storage temperature to be uniform,

$$SC^s\, dT_s/dt = J_m^c C_p^c (T_{c,out} - T_{c,in}) - J_m^L C_p^L (T_{L,in} - T_{L,out}) - h_s(\overline{T}_s - T_0). \qquad (16.21)$$

Here S is the storage volume (or mass), C^s is the heat capacity on a volume (or mass) basis of the material used for storage (assuming for simplicity storage in the form of sensible heat), superscripts c and L distinguish the flows in the collector and the load circuits, and finally, T_0 is the temperature of the surroundings of the storage volume (e.g. soil in the case of underground storage), to which the storage is losing energy through its walls. (16.21) must be supplemented by (16.19) and a similar equation for the load area, e.g. of the form

$$LC^L\, dT_L/dt = J_m^L C_p^L (T_{L,in} - T_{L,out}) - h_L(\overline{T}_L - T_0), \qquad (16.22)$$

assuming that the load area is a building of volume (or mass) L and average temperature $\overline{T}_L$, which may lose energy to surroundings of ambient air temperature T_a. If the load is in the form of water heating, it may be described by a term of the same form as the first one on the right-hand side of (16.22), but with $T_{L,out}$ being the water inlet temperature and $T_{L,in}$ being the water temperature of the water supplied by the system for usage.

Together with the four heat exchanger equations (6.2) with h_c, h_{cs}, h_{sL} and h_L (see Fig. 16.9), (16.19), (16.21) and (16.22) constitute seven equations for determining the seven temperatures appearing in Fig. 16.9. If the load circuit is not activated, the four equations for $T_{c,in}$, $T_{c,out}$, $\overline{T}_s$ and $\overline{T}_c$ remain. Instead of the heat exchanger equation (6.2) for the collector, which for $h_c \to \infty$ gives $T_{c,out} = \overline{T}_c$ owing to the assumption that the reservoir of temperature $\overline{T}_c$ is large compared with the amounts of heat exchanged, relations such as

(16.18) may be used, or a more elaborate calculation taking into account the geometry of the collector and fluid pipes or channels may be performed.

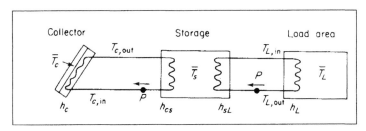

Figure 16.9. Solar heating system with storage. h denotes a heat exchanger, P denotes a circulation pump, and the T's are temperatures.

The efficiency of the collector itself may be defined as

$$\eta_c = J_m{}^c C_p{}^c (T_{c,out} - T_{c,in}) / E_{s,\gamma}{}^{sw}, \tag{16.23}$$

where the numerator may be obtained from (16.19) and where the denominator is the total short-wavelength radiation flux on the collector surface characterised by its tilt and azimuth angles s and γ relative to a horizontal plane (i.e. direct and scattered/diffuse radiation, plus any radiation reflected from other surfaces onto the plane considered). Only in case the heat capacity of the collector is neglected, or if the collector is operating at a constant temperature, can the collector heat output (to the fluid) be taken as the collector gain $E_c{}^{gain}(T_c)$ from (16.15) directly.

If an energy store is used in connection with the collector, the relevant efficiency is that of delivering the heat to the storage, which is formally the same as (16.23). In practice, however, the introduction of an energy storage may completely alter the efficiency by changing $T_{c,in}$, which again causes a change in $T_{c,out}$. When the storage temperature, T_s, is low, the average collector plate temperature $\overline{T_c}$ will be fairly low and the loss terms correspondingly small. Since $T_{c,in}$ will always be higher than T_s (but not always higher than $\overline{T_s}$, since often the fluid circuit is made to leave the storage at a place of low temperature, if T_s is not uniform), the accumulation of stored energy and associated increases in T_s and $T_{c,in}$ will cause the collector loss terms to increase, implying a diminishing efficiency. Examples of this behaviour are apparent from numerous simulations of complete solar heating systems (see e.g. Sørensen, 2004).

Another important factor in determining the efficiency of energy collection is the rate of fluid circulation through the collector circuit, $J_m{}^c$. Figure 16.10 gives an example of the performance of the same system with two different values of $J_m{}^c$. With the low fluid velocity, the exit temperature from the collector, $T_{c,out}$, is about 45°C above the inlet temperature, but the energy transferred to the storage is much less than in the case of a higher fluid velocity, causing a smaller difference between $T_{c,out}$ and $T_{c,in}$, but larger net collector gain, owing to the smaller loss terms. The optimum fluid flow rate for a given heat exchange capability h_{cs} (assuming the heat transfer at the collector to be nearly perfect) may be found by calculating the collection efficiency as a function of J_m^c or, as a more generally applicable criterion, by calculating the coverage of the actual energy demand that is sought covered by the solar heat system. If a given solar heat system is considered, the percentage of the total heat load over an extended period, such as a year, may be evaluated as function of J_m^c, as done in Fig. 16.11.

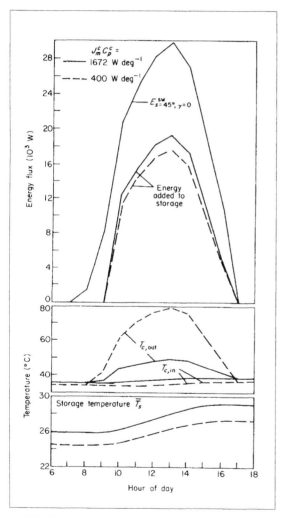

Figure 16.10. Single-day performance of a solar heating system, based on a simulation model using the Danish Reference Year (latitude 56°N). The day is 8 February, the collector area $A = 40$ m^2, and the storage volume $S = 25$ m^3 (water equivalent). The collector has one glass cover layer and a selective surface ($\varepsilon_c^{lw} = 0.11$), and the heat transfer coefficient $h_{cs} = 4000$ W K^{-1}. Two different flow rates are considered for the collector circuit, $J_m^c C_p^c = 1672$ or 400 W K^{-1} corresponding to 0.4 or 0.096 kg of water per second.

If 100% coverage is aimed at, the parameter to be optimised may be taken as the minimum size (e.g. collector area A), which will allow the full load to be covered at any time. The example shown in Fig. 16.11 relates to the heating and hot water requirements of a one-family dwelling at 56°N latitude. For each value of the heat exchange coefficient h_{cs} between the collector circuit and the storage, the optimum flow rate is determined. With increasing h_{cs}, the region of "acceptable" values of J_m^c becomes larger, the increased J_m^c indicated for large h_{cs} does not significantly improve the performance of the system, and it will soon become uneconomical due to the energy spent in pumping the fluid through the circuit (the energy required for the pump is negligible for low fluid speeds, but the pipe resistance increases with the fluid speed, and the resistance through the heat exchanger increases when h_{cs} does). The results of Fig. 16.11 may then be interpreted as suggesting a modest increase in fluid velocity, from about 0.1 to about 0.2 kg (or litre) s^{-1}, when the heat exchange coefficient is increased from 400 to 4000 W K^{-1} (for the particular system considered).

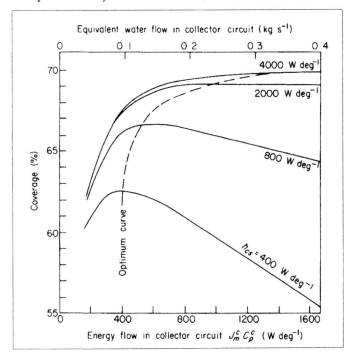

Figure 16.11. Annual average coverage of heating and hot water load for a one-family dwelling (average load 1.91 kW) under conditions set by the Danish Reference Year (56°N) as a function of water flow in collector circuit and heat exchanger transfer coefficient. The collector is as described in the caption to Fig. 16.10 and placed on a south-facing roof with tilt angle s = 45°.

The optimisation problem outlined in this example is typical of the approach to the concept of an "efficiency" for the total solar heat system, including storage and load sections. Because the time distribution of the load is, in general, barely correlated with that of the solar energy collection (in fact, it is more "anti-correlated" with it), an instantaneous efficiency defined as the ratio between the amount of energy delivered to the load area and the amount of incident solar energy is meaningless. A meaningful efficiency may be defined as the corresponding ratio of average values, taken over a sufficient length of time to accommodate the gross periodicity of solar radiation and of load, i.e. normally a year,

$$\bar{\eta}_{system} = \bar{c} / \bar{E}_{s,\gamma}^{sw}, \tag{16.24}$$

where c is the average load covered by the solar energy system and $\bar{E}_{s,\gamma}^{sw}$ is the average incident short-wavelength radiation on the surface of the collector (its cover if any). Conventionally, the long-wavelength or other heat contributions to the incident energy flux are not included in the efficiency calculations (16.24) or (16.23).

Concentrating collectors and tracking systems

Various advanced versions of the flat-plate collector are conceivable in order to increase the amount of absorbed energy and decrease the amount of heat lost. The incident average flux may be increased by replacing the fixed installation (of tilt and azimuth angles s and γ) by a movable one, which can be made to track the direction towards the Sun. Fully tracking (i.e. in both height and azimuth angle) devices may be mounted in a manner similar to the one used for (star-)tracking astronomical instruments, and incompletely tracking systems (e.g. only following the azimuth angle of the Sun) might combine a simpler design with some improvement in energy gain. The gain made possible by tracking the Sun is not from scattered and (diffusely) rejected fluxes as they do not change much on average by going from a fixed to a tracking collector, but from the direct part, which becomes replaced by the normal incidence flux S_N (i.e. losing a factor of $\cos\theta$, with θ being the angle between the direction to the Sun and the normal to the collector plane). For a flat-plate collector, the maximum average gain from using a fully tracking system (less than a factor 2) would rarely justify the extra cost of the tracking system, which, at least at present, exceeds the cost of doubling the collector area.

Other possible improvements of the flat-plate collector system include coating the cover layers to reduce reflection losses on the outside and increase reflection (back on to the absorber) on the inside surfaces. The mainly convective heat loss E_{top}^{sens} may be reduced by installation of shields that diminish the wind velocity outside the top cover layer. The strong dependence of convection losses on wind speed is clear from (16.12). The convective losses may be more radically diminished by evacuating the space between the absorber plate and the cover system. In vacuum, heat transport can take place only by radiation. The evacuation of the substantial space between absorber and cover typical of collector designs of the type shown in Fig. 16.4 is not practical, and systems using evacuation-type insulation would be built in a different way, e.g. with units of evacuated cylinders or spheres containing the absorbers and possibly concentrating equipment.

Focusing, or more generally concentrating, devices constitute a separate class of solar collectors. They are necessary if very high temperatures are required, but may, in principle, also be considered for heating purposes involving modest temperature rises over the ambient. For a flat-plate collector without any of the special loss-reducing features mentioned above, the net gain is a rapidly decreasing function of the plate temperature, as seen in Fig. 16.7. Since the losses are proportional to the absorber area, there is a special interest in reducing this area relative to the collector area A determining the incident radiation energy.

One extreme is a point-focusing device, such as the parabolic reflectors in Fig. 16.12 or the lenses in Fig. 16.14. If the absorber is placed in the focus on the symmetry axis and its dimension corresponds to that of the image of the Sun when the Sun is in the direction of the symmetry axis ("optical axis"), then direct radiation will be accepted, but only as long as the optical axis is made to track the Sun. Imperfections and, in the case of lenses, the wavelength dependence of the refraction index may further enlarge the image dimensions. The absorber dimension would have to be chosen so that it covered the bulk of the image under most conditions.

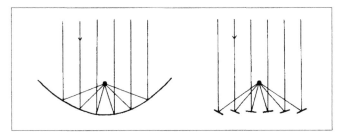

Figure 16.12. Parabolic- and Fresnel-type reflectors, e.g. for use in heliostats and tower absorbers, respectively.

The total short-wavelength radiation received by the absorber, E_c^{sw}, may be written in the same form as for the flat-plate collector, (16.3) or (16.5), but with a transmission–absorption product, $P^{t.a.}(\Omega)$, suitable for the focusing geometry rather than (16.4). If reflections and re-absorptions at the absorber surface are neglected, the transmission–absorption product for a focusing device may be written as the product of the absorptance α^{sw} of the absorber surface and a transmission function $t^{sw}(\Omega)$, which expresses the fraction of the incident short-wavelength radiation from the direction Ω that reaches the absorber surface (directly or after one or more reflections or refractions within the device),

$$P^{t.a.}(\Omega) = \alpha^{sw} t^{sw}(\Omega). \tag{16.25}$$

For the parabolic reflector (Fig. 16.12) and the lens (Fig. 16.14), the idealised transmission function is zero, unless the absorber is placed exactly in the focal point corresponding to incoming radiation from the direction Ω.

Thus, either the collector has to fully track the Sun or the absorber has to be moved to the focal point corresponding to a given direction to the Sun. For scattered radiation, which by definition does not come from the direction of the Sun (the same being normally the case for reflected radiation), $t^{sw}(\Omega)$ is zero, and such radiation will not be available for these devices. As a compromise, the acceptance angle may be increased at the expense of a high concentration ratio. Some scattered and reflected radiation will be accepted, and complete tracking may not be necessary.

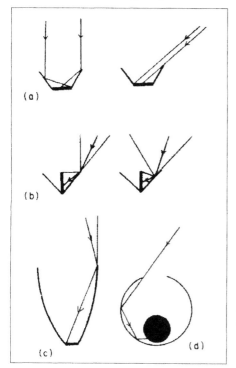

Figure 16.13. Examples of concentrating collectors.

    Figure 16.13 gives a number of examples of such devices. The inclined "booster" mirrors along the rim of a flat-plate collector (Fig. 16.13a) result in a modest increase in concentration for incident angles not too far from the normal to the absorber plate, but a decrease (due to shadows) for larger incident angles. The V- or cone-shaped collector with the absorber in the shape of a wall or an upright line (cylinder) in the middle (Fig. 16.13b) has an acceptance angle depending on the distance from the convergence point of the cone. At this point, the acceptance angle equals the opening angle, but at points further out along the cone sides the acceptance angle is smaller (depending on the length of the absorber line, or wall height) relative to that of the cone. A fairly high concentration can be achieved with a cusp construction (Fig. 16.13c), comprising two parabolic sections, each of which has its focus at the point of intersection between the opposite parabolic section and the absorber. Figure 16.13d shows a "trapping" device consisting of a cylinder (or sphere) with a window, and an absorber displaced from the centre. The dimensions of the absorber are such that a large fraction of the radiation passing through the window will be absorbed after (multiple) reflections on the reflecting inside walls of the cylinder. Also, for the long-wavelength radiation emitted by the absorber, the reflecting walls will lead a substantial fraction back to the absorber, so that the loss-to-gain ratio is improved, even though there is no actual concentration (if the dimension of the absorber is as large as that of the window).

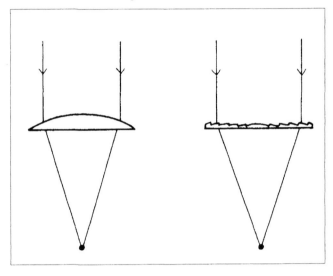

Figure 16.14. Simple Fresnel-type lenses.

Energy collection from focusing systems

A survey of a number of optically focusing systems for solar energy collection may be found in Meinel and Meinel (1976).

In the absence of cover systems, the long-wavelength emission from a focusing system may be written in the form (16.6),

$$- E_c^{lw} = A_a \varepsilon^{lw} \sigma (T_c^{4} - T_e^{4}),$$ (16.26)

where A_a is the absorber area.

Similarly, the sensible heat exchange with the surroundings of ambient temperature T_a may be given in analogy to (16.7) and (16.8),

$$E_c^{sens} = - A_a U (T_c - T_a),$$ (16.27)

where U may depend on wind speed in analogy to (16.9). The collector gain is now given by $E_c^{gain} = E_c^{sw} + E_c^{lw} + E_c^{sens}$, where the first term is proportional to A, while the two last terms are proportional to A_a. Assuming now a fluid flow J_m^c through the absorber, the expression for determination of the amount of energy extracted becomes [cf. (16.19)]

$$J_m^c C_p^c (T_{c,out} - T_{c,in}) = E_c^{gain}(\overline{T}_c) - A_a C' \, d\overline{T}_c / dt$$

$$= E_c^{sw} + E_c^{lw}(\overline{T}_c) + E_c^{sens}(\overline{T}_c) - A_a C' \, d\overline{T}_c / dt), \qquad (16.28)$$

where $A_a C'$ is the heat capacity of the absorber and the relation between $\overline{T}_c$ and $T_{c,out}$ may be of the form (16.20) or (16.22). Inserting (16.5) with use of (16.25) to (16.27) into (16.28), it is apparent that the main additional parameters specifying the focusing system are the area concentration ratio,

$$X = A/A_a, \qquad (16.29)$$

and the energy flux concentration ratio,

$$C^{flux} = A t^{SW}(\Omega) / A_i, \qquad (16.30)$$

where t^{SW} is the transmission function appearing in (16.25) and A_i is the area of the actual image (which may be different from A_a). Knowledge of A_i is important for determining the proper size of the absorber, but it does not appear in the temperature equation (16.28), except indirectly through calculation of the transmission function t^{SW}.

For a stalled collector, $J_m^c = 0$ and d $\overline{T}_c / dt = 0$ give the maximum absorber temperature $T_{c,max}$. If this temperature is high, and the wind speed is low, the convective heat loss (16.29) is negligible in comparison with (16.26), yielding

$$\sigma(T_{c,max}^4 - T_c^4) = X D \, \alpha^{sw} \varepsilon^{lw},$$

where D is the incident short-wavelength flux in the direction of the Sun [cf. (16.5)], assuming the device accepts only the flux from this direction and assuming perfect transmission to the absorber ($t^{SW} = 1$). With the same conditions, the best performance of an operating collector is obtained from (16.30), assuming a steady-state situation. The left-hand side of (16.30) is the amount of energy extracted from the collector per unit of time, E^{extr}:

$$E^{extr} = A \alpha^{SW} D - A_a \varepsilon^{lw} \sigma (\overline{T}_c^4 - T_c^4), \qquad (16.31)$$

and the efficiency corresponding to the idealised assumptions

$$\eta_c^{ideal} = \frac{E^{extr}}{AD} = \alpha^{SW} \left(1 - \frac{\varepsilon^{lw} \sigma}{X \alpha^{SW} D} (\overline{T}_c^4 - T_e^4) \right). \qquad (16.32)$$

This relation is illustrated in Fig. 16.15, assuming $D = 800$ W m^{-2}, $T_c = 300$ K and $\alpha^{SW} = 0.9$ independent of absorber temperature. It is clear that the use of a selective absorber surface (large ratio $\alpha^{SW}/\varepsilon^{lw}$) allows a higher efficiency to be obtained at the same temperature and that the efficiency drops rapidly with decreasing light intensity D. In case the total short-wavelength flux on the collector is substantially different from the direct component, the actual efficiency relative to the incident flux $E_{s,\gamma}^{SW}$,

$$\eta_c = E^{extr} / E_{s,\gamma}^{SW},$$

may become far less than the "ideal" efficiency (16.32).

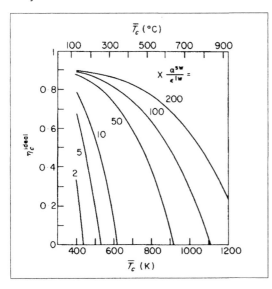

Figure 16.15. Ideal collector efficiency for concentrating collectors, evaluated for a direct radiation incident flux of $D = 800$ W m^{-2}, an environmental temperature $T_e = 300$ K and $\alpha^{SW} = 0.9$.

SOLAR THERMAL ELECTRICITY GENERATORS

CHAPTER 17

Conversion of solar radiation into electric energy may be achieved either in two steps by, for example, first converting radiation to heat, and then heat into electricity, using one of the methods described in Chapters 2 to 6, or alternatively by direct conversion of radiation into electricity, using the photovoltaic conversion scheme discussed in Chapter 14. Two-step conversion using chemical energy rather than heat as the intermediate energy form is also possible, for instance, by means of the photogalvanic conversion scheme (Sørensen, 2005). The present chapter gives examples of the layout of solar radiation-to-heat-to-electricity systems and discusses their performance.

Photo-thermoelectric converters

A two-step conversion device may be of the general form shown in Fig. 17.1. The solar collector may be of the flat-plate type or may perform a certain measure of concentration, requiring partial or full tracking of the Sun. The second step is performed by a thermodynamic engine cycle, for example, a Rankine cycle with expansion through a turbine as indicated in the figure. Owing to the vaporisation of the engine-cycle working fluid in the heat exchanger linking the two conversion steps, the heat exchange performance does not follow the simple description leading to (6.2). The temperature behaviour in the heat exchanger is more like the type shown in Fig. 17.2. The fluid in the collector circuit experiences a uniform temperature decrease as a function of the distance travelled through the heat exchanger, from x_1 to x_2. The working fluid entering the heat exchanger at x_2 is heated to boiling point, after which further heat exchange is used to evaporate the working fluid (and perhaps superheat the gas), so that the temperature curve becomes nearly flat after a certain point.

The thermodynamic engine cycle (the right-hand side of Fig. 17.1) can be described by the method outlined in Chapter 2, with the efficiency of heat to electricity conversion given by (2.19) [limited by the ideal Carnot process efficiency (2.4)],

$$\eta_w = E^{electric} / J_{Q,in}{}^w \leq (T_{w,in} - T_{w,out}) / T_{w,in},$$

where $E^{electric} = - J_q F_q$ is the electric power output [cf. (2.14)] and

$$J_{Q,in}{}^w = J_m{}^c C_p{}^c (T_{c,out} - T_{c,in})$$

is the rate of heat transfer from the collector circuit to working fluid circuit in the heat exchanger. The right-hand side is given by (16.28) for a concentrating solar collector and by (16.19) and (16.15) for a flat-plate collector. The overall conversion efficiency is the product of the efficiency η_c of the collector system and η_w,

$$\eta = \eta_c \eta_w.$$

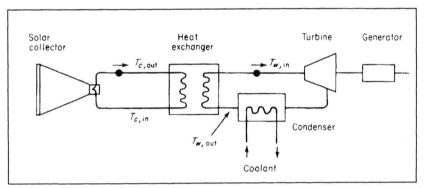

Figure 17.1. Photo-thermoelectric generator shown as based on concentrating solar collectors.

The determination of the four temperatures $T_{c,in}$, $T_{c,out}$, $T_{w,in}$ and $T_{w,out}$ (cf. Fig. 17.1) requires an equation for the collector performance, for the heat transfer to the collector fluid, for the heat transfer in the heat exchanger and for the processes involving the working fluid of the thermodynamic cycle. If the collector performance is assumed to depend only on an average collector temperature, $\overline{T}_c$ [e.g. given by (16.18)], and if $T_{w,in}$ is considered to be roughly equal to $\overline{T}_c$ (Fig. 17.2 indicates that this may be a fair first-order approximation), then the upper limit for η_w is approximately

$$\eta_w \lesssim (\overline{T}_c - T_{w,out}) / \overline{T}_c.$$

Taking $T_{w,out}$ as 300 K (determined by the coolant flow through the condenser in Fig. 17.1), and replacing η_c by the idealised value (16.32), the overall efficiency may be estimated as shown in Fig. 17.3, as a function of $\overline{T}_c$. The assumptions for $\eta_c{}^{ideal}$ are as in Fig. 16.15: no convective losses (this may be important for flat-plate collectors or collectors with a modest concentration factor) and an incident radiation flux of 800 W m^{-2} reaching the absorber (for strongly focusing collectors, this also has to be direct radiation). This is multiplied by η_w^{ideal} (the Carnot limit) to yield η_{total}^{ideal} depicted in Fig. 17.3, with the further assumptions regarding the temperature averaging and temperature drops in the heat exchanger that allowed the introduction of $\overline{T}_c$ as the principal variable both in η_w^{ideal} and in η_c^{ideal}.

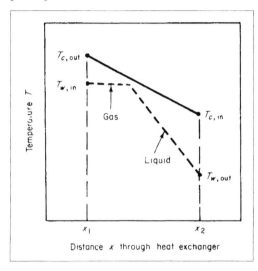

Figure 17.2. Temperature behaviour in heat exchanger for two-phase working fluid of temperature T_w and the fluid in the collector circuit of temperature T_c (based on Athey, 1976).

In realistic cases, the radiation reaching the absorber of a focusing collector is perhaps half of the total incident flux $E_{s,\gamma}^{SW}$ and η_w is maybe 60% of the Carnot value, i.e.

$$\eta_{total} \approx 0.3\ \eta_{total}^{ideal}.$$

This estimate may also be valid for small concentration values (or rather small values of the parameter $X\alpha^{sw}/\varepsilon^{lw}$ characterising the curves in Figs. 17.3 and 16.15), since the increased fraction of $E_{s,\gamma}^{sw}$ being absorbed is compensated for by high convective heat losses. Thus, the values of η_{total}^{ideal} between 6 and 48%, obtained in Fig. 17.3 for suitable choices of $\overline{T}_c$ (this can be adjusted by altering the fluid flow rate J_m^c), may correspond to realistic efficiencies of 2–15%, for values of $X\alpha^{sw}/\varepsilon^{lw}$ increasing from 1 to 200.

The shape of the curves in Fig. 17.3 is brought about by the increase in η_w^{ideal} as a function of $\overline{T}_c$, counteracted by the accelerated decrease in η_c^{ideal} as a function of $\overline{T}_c$, which is seen in Fig. 16.15.

Photo-thermoelectric conversion systems based on flat-plate collectors have been discussed, for example, by Athey (1976); systems based on solar collectors of the type shown in Fig. 16.13d have been considered by Meinel and Meinel (1972). They estimate that a high temperature on the absorber can be obtained by use of selective surfaces, evacuated tubes (e.g. made of glass with a reflecting inner surface except for the window shown in Fig. 16.13d), and a crude Fresnel lens (see Fig. 16.14) concentrating the incoming radiation on the tube window. Molten sodium is suggested as the collector fluid. Finally, fully tracking systems based on the concept shown on the right-hand side of Fig. 16.12 ("power towers") have been suggested, e.g. by Teplyakov and Aparisi (1976) and by Hildebrandt and Vant-Hull (1977).

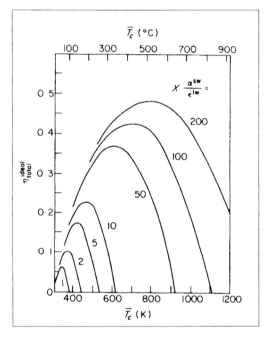

Figure 17.3. Ideal overall efficiency of photo-thermoelectric converter, based on the assumptions underlying Figs. 16.15 and 17.2 and with the Rankine cycle efficiency replaced by an ideal Carnot cycle efficiency.

CHAPTER 18

SOLAR COOLING AND OTHER APPLICATIONS

A variety of applications of solar energy thermal conversion have been considered, in addition to heat use. Living comfort and food preservation require cooling in many places. As in the case of heating, the desired temperature difference from the ambient is usually only about 10–30°C. Passive systems such as those depicted in Figs. 16.1 and 16.2 may supply adequate cooling in the climatic regions for which they were designed. Radiative cooling is very dependent on the clearness of the night sky (Sørensen, 2004). In desert regions with a clear night sky, the difference between the winter ambient temperature and the effective temperature of the night sky has been used to freeze water. In Iran, the ambient winter temperature is typically a few degrees Celsius above the freezing point of water, but the temperature T_e of the night sky is below 0°C. In previous centuries, ice (for use in the palaces of the rich) was produced from shallow ponds surrounded by walls to screen out the daytime winter Sun, and the ice produced during the night in such natural icemakers ("yakhchal"; see Bahadori, 1977) could be stored for months in deep (say, 15 m) underground containers.

Many places characterised by hot days and cool nights have taken advantage of the heat capacity of thick walls or other structures to smooth out the diurnal temperature variations. If additional cooling is required, active solar cooling systems may be employed, and a "cold storage" may be introduced in order to cover the cooling need independent of the variations in solar radiation, in analogy to the "hot storage" of solar heating systems.

The solar cooling system may consist of flat-plate collectors delivering the absorbed energy to a "hot storage", which is then used to drive an absorption cooling cycle (Fig. 18.1), drawing heat from the "cool storage", to which the load areas are connected (see e.g. Wilbur and Mancini, 1976). In principle, only one kind of storage is necessary, but with both hot and cold storage the system can simultaneously cover both heating needs (e.g. hot water) and cooling needs (e.g. air conditioning).

The absorption cooling cycle (Fig. 18.1) is achieved by means of an absorbent–refrigerant mix, such as LiBr–H_2O or H_2O–NH_3. The lithium–bromide–water mix is more suitable for flat-plate solar collector systems, giving higher efficiency than the water–ammonia mix for the temperatures characteristic of flat-plate collectors. LiBr is hygroscopic, i.e. it can mix with water in any ratio. The solar heat is used in a "generator" to vaporise some water from the mix. This vapour is led to a condenser unit, using a coolant flow, and is then expanded to regain a gaseous phase, whereby it draws heat from the area to be cooled, and is returned to an "absorber" unit. Here it becomes absorbed in the LiBr–H_2O mix with the help of a coolant flow. The coolant inlet temperature would usually be the ambient one, and the same coolant passes through the absorber and the condenser. The coolant exit temperature would then be above ambient, and the coolant cycle could be made closed by exchanging the excess heat with the surroundings (e.g. in a "cooling tower"). The refrigerant-rich mix in the absorber is pumped back to the generator and is made up for by recycling refrigerant-poor mix from the generator to the absorber (e.g. entering through a spray system). In order not to waste solar heat or coolant flow, these two streams of absorbent–refrigerant mix are made to exchange heat in a heat exchanger. The usefulness of a given absorbent–refrigerant pair is determined by the temperature dependence of vaporisation and absorption processes.

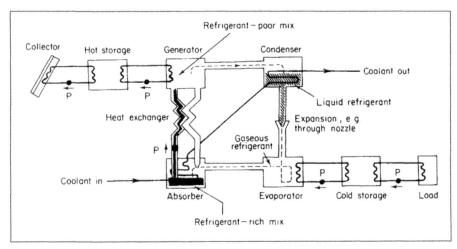

Figure 18.1. Solar absorption cooling system using a mix of absorbent (e.g. LiBr) and refrigerant (e.g. H_2O). P denotes pump. The coolant may be recycled after having rejected heat to the environment in a cooling tower.

In dry climates a very simple method of cooling is to spray water into an air-stream (evaporative cooling). If the humidity of the air should remain unchanged, the air has first to be dried (spending energy) and then cooled by evaporating water into it, until the original humidity is again reached.

In principle, cooling by means of solar energy may also be achieved by first converting the solar radiation to electricity by one of the methods described in Chapters 14, 15 or 17 and then using the electricity to drive a thermo-dynamic cooling cycle, such as the Rankine cycle in Fig. 3.1.

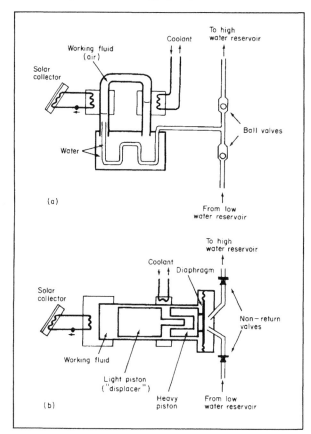

Figure 18.2. Solar water pumps based on Stirling cycles. In (a), air and water should not mix. The ball valves produce cyclic operation. In (b), the light and heavy pistons should be out of phase by a half cycle (based on West, 1974; Beale 1976).

The same applies if the desired energy form is work, as in the case of pumping water from a lower to a higher reservoir (e.g. for irrigation of agricultural land). In practice, however, the thermodynamic cycles discussed in Chapter 3 are used directly to make a pump produce mechanical work on the basis of the heat obtained from a solar collector. Figure 18.2 shows two types of devices based on the Stirling cycle, using air or another gas as a working fluid (cf. Fig. 3.1). In the upper part of the figure, two ball valves ensure an oscillatory, pumping water motion, maintained by the tendency of the temperature gradient in the working fluid (air) to move air from the cold to the hot region. The lower part of the figure shows a "free piston

Stirling engine" in which the oscillatory behaviour is brought about by the presence of two pistons of different mass, the heavy one being delayed by half an oscillatory period with respect to the lighter one. The actual water pumping is made by the "membrane-movements" of a diaphragm, but if the power level is sufficient the diaphragm may be replaced by a piston-type pump. For Stirling cycle operation (cf. Chapter 5), the working fluid is taken as a gas. This is an efficient cycle at higher temperatures (e.g. with use of a focusing solar collector), but if the heat provided by the (flat-plate) solar collector is less than about 100°C, a larger efficiency may be obtained by using a two-phase working fluid (corresponding to one of the Rankine cycles shown in Fig. 3.1).

If focusing solar collectors are used, the diaphragm may be replaced by a more efficient piston pump.

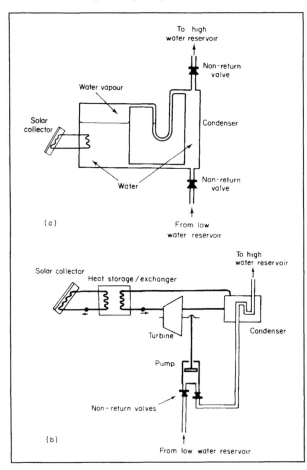

Figure 18.3. Solar water pumps based on Rankine cycles. In (a), the formation of vapour, its transfer through the U-tube and condensation produces a cyclic pressure–suction variation, which draws water from the lower reservoir and delivers water to the high reservoir (based on Boldt, 1978). In (b), a turbine-type pump is shown, the working fluid of which may be an organic compound (cf. Meinel and Meinel, 1976).

Two Rankine-type solar pumps, based on a fluid-to-gas and gas-to-fluid cycle, are outlined in Fig. 18.3. The one shown in the upper part of the figure is based on a cyclic process. The addition of heat evaporates water in the container on the left, causing water to be pumped through the upper one-way valve. When the vapour reaches the bottom of the U-tube, all of it moves to the container on the right and condenses. New water is drawn from the bottom reservoir, and the pumping cycle starts all over again. This pump is intended for wells of shallow depth, below 10 m (Boldt, 1978).

The lower part of Fig. 18.3 shows a pump operated by a conventional Rankine engine, expanding the working fluid to gas phase passing through a turbine, and then condensing the gas using the pumped water as coolant, before returning the working fluid to the heat exchanger receiving solar absorbed heat. For all the pumps based on a single thermodynamic cycle, the maximum efficiency (which cannot be reached in a finite time) is given by (2.4) and the actual efficiency is given by an expression of the form (2.19).

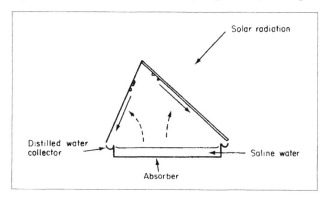

Figure 18.4. Solar still.

Distillation of saline water (e.g. sea water) or impure well water may be obtained by a solar still, a simple example of which is shown in Fig. 18.4. The side of the cover system facing the Sun is transparent, whereas the part sloping in the opposite direction is highly reflective and thus remains at ambient temperature. It therefore provides a cold surface for condensation of water evaporated from the saline water surface (kept at elevated temperature due to the solar collector part of the system). In more advanced systems, some of the heat released by condensation (about 2.3×10^6 J kg^{-1}, which is also the energy required for the vaporisation) is recovered and used to heat more saline water (Talbert et al., 1970).

V. ELECTROCHEMICAL ENERGY CONVERSION PROCESSES

FUEL CELLS

19.1 Electrochemical conversion

Electrochemical energy conversion is the direct conversion of chemical energy, i.e. free energy of the form (2.8) into electrical power or vice versa. A device that converts chemical energy into electric energy is called a *fuel cell* (if the free energy-containing substance is stored within the device rather than flowing into the device, the name "primary battery" is sometimes used). A device that accomplishes the inverse conversion (e.g. electrolysis of water into hydrogen and oxygen) may be called a *driven cell*. The energy input for a driven cell need not be electricity, but could be solar radiation, for example, in which case the process would be photochemical rather than electrochemical. If the same device is used for conversion in both directions, or if the free energy-containing substance is regenerated outside the cell (energy addition required) and recycled through the cell, it may be called a *regenerative* or *reversible fuel cell* and, if the free energy-containing substance is stored inside the device, a *regenerative* or *secondary battery*. All fuel cell types are in principle capable of operation in both directions, although a particular cell may have been optimised for a particular process.

The basic ingredients of an electrochemical device are two electrodes (sometimes called anode and cathode) and an intermediate electrolyte layer capable of transferring positive ions from the negative to the positive electrode (or negative ions in the opposite direction), while a corresponding flow of electrons in an external circuit from the negative to the positive electrode provides the desired power. Use has been made of solid electrodes and fluid electrolytes (solutions), as well as fluid electrodes (e.g. in high-temperature batteries) and solid electrolytes (such as ion-conducting semiconductors). A more detailed treatise of fuel cells may be found in Sørensen (2005).

19.2 Fuel cells

The difference in electric potential, $\Delta\phi_{ext}$, between the electrodes (cf. the schematic illustration in Fig. 19.1) corresponds to an energy difference $e\Delta\phi_{ext}$ for each electron. The total number of electrons that could traverse the external circuit may be expressed as the product of the number of moles of electrons, n_e, and Avogadro's constant N_A, so the maximum amount of energy emerging as electrical work is

$$\Delta W^{(elec)} = n_e\, N_A\, e\, \Delta\phi_{ext}, \tag{19.1}$$

where $\mathscr{F} = N_A e = 96\,400$ C mol^{-1} (Faraday's constant) is sometimes introduced. This energy must correspond to a loss (conversion) of free energy,

$$-\Delta G = \Delta W^{(elec)} = n_e\mathscr{F}\Delta\phi_{ext}, \tag{19.2}$$

which constitutes the total loss of free energy from the "fuel" for an ideal fuel cell. This expression may also be derived from (2.8), using (2.2) and $\Delta Q = T\,\Delta S$, because the ideal process is reversible, and $\Delta W = -P\,\Delta V + \Delta W^{(elec)}$.

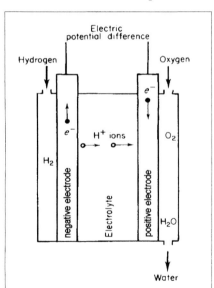

Figure 19.1. Schematic picture of a hydrogen–oxygen fuel cell. The electrodes are in this case porous, so that the fuel gases may diffuse through them.

Figure 19.1 shows an example of a fuel cell, based on the free energy change $\Delta G = -7.9 \times 10^{-19}$ J for the reaction

$$2H_2 + O_2 \rightarrow 2H_2O.$$

Hydrogen gas is led to the negative electrode, which may consist of a porous material, allowing H^+ ions to diffuse into the electrolyte, while the electrons enter the electrode material and may flow through the external circuit. If a catalyst (e.g. a platinum film on the electrode surface) is present, the reaction at the negative electrode

$$2H_2 \rightarrow 4H^+ + 4e^- \tag{19.3}$$

may proceed at a much enhanced rate (see e.g. Bockris and Shrinivasan, 1969). Gaseous oxygen (or oxygen-containing air) is similarly led to the positive electrode, where a more complex reaction takes place, the net result of which is

$$O_2 + 4H^+ + 4e^- \rightarrow 2H_2O. \tag{19.4}$$

This reaction may be built up by simpler reactions with only two components, such as oxygen first picking up electrons or first associating with a hydrogen ion. Also at the positive electrode, the reaction rate can be stimulated by a catalyst. Instead of the porous material electrodes, which allow direct contact between the input gases and the electrolyte, membranes can be used (cf. also Bockris and Shrinivasan, 1969) like those found in biological material, i.e. membranes that allow H^+ to diffuse through but not H_2, etc.

The drop in free energy (19.2) is usually considered to be mainly associated with the reaction (19.4), expressing G in terms of a chemical potential, e.g. of the H^+ ions dissolved in the electrolyte. Writing the chemical potential μ as Faraday's constant times a potential ϕ, the free energy for n moles of hydrogen ions is

$$G = n\mu = n\mathscr{F}\phi = nN_A e\phi. \tag{19.5}$$

When the hydrogen ions "disappear" at the positive electrode according to the reaction (19.4), this chemical free energy is converted into the electrical energy (19.1) or (19.2), and since the numbers of electrons and hydrogen ions in (4.73) are equal, $n = n_e$, the chemical potential μ is given by

$$\mu = \mathscr{F}\phi = \mathscr{F}\Delta\phi_{ext}. \tag{19.6}$$

Here ϕ is the quantity usually referred to as the electromotive force (e.m.f.) of the cell, or "standard reversible potential" of the cell, if taken at standard atmospheric pressure and temperature. From the value of ΔG quoted above, corresponding to -2.37×10^5 J per mole of H_2O formed, the cell electromotoric force (e.m.f.) becomes

$$\phi = -\Delta G / n\mathscr{F} = 1.23 \text{ V}, \tag{19.7}$$

with $n = 2$ since there are two H^+ ions for each molecule of H_2O formed. In the case of multiple constituents (e.g. in a solution), the chemical potential (19.6) may be parametrised in the form (Maron and Prutton, 1959)

$$\mu_i = \mu_i^0 + \mathscr{R} T \log (f_i x_i), \tag{19.8}$$

where i labels the constituents of mole fraction x_i and f_i are empirical parameters. The cell e.m.f. may thus be expressed in terms of the properties of the reactants and the electrolyte, including the empirical activity coefficients f_i appearing in (19.8), as a result of generalising the expression obtained from the definition of the free energy, (2.6), and assuming P, V and T to be related by the ideal gas law, $PV = \mathscr{R}T$, valid for one mole of an ideal gas (cf. e.g. Angrist, 1976).

The efficiency of a fuel cell is the ratio between the electrical power output (19.1) and the total energy drawn from the fuel. However, it is possible to exchange heat with the surroundings, and the energy lost from the fuel may thus be different from ΔG. For an ideal (reversible) process, the heat added to the system is

$$\Delta Q = T \, \Delta S = \Delta H - \Delta G,$$

and the efficiency of the ideal process thus becomes

$$\eta^{ideal} = -\Delta G / (-\Delta G - \Delta Q) = \Delta G / \Delta H. \tag{19.9}$$

For the hydrogen–oxygen fuel cell considered above, the enthalpy change during the two processes (19.3) and (194) is $\Delta H = -9.5 \times 10^{-19}$ J or -2.86×10^5 J per mole of H_2O formed, and the ideal efficiency is

$$\eta^{ideal} = 0.83.$$

There are reactions with positive entropy change, such as $2C + O_2 \rightarrow 2CO$, which may be used to cool the surroundings and at the same time create electric power with an ideal efficiency above one (1.24 for CO formation).

In actual fuel cells, a number of factors tend to diminish the power output. They may be expressed in terms of "expenditure" of cell potential fractions on processes not contributing to the external potential,

$$\Delta \phi_{ext} = \phi - \phi_1 - \phi_2 - \phi_3 - \ldots,$$

where each of the terms $-\phi_i$ corresponds to a specific loss mechanism. Examples of loss mechanisms are blocking of pores in the porous electrodes, e.g. by piling up of the water formed at the positive electrode in the process (19.4), internal resistance of the cell (heat loss) and the building up of potential barriers at or near the electrolyte–electrode interfaces. Most of these mechanisms limit the reaction rates and thus tend to place a limit on the current of ions that may flow through the cell. There will be a limiting current, I_L, beyond which it will not be possible to draw any more ions through the electrolyte, because of either the finite diffusion constant in the electrolyte, if the ion transport is dominated by diffusion, or the finite effective surface of the electrodes at which the ions are formed. Figure 19.2 illustrates the change in $\Delta \phi_{ext}$ as a function of current,

$$I = \Delta\phi_{ext} R_{ext},$$

expressed as the difference between potential functions at each of the electrodes, $\Delta\phi_{ext} = \phi_c - \phi_a$. This representation gives some indication of whether the loss mechanisms are connected with the positive or negative electrode processes, and it is seen that the largest fraction of the losses is connected with the more complex positive electrode reactions in this example. For other types of fuel cells it may be negative ions that travel through the electrolyte, with corresponding changes in characteristics.

It follows from diagrams of the type shown in Fig. 19.2 that there will be an optimum current, usually lower than I_L, for which the power output will be maximum,

$$\max(E) = I^{opt}\,\Delta\phi_{ext}{}^{opt}.$$

Dividing by the rate at which fuel energy ΔH is added to the system in a steady-state situation maintaining the current I^{opt}, one obtains the actual efficiency of conversion maximum,

$$\max(\eta) = I^{opt}\,\Delta\phi_{ext}{}^{opt}\,/\,(\mathrm{d}H/\mathrm{d}t). \tag{19.10}$$

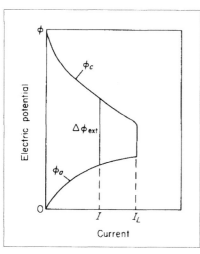

Figure 19.2. Fuel cell negative electrode potential ϕ_a and positive electrode potential ϕ_c as a function of current. The main cause of the diminishing potential difference $\Delta\phi_{ext}$ for increasing current is at first incomplete electrocatalysis at the electrodes, for larger currents also ohmic losses in the electrolyte solution, and finally a lack of ion transport (based on Bockris and Shrinivasan, 1969).

The potential losses within the cell may be described in terms of an internal resistance, the existence of which implies a non-zero energy dissipation, if energy is to be extracted in a finite time. This is the same fundamental type of loss as that encountered in the case of solar cells (14.16) and in the general thermodynamic theory described in Chapter 2.

19.3 Fuel cell technology

The idea of converting fuel into electricity by an electrode–electrolyte system originated in the 19th century (Grove, 1839). The basic principle behind a hydrogen–oxygen fuel cell was described in Section 19.2. The first practical applications were in powering space vehicles, starting during the 1970s. A range of different fuel cell types has been developed, with applications at different temperature in stationary or mobile environments as the main technology determinant. A more complete description of the possibilities may be found in Sørensen (2005).

An early development for stationary power applications is the phosphoric acid cells, which use porous carbon electrodes with a platinum catalyst and phosphoric acid as electrolyte and feed hydrogen to the negative electrode, with electrode reactions given by (19.3) and (19.4). The operating temperature is in the range 175–200°C, and water has to be continuously removed.

Another early fuel cell development, the alkaline cells, uses KOH as electrolyte and have electrode reactions of the form

$$H_2 + 2OH^- \rightarrow 2H_2O + 2e^-,$$
$$\tfrac{1}{2} O_2 + H_2O + 2e^- \rightarrow 2OH^-.$$

These cells operate in the temperature range 70–100°C, but specific catalysts require maintenance of fairly narrow temperature regimes. Also, the hydrogen fuel must have a high purity and notably not contain any CO_2. Alkaline fuel cells have been used extensively on spacecraft and have recently been proposed for road vehicles (Hoffmann, 1998a). Their relative complexity and use of corrosive compounds requiring special care in handling would seem to constitute a barrier for them to become acceptable for general-purpose use.

The third low-temperature fuel cell type in commercial use is the proton exchange membrane (PEM) cell. It has been developed over a fairly short period of time and is considered to hold the greatest promise for economic application in the transportation sector. It contains a solid polymer membrane sandwiched between two gas diffusion layers and electrodes. The membrane material may be polyperfluorosulphonic acid. A platinum or Pt–Ru alloy catalyst is used to break hydrogen molecules into atoms at the negative electrode, and the hydrogen atoms are then capable of penetrating the membrane and reaching the positive electrode, where they combine with oxygen to form water, again with the help of a platinum catalyst. The electrode reactions are again (19.3) and (19.4), and the operating temperature is 50–100°C (Wurster, 1997). Figure 19.3 shows a typical layout of an individual cell. Several of these are then stacked on top of each other. This modularity implies that PEM fuel cells can be used for applications requiring little power (1 kW or less). PEM cell stacks are dominating the current wealth of demonstration projects in road transportation, portable power and special applications. The efficiency of conversion for the small systems is between 40% and 50%, but a 50 kW system has recently achieved an efficiency in the laboratory near 60%. As indicated in Fig. 19.4, an advantage of particular importance for automotive applications is the high efficiency at part loads, which alone gives a factor of two improvement over current internal combustion engines.

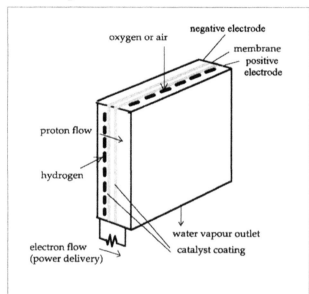

Figure 19.3. Layout of a PEM fuel cell layer, several of which may be stacked.

For use in automobiles, compressed and liquefied hydrogen are limited by low energy density and safety precautions for containers and, first of all, by the requirement of a new infrastructure for fuelling. The hydrogen storage problem, which until recently limited fuel cell projects to large vehicles such as buses, may be solved by use of metal hydride or carbon nanofibre stores (see Chapter 34), while usage in small passenger vehicles may get away with compressed hydrogen containers, as long as the overall efficiency of the vehicle is high. In order to avoid having to make large changes to the current gasoline and diesel fuel filling stations, it was for a while proposed to use methanol as the fuel distributed to the vehicle fuel tank. The energy density of methanol is 4.4 kWh litre^{-1}, which is half that of gasoline. This is quite acceptable owing to the higher efficiency of conversion. Hydrogen is then formed onboard by a methanol reformer, before being fed to the fuel cell to produce the electric power for an electric motor. The set-up is illustrated in Fig. 19.5. Prototype vehicles with this set-up have been tested (cf. Takahashi, 1998; Brown, 1998), but commercialisation has more or less stopped, because the downsized methanol reformer systems had too many operating problems that turned out to be difficult to avoid at a reasonable cost.

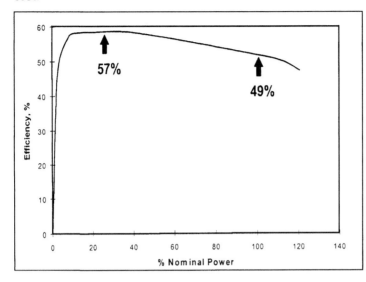

Figure 19.4. Expected part-load efficiencies for a 50-kW PEM fuel cell, projected from measurements involving 10–20 cell test stacks (Patil, 1998).

Methanol, CH_3OH, may even be used directly in a fuel cell, without the extra step of reforming to H_2. PEM fuel cells accepting methanol as a feedstock have been developed (Sørensen, 2005). They have lower efficiency than hydrogen PEM cells and are mainly aimed at the portable power market. As methanol can be produced from biomass, hydrogen may in this way be eliminated from the energy system. On the other hand, handling of surplus production of wind or photovoltaic power might still conveniently involve hydrogen as an intermediate energy carrier, as it may have direct uses and thus may improve the system efficiency by avoiding the losses in methanol production. The electric power to hydrogen conversion efficiency is over 65% in high-pressure alkaline electrolysis plants (Wagner et al., 1998), and the efficiency of the further hydrogen (plus CO or CO_2 and a catalyst) to methanol conversion is around 70%. Also, the efficiency of producing methanol from biomass is about 45%, whereas higher efficiencies are obtained if the feedstock is methane (natural gas) (Jensen and Sørensen, 1984; Nielsen and Sørensen, 1998).

For stationary applications, fuel cells of higher efficiency may be constructed, aimed at processes operating at higher temperatures. One line of research has been molten carbonate fuel cells, with electrode reactions

$$H_2 + CO_3^{2-} \to CO_2 + H_2O + 2e^-,$$
$$\tfrac{1}{2} O_2 + CO_2 + 2e^- \to CO_3^{2-}.$$

The electrolyte is a molten alkaline carbonate mixture retained within a porous aluminate matrix. The carbonate ions formed at the positive electrode travel through the matrix and combine with hydrogen at the negative electrode at an operating temperature of about 650°C. The process was originally aimed at hydrogen supplied from coal gasification or natural gas conversion. Test of a new 250 kW system is taking place at a power utility company in Germany (Hoffmann, 1998b). The expected conversion efficiency is about 55%, but some of the high-temperature heat may additionally be utilised. Earlier molten carbonate fuel cell tests have encountered severe problems with corrosion.

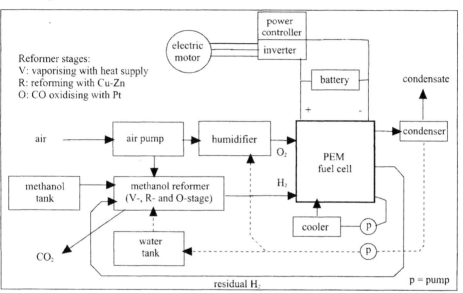

Figure 19.5. Layout of the power system for a methanol-to-hydrogen powered vehicle with fuel cell conversion and an electric motor. The power controller allows shift from direct drive to battery charging.

Considerable efforts are dedicated to the solid electrolyte cells. Solid oxide fuel cells (SOFC) use zirconia as the electrolyte layer to conduct oxygen ions formed at the positive electrode. Electrode reactions are

$$H_2 + O^{2-} \rightarrow H_2O + 2e^-,$$
$$\tfrac{1}{2} O_2 + 2e^- \rightarrow O^{2-}.$$

The reaction temperature is 700–1000°C, which makes it possible to auto-reform several fuels other than hydrogen within the system. On the other hand, lower temperatures are desirable, due to lower corrosion problems (and dreams of automotive applications), and may be achieved at some expense of efficiency by using a thin electrolyte layer (about 10 µm) of yttrium-stabilised zirconia sprayed onto the negative electrode as a ceramic powder (Kahn, 1996). A number of prototype plants (in the 100-kW size range) are in operation. Current conversion efficiency is about 55% but could reach 70–80% in the future (Hoffmann, 1998b).

Particularly for vehicle applications of hydrogen-based technologies using compressed hydrogen as a fuel, efforts are needed to ensure a high level of safety in collisions and other accidents. Because of the difference between the physical properties of hydrogen fuel and the hydrocarbon fuels currently in use (higher diffusion speed, flammability and explosivity over wider ranges of mixtures with air), there is a need for new safety-related studies, particularly if hydrogen is stored on board in containers at pressures above the currently common 20–30 MPa. A few such studies have already been made (Brehm and Mayinger, 1989).

Reverse operation of fuel cells

A driven-cell conversion based on the dissociation of water may be accomplished by any fuel cell. This process is in current practice using alkaline fuel cells and goes by the name of *electrolysis*. The expression (19.10) indicates that the efficiency of an ideal fuel cell in reverse operation may theoretically reach 1.20, implying that the electrolysis process draws heat from the surroundings.

By combining a fuel cell with an electrolysis unit, a regenerative system has been obtained, and if hydrogen and oxygen can be stored, an energy storage system or a "battery" results. The electric energy required for the electrolysis need not come from the fuel cell, but may be the result of an intermittent energy conversion process (e.g. wind turbine, solar photovoltaic cell, etc.). The combined system may be two fuel cells, each optimised for one of the process directions, or it may be a single fuel cell with decent performance in both directions. The problem in such dual systems is typically to raise the electrolysis efficiency above 50%, while the power production efficiency remains around 50%.

Direct application of radiant energy to the electrodes of an electrochemical cell has been suggested, aiming at maintaining a driven-cell process (e.g. dissociation of water) without having to supply electric energy. The electrodes could be made of suitable p- and n-type semiconductors, and the presence of photo-induced electron–hole excitations might modify the electrode potentials ϕ_a and ϕ_c in such a way that the driven-cell reactions become thermodynamically allowed. Owing to the electrochemical losses discussed in Section 19.1, additional energy would have to be provided by the solar radiation. If the radiation-induced electrode processes have a low efficiency, the overall efficiency may still be higher for photovoltaic conversion of solar energy into electricity followed by conventional electrolysis (Manassen *et al.*, 1976). In recent years, several such photoelectrochemical devices have been constructed and are undergoing tests.

OTHER ELECTROCHEMICAL ENERGY CONVERSION

CHAPTER

20

As mentioned in Section 19.1, batteries are basically performing the same tasks as fuel cells, but with internal storage of the ingredients. They will be treated as part of the storage coverage in Chapter 33. There are many other suggestions for using electrochemistry in connection with energy conversion and this chapter just gives one fairly speculative example based on ocean variations in salinity.

20.1 Salinity differences

Useful chemical energy may be defined as energy that can be released through exotermic chemical reactions. In general, chemical energy is associated with chemical bindings of electrons, atoms and molecules. The bindings may involve overlapping electron wavefunctions of two or more atoms, attraction between ionised atoms or molecules, and long-range electromagnetic fields created by the motion of charged particles (notably electrons). In all cases, the physical interaction involved is the Coulomb force. Examples of chemical energy connected with molecular binding structure are bio-energy sources, including fossil fuels (cf. Part VI below).

The organisation of atoms or molecules in regular lattice structures represents another manifestation of chemical bindings. Some substances possess different crystalline forms, which may exist under given external conditions. In addition to the possibility of different solid phases, phase changes associated with transitions among solid, liquid and gas phases all represent different levels of latent energy. Conversion between such states can be considered as energy storage and retrieval, and several examples will be mentioned in Chapters 29 and 34.

Solutions represent another form of chemical energy, relative to the pure solvent. The free energy of a substance with components $i = 1, 2,...,$ there being n_i mol of the ith component, may be written

$$G = \sum_i n_i \mu_i,$$ (20.1)

where μ_i is called the "chemical potential" of component i. For a solution, μ_i can be expressed in the form [cf. (19.8)]

$$\mu_i = \mu_i^0 + \mathcal{R} T \log (f_i x_i),$$ (20.2)

where $\mathcal{R}$ is the gas constant (8.3 J K^{-1} mol^{-1}), T is the temperature (K) and $x_i = n_i/(\Sigma_j n_j)$ the mole fraction. μ_i^0 is the chemical potential that would correspond to $x_i = 1$ at the given pressure P and temperature T, and f_i is the "activity coefficient", an empirical constant that approaches unity for "ideal solutions", an example of which is the solvent of a very dilute solution (whereas, in general, f_i cannot be expected to approach unity for the dissolved component of a dilute solution).

It follows from (20.1) and (20.2) that a solution represents a lower chemical energy than the pure solvent. The most common solution present in large amounts on the Earth is saline ocean water. Relative to this, pure or fresh water such as river run-off represents an elevated energy level. In addition, there are marked salinity differences within the oceans (Sørensen, 2004).

Taking the average ocean salinity as about 33×10^{-3} (mass fraction), and regarding this entirely as ionised NaCl, $n_{Na+} = n_{Cl-}$ becomes about 0.56×10^3 mol and $n_{water} = 53.7 \times 10^3$ mol, considering a volume of 1 m^3. The chemical potential of ocean water, μ, relative to that of fresh water, μ^0, is then from (20.2)

$$\mu - \mu^0 = \mathcal{R} T \log x_{water} \approx -2\mathcal{R} T n_{Na+} / n_{water}.$$

Consider now a membrane that is permeable for pure water but impermeable for salt (i.e. for Na$^+$ and Cl$^-$ ions) as indicated in Fig. 20.1. On one side of the membrane, there is pure (fresh) water, and on the other side saline (ocean) water. Fresh water will flow through the membrane, trying to equalise the chemical potentials μ^0 and μ initially prevailing on each side. If the ocean can be considered as infinite and being rapidly mixed, then n_{Na+} will remain fixed, also in the vicinity of the membrane. In this case each m^3 of fresh water penetrating the membrane and becoming mixed will release an amount of energy, which from (20.1) is

$$\delta G = \sum_i (n_i \delta \mu_i + \mu_i \delta n_i) \approx n_{water} (\mu^0 - \mu) \approx 2 \mathcal{R} T n_{Na+}.$$ (20.3)

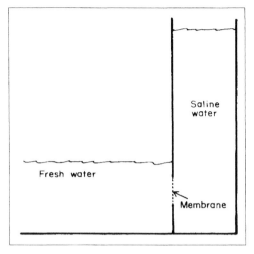

Figure 20.1. Schematic picture of an osmotic pump. In order to mix the fresh water penetrating the semi-permeable membrane in the direction towards the right and to maintain the salinity in the salt water compartment, new saline water would have to be pumped into the salt water compartment, and water motion near the membrane would have to be ensured.

For a temperature $T \approx 285$ K (considered fixed), $\delta G \approx 2.65 \times 10^6$ J. The power corresponding to a fresh-water flow of 1 m^3 s^{-1} is thus 2.65×10^6 W (cf. Norman, 1974). The worldwide run-off of about 4×10^{13} m^3 y^{-1} would thus correspond to an average power of around 3×10^{12} W.

The arrangement schematically shown in Fig. 20.1 is called an osmotic pump. The flow of pure water into the tube will ideally raise the water level in the tube, until the pressure of the water head balances the force derived from the difference in chemical energy. The magnitude of this "osmotic pressure", P^{osm}, relative to the atmospheric pressure P_0 on the fresh-water surface, is found from the thermodynamic relation

$$V \, dP - S \, dT = \Sigma_i n_i \, d\mu_i,$$

where V is the volume, S is the entropy and T is the temperature. Assuming that the process will not change the temperature (i.e. considering the ocean a large reservoir of fixed temperature), insertion of (20.3) yields

$$P^{osm} = \delta P \approx n_{water} \, V^{-1} \, \delta\mu_{water} \approx 2\mathcal{R} \, T \, n_{Na+} \, V^{-1}. \tag{20.4}$$

Inserting the numerical values of the example above, $P^{osm} = 2.65 \times 10^6$ N m^{-2}, corresponding to a water-head some 250 m above the fresh-water surface. If the assumption of fixed mole fraction of salt in the tube is to be realised, it would presumably be necessary to pump saline water into the tube. The energy spent for pumping, however, would be mostly recoverable, since it also adds to the height of the water-head, which may be used to generate electricity as in a hydropower plant.

An alternative way of releasing the free energy difference between solutions and pure solvents is possible when the dissolved substance is ionised (the solution is then called electrolytic). In this case direct conversion to electricity is possible, as further discussed in the following section.

20.2 Conversion of salinity gradient resources

As discussed in Section 20.1, a salinity difference, such as the one existing between fresh (river) and saline (ocean) water, may be used to drive an osmotic pump, and the elevated water may in turn be used to generate electricity by an ordinary turbine (cf. Chapter 11).

An alternative method, aiming directly at electricity production, takes advantage of the fact that the salts of saline water are ionised to a certain degree and thus may be used to derive an electrochemical cell of the type discussed in Sector 19.2 (Pattie, 1954; Weinstein and Leitz, 1976).

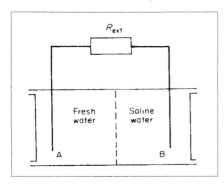

Figure 20.2. Schematic picture of dialytic battery.

This electrochemical cell, which may be called a "dialytic battery" since it is formally the same as the dialysis apparatus producing saline water with electric power input, is shown schematically in Fig. 20.2. In this case, the membrane allows one kind of ion to pass and thereby reach the corresponding electrode (in contrast to the osmotic pump in Fig. 20.1, where water could penetrate the membrane but ions could not). The free energy difference between the state with free Na^+ and Cl^- ions (assuming complete ionisation), and the state with no such ions is precisely the difference between saline and fresh water, calculated in Section 20.1 and approximately given by (20.3). Assuming further that each Na^+ ion reaching the electrode A neutralises one electron, and that correspondingly each Cl^- ion at the electrode B gives rise to one electron (which may travel in the external circuit), then the electromotoric force, i.e. the electric potential ϕ of the cell, is given in analogy to (19.5), with the number of positive ions per mole equal to n_{Na^+}, and ϕ is related to the change in free energy (20.3) just as (19.5) was related to (19.2),

$$n_e \mathscr{F} \phi = n_{Na^+} \mathscr{F} \phi = \delta G \approx 2 \mathscr{R} T \, n_{Na^+},$$

or

$$\max \left(\Delta \phi_{ext} \right) = \phi \approx 2 \mathscr{R} \, T/\mathscr{F}. \tag{20.5}$$

Inserting the numerical values given in Section 20.1, and $T = 300$ K, the cell electromotive force becomes $\phi \approx 0.05$ V. The actual value of the external potential, $\Delta \phi_{ext}$, may be reduced as a result of various losses, as described in Section 19.2 (e.g. Fig. 19.2). ϕ is also altered if the "fresh" water has a finite amount of ions or if ions other than Na^+ and Cl^- are present.

If the load resistance is R_{ext}, the power delivered by the cell is given by the current $I = \Delta \phi_{ext} \, R_{ext}^{-1}$,

$$E = I \Delta \phi_{ext},$$

and as usual the maximum power corresponds to an external potential difference $\Delta \phi_{ext}$, which is smaller than the open circuit value (20.5). The internal losses may also be represented by an internal resistance R_{int} defined by

$$\Delta \phi_{ext} = \phi - I R_{int} = \phi - \Delta \phi_{ext} \left(R_{int}/R_{ext} \right).$$

Thus, the power output may also be written

$$E = \phi^2 R_{ext}/(R_{int} + R_{ext})^2.$$

R_{int} depends on electrode properties, as well as on n_{Na^+} and n_{Cl^-} and their variations across the cell.

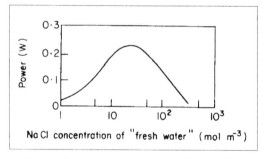

No Cl concentration of "fresh water" (mol m^{-3})

Figure 20.3. Measured performance of dialytic battery, containing about 30 membrane pairs in a stack. The external load was $R_{ext} = 10 \, \Omega$ (close to power optimisation), and electrode losses had been compensated for by insertion of a balancing power source (based on Weinstein and Leitz, 1976).

Several anode-membrane plus cathode-membrane units may be stacked beside each other in order to obtain a sufficiently large current (necessary because the total internal resistance has a tendency to increase strongly, if the current is below a certain value). Figure 20.3 shows the results of a small-scale experiment (Weinstein and Leitz, 1976) for a stack of about 30 membrane-electrode pairs and an external load, R_{ext}, close to the one giving maximum power output. The power is shown, however, as a function of the salinity of the fresh-water, and it is seen that the largest power output is obtained for a fresh water salinity which is not zero but 3–4% of the sea water salinity. The reason is that although the electromotive force (20.5) diminishes with increasing fresh-water salinity, this is at first more than compensated for by the improved conductivity of the solution (decrease in R_{int}), when ions are also initially present in the fresh-water compartment.

Needless to say, small-scale experiments of the kind described above are a long way from a demonstration of viability of the salinity gradient conversion devices on a large scale, and it is by no means clear whether the dialysis battery concept will be more or less viable than power turbines based on osmotic pumps. The former seems likely to demand a larger membrane area than the latter, but no complete evaluation of design criteria has been made in either case.

BENT SØRENSEN - 2000

VI.   BIOENERGY CONVERSION
PROCESSES

CHAPTER 21

COMBUSTION

21.1 Fossil biofuels

Traditional uses of organic fuels as energy carriers comprise the use of fresh biomass, i.e. storage and combustion of wood fuels, straw and other plant residues, and in recent centuries including storage and combustion of fossilised biomass, i.e. fuels derived from coal, oil and natural gas. While the energy density of living biomass is in the range of 10–30 MJ per kg of dry weight, the several million years of fossilation processes typically increase the energy density by a factor of two (see Fig. 21.1). The highest energy density is found for oil, where the average density of crude oil is 42 MJ kg^{-1}.

Source	Reserves (EJ)	Resources (> 50% prob.)	Occurrence (speculative)	1990 use (EJ)	Accum. use 1860–1990
Coal				91	5 203
hard coal	15 000	100 000	900 000		
brown coal/lignite	4 000	30 000	90 000		
peat	0	4 000	4 000		
Oil				128	3 343
conventional	5 000	3 000	13 000		
unconventional	7 000	9 000	20 000		
Natural gas				71	1 703
conventional	5 000	5 000	15 000		
unconventional	7 000	20 000	25 000		
in hydrates	0	0	100 000		

Table 21.1. Fossil reserves, additional resources and consumption in EJ (UN, 1981; Jensen and Sørensen, 1984; Nakicenovic et al., 1996).

The known reserves of fossil fuels that may be economically extracted today, and estimates of further resources exploitable at higher costs, are indicated in Table 21.1. The finiteness of such resources is, of course, together with the environmental impact issues, the reason for turning to renewable energy sources, including renewable usage of biomass resources.

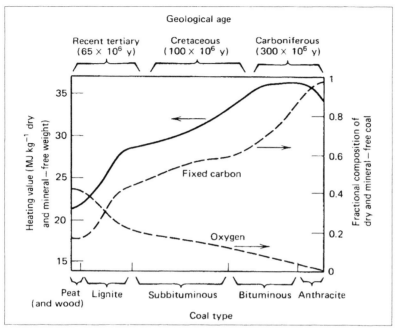

Figure 21.1. Coal to peat classification and selected properties. The dashed lines indicate the fractional content of oxygen and of fixed carbon. The remaining fraction consists of volatile matter (based on US DoE, 1979).

However, fresh biomass is a large potential source of renewable energy that in addition to use for combustion may be converted into a number of liquid and gaseous biofuels. This may be achieved by thermochemical or by biochemical methods, as described below.

Biomass is not just stored energy; it is also a store of nutrients and a potential raw material for a number of industries. Therefore, bio-energy is a topic that cannot be separated from food production, timber industries (serving construction purposes, paper and pulp industries, etc.), and organic feedstock-dependent industries (chemical and biochemical industries, notably). Furthermore, biomass is derived from plant growth and animal husbandry, linking the energy aspect to agriculture, livestock, silviculture, aquaculture, and quite generally to the management of the global ecological system. Thus, procuring and utilising organic fuels constitute a profound interference with the natural biosphere, and the understanding of the range of impacts as well as the elimination of unacceptable ones should be an integral part of any scheme for diversion of organic fuels to human society.

21.2 Heat production from biomass combustion

Heat may be obtained from biological materials by burning, eventually with the purpose of further conversion. Efficient burning usually requires the reduction of water content, for example, by drying in the Sun. The heat produced by burning cow dung is about 1.5×10^7 J per kg of dry matter, but initially only about 10% is dry matter, so the vaporisation of 9 kg of water implies an energy requirement of 2.2×10^7 J, i.e. the burning process is a net energy producer only if substantial sun-drying is possible. Firewood and other biomass sources constitute stores of energy, since the drying may be performed during summer, such that these biological fuels can be used during winter periods when the heating requirement may be large and the possibility of sun-drying may not exist. The heat produced by burning 1 kg of dry wood or sawmill scrap is about 1.25×10^7 J (1 kg in these cases corresponds to a volume of approximately 1.5×10^{-3} m^3), and the heat from burning 1 kg of straw (assumed water content 15%) is about 1.5×10^7 J (Eckert, 1960). The boilers used for firing with wood or straw have efficiencies that are often considerably lower than those of oil or gas burners. Typical efficiencies are in the range 0.5–0.6 for the best boilers. The rest of the enthalpy is lost, mostly in the form of vapour and heat leaving the chimney (or other smoke exit), and therefore is not available at the load area.

Combustion is the oxidation of carbon-containing material in the presence of sufficient oxygen to complete the process

$$C + O_2 \rightarrow CO_2.$$

Wood and other biomass is burned for cooking, for space heating, and for a number of specialised purposes, such as provision of process steam and electricity generation. In rural areas of many Third World countries a device consisting of three stones for outdoor combustion of twigs is still the most common. In the absence of wind, up to about 5% of the heat energy may reach the contents of the pot resting on top of the stones. In some countries, indoor cooking on simple chulas is common. A chula is a combustion chamber with a place for one or more pots or pans on top, resting in such a way that the combustion gases will pass along the outer sides of the cooking pot and leave the room through any opening. The indoor air quality is extremely poor when such chulas are in use, and village women in India using chulas for several hours each day are reported to suffer from severe cases of eye irritation and diseases of the respiratory system (Vohra, 1982).

Earlier, most cooking in Europe and its colonies was done on stoves made of cast iron. These stoves, usually called European stoves, had controlled air intake and both primary and secondary air inlets, chimneys with regulation of gas passage, and several cooking places with ring systems allowing the pots to fit tightly in holes, with a part of the pot indented into the hot gas stream. The efficiency would be up to about 25%, counted as energy delivered to the pots divided by wood energy spent, but such efficiencies would only be reached if all holes were in use and if the different temperatures prevailing at different boiler holes could be made useful, including afterheat. In many cases the average efficiency would hardly have exceeded 10%, but in many of the areas in question the heat lost to the room could also be considered as useful, in which case close to 50% efficiency (useful energy divided by input) could be reached. Today, copies of the European stove are being introduced in several regions of the Third World, with use of local materials such as clay and sand–clay mixtures instead of cast iron.

Wood-burning stoves and furnaces for space heating have conversion efficiencies from below or about 10% (open furnace with vertical chimney) up to 50% (oven with two controlled air inlets and a labyrinth-shaped effluent gas route leading to a tall chimney). Industrial burners and stokers (for burning wood scrap) typically reach efficiencies of about 60%. Higher efficiencies require a very uniform fuel without variations in water content or density.

Most biological material is not uniform, and some pre-treatment can often improve both the transportation and storage processes and the combustion (or other final use). Irregular biomass (e.g. twigs) can be chopped or cut to provide unit sizes fitting the containers and burning spaces provided. Furthermore, compressing and pelletising the fuel can make it considerably more versatile. For some time, straw compressors and pelletisers have been available, so that the bulky straw bundles can be transformed into a fuel with volume densities approaching that of coal. Other biomass material can conceivably be pelletised with advantage, including wood scrap, mixed biomass residues, and even aquatic plant material (Anonymous, 1980). Portable pelletisers are available (e.g. in Denmark) which allow straw to be compressed in the fields so that longer transport becomes economically feasible and so that even long-term storage (seasonal) of straw residues becomes attractive.

A commonly practiced conversion step is from wood to charcoal. Charcoal is easier to store and to transport. Furthermore, charcoal combustion – for example, for cooking – produces less visible smoke than direct wood burning and is typically so much more efficient than wood burning that, despite wood-to-charcoal conversion losses, less primary energy is used to cook a given meal with charcoal than with wood.

Particularly in the rich countries, a considerable source of biomass energy is urban refuse, which contains residues from food preparation and discarded consumer goods from households, as well as organic scrap material from commerce and industry. Large-scale incineration of urban refuse has become an important source of heat, particularly in Western Europe, where it is used mainly for district heating (deRenzo, 1978).

For steam generation purposes, combustion is performed in the presence of an abundant water source ("waterwall incineration"). In order to improve pollutant control, fluidised bed combustion techniques may be utilised (Cheremisinoff et al., 1980). The bed consists of fine-grain material, for example, sand, mixed with material to be burned (particularly suited is sawdust, but any other biomass including wood can be accepted if finely chopped). The gaseous effluents from combustion, plus air, fluidise the bed as they pass through it under agitation. The water content of the material in the bed may be high (in which case steam production entails). Combustion temperatures are lower than for flame burning, and this partly makes ash removal easier and partly reduces tar formation and salt vaporisation. As a result, the reactor life is extended and air pollution can better be controlled.

In general, the environmental impacts of biomass utilisation through combustion may be substantial and comparable to, though not entirely of the same nature as, the impacts from coal and oil combustion (see Table 21.2). In addition, ashes will have to be disposed of. For boiler combustion, the sulphur dioxide emissions are typically much smaller than those for oil and coal combustion, which would give 15–60 kg t^{-1} in the absence of flue gas cleaning. If ash is re-injected into the wood burner, higher sulphur values appear, but these values are still below the fossil fuel emissions in the absence of sulphur removal efforts.

Substance emitted	Emissions (kg/10^3kg)
Particulates	12.5–15.0
Organic compounds[a]	1.0
Sulphur dioxide	0–1.5[b]
Nitrogen oxides	5.0
Carbon monoxide	1.0

Table 21.2. Uncontrolled emissions from biomass combustion in boilers (kg per ton of fuel, based on woody biomass; US EPA, 1980).

[a] Hydrocarbons including methane and traces of benzo(a)pyrine.
[b] Upper limit is found for bark combustion. Ten times higher values are reported in cases where combustion ashes are re-injected.

Particulates are not normally regulated in home boilers, but for power plants and industrial boilers, electrostatic filters are employed with particulate removal up to over 99%. Compared to coal burning without particle removal, wood emissions are 5–10 times lower. When starting a wood boiler there is an initial period of very visible smoke emission, consisting of water vapour and high levels of both particulate and gaseous emissions. After reaching operating temperatures, wood burns virtually without visible smoke. When stack temperatures are below 60°C, again during start-up and incorrect burning practice, severe soot problems arise.

Nitrogen oxides are typically 2–3 times lower for biomass burning than for coal burning (without removal, per kilogram of fuel), often leading to similar emissions if taken per unit of energy delivered.

Particular concern should be directed at the organic compound emissions from biomass burning. In particular, benzo(a)pyrene emissions are found to be up to 50 times higher than for fossil fuel combustion, and the measured concentrations of benzo(a)pyrene in village houses in Northern India (1.3–9.3 $\times$ 10^{-9} kg m^{-3}), where primitive wood-burning chulas are used for several (6–8) hours every day, exceed the German standards of 10^{-11} kg m^{-3} by 2–3 orders of magnitude (Vohra, 1982). However, boilers with higher combustion temperatures largely avoid this problem, as indicated in Table 21.2. In areas such as New England or Denmark, where wood stoves are used in suburban areas for "cosiness" a few hours during some evenings, these account for 85% of the total air pollution from energy use, but only for a few percent of the total energy use (Information, 2007).

The lowest emissions are achieved if batches of biomass are burned at optimal conditions, rather than regulating the boiler up and down according to heating load. Therefore, wood heating systems consisting of a boiler and a heat storage unit (gravel, water) with several hours of load capacity will lead to the smallest environmental problems (Hermes and Lew, 1982). This example shows that there can be situations where energy storage would be introduced entirely for environmental reasons.

Finally, it should be mentioned that occupational hazards arise during tree felling and handling. The accident rate is high in many parts of the world, and safer working conditions for forest workers are imperative if wood is to be used sensibly for fuel applications.

Finally, concerning carbon dioxide, which accumulates in the atmosphere as a consequence of rapid combustion of fossil fuels, it should be kept in mind that the carbon dioxide emissions during biomass combustion are balanced in magnitude by the net carbon dioxide assimilation by the plants, so that the atmospheric CO_2 content is not affected, at least by the use of biomass crops in fast rotation. However, the lag time for trees may be decades or centuries, and in such cases, the temporary carbon dioxide imbalance may contribute to climatic alterations.

Composting

An alternative way of deriving some heat from biomass is composting. Primary organic materials form the basis for a number of energy conversion processes other than burning. Since these produce liquid or gaseous fuels, plus associated heat, they will be dealt with in the following sections on fuel production. However, conversion aiming directly at heat production has also been utilised, with non-combustion processes based on manure from livestock animals and in some cases on primary biomass residues.

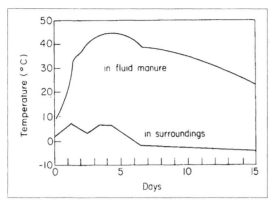

Figure 21.2. Temperature development in composting device based on liquid manure from pigs and poultry and with a blower providing constant air supply. Temperature of surroundings is also indicated (based on Popel, 1970).

Two forms of composting are in use: one based on fluid manure (less than 10% dry matter), and the other based on solid manure (50–80% dry matter). The chemical process is in both cases a bacterial decomposition under aerobic conditions, i.e. the compost heap or container has to be ventilated in order to provide a continuous supply of oxygen for the bacterial processes. The bacteria required for the process (of which lactic acid producers constitute the largest fraction; cf. McCoy, 1967) are normally all present in manure, unless antibiotic residues that kill bacteria are retained after some veterinary treatment. The processes involved in composting are very complex, but it is believed that decomposition of carbohydrates is responsible for most of the heat production (Popel, 1970]. A fraction of the carbon from the organic material is found in new-bred micro-organisms.

A device for treating fluid manure may consist of a container with a blower injecting fresh air into the fluid in such a way that it becomes well distributed over the fluid volume. An exit air channel carries the heated airflow to, say, a heat exchanger. Figure 21.2 shows an example of the temperature of the liquid manure, along with the temperature outside the container walls, as a function of time. The amount of energy required for the air blower is typically around 50% of the heat energy output, and is in the form of high-quality mechanical energy (e.g. from an electrically driven rotor). Thus, the simple efficiency may be around 50%, but the second law efficiency (2.20) may be quite low.

Heat production from solid manure follows a similar pattern. The temperature in the middle of a manure heap ("dunghill") may be higher than that of liquid manure, owing to the low heat capacity of the solid manure (see Fig. 21.3). Air may be supplied by blowers placed in the bottom of the heap, and in order to maintain air passage space inside the heap and remove moisture, occasional re-stacking of the heap is required. A certain degree of variation in air supply can be allowed, so that the blower may be driven by a renewable energy converter, for example, a windmill, without storage or back-up power. With frequent re-stacking, air supply by blowing is not required, but the required amount of mechanical energy input per unit of heat extraction is probably as high as for liquid manure. In addition, the heat extraction process is more difficult, demanding, for example, that heat exchangers be built into the heap itself (water pipes are one possibility). If an insufficient amount of air is provided, the composting process will stop before the maximum heat has been obtained.

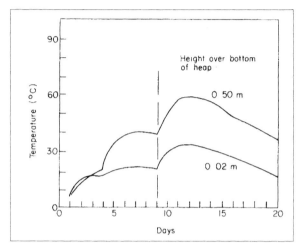

Figure 21.3. Temperature development at different locations within a solid manure composting heap (total height 0.8 m). After nine days, air is added from below (based on Olsen, 1975).

Owing to the bacteriological decomposition of most of the large organic molecules present, the final composting product has considerable value as fertiliser.

Metabolic heat

Metabolic heat from the life processes of animals can also be used by man, in addition to the heating of human habitats by man's own metabolic heat. A livestock shed or barn produces enough heat, under most conditions of winter occupancy, to cover the heating requirements of adjacent farm build-ings, in addition to providing adequate temperature levels for the animals. One reason for this is that livestock barns must have a high rate of ventila-tion in order to remove dust (e.g. from the animal's skin, fur, hair or feath-ers) and water vapour. For this reason, utilisation of barn heat may not re-quire extra feeding of the animals, but may simply consist of heat recovery from air that for other reasons has to be removed. Utilisation for heating a nearby residence building often requires a heat pump (Chapter 6), because the temperature of the ventilation air is usually lower than that required at the load area, so that a simple heat exchanger would not work.

In temperate climates, the average temperature in a livestock shed or barn may be about 15°C during winter. If young animals are present, the required temperature is higher. With an outside temperature of 0°C and no particular insulation of walls, the net heat production of such barns or sheds is positive when the occupants are fully grown animals, but negative if the occupants are young individuals and very much so if newborn animals are present (Olsen, 1975). In chicken or pig farms, the need for heat input may be avoided by having populations of mixed age or heat exchange between compartments for young and adult animals. The best candidates for heat extraction to other applications might then be dairy farms.

A dairy cow transfers about 25% of the chemical energy in the fodder to the milk and a similar amount to the manure (Claesson, 1974). If the weight of the cow remains constant, the rest is converted to heat and is rejected as heat radiation, convection or latent heat of water vaporisation. The distribution of the heat production in sensible and latent forms of heat is indicated in Fig. 21.4. It is seen to be strongly dependent on the air temperature in the barn. At 15°C, about two-thirds of the heat production is in the form of sensible heat. Heat transfer to a heat pump circuit may take place from the ventilation air exit. Water condensation on the heat exchanger surface involved may help to prevent dust particles from accumulating on the heat exchanger surface.

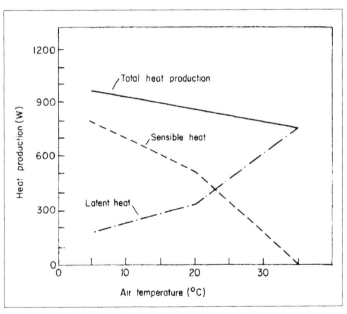

Figure 21.4. Average heat production and form of heat for a "standard" cow (the heat production of typical cows of "red" or "black-spotted" races is about 20% higher, while that of typical "Jersey" cows is about 30% lower) (based on Petersen, 1972).

BIOLOGICAL CONVERSION INTO GASEOUS FUELS

A number of conversion processes aim at converting one fuel into another, which is considered more versatile. Examples are charcoal production from wood (efficiency of conversion about 50%), liquefaction and gasification (Squires, 1974). Figure 22.1 gives an overview of non-food uses of biomass.

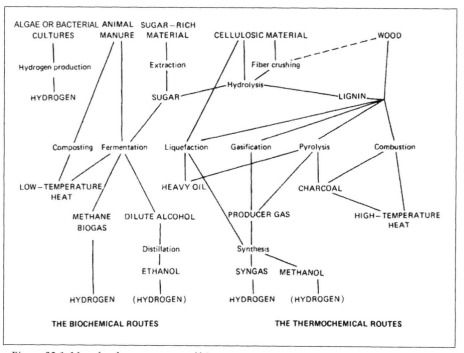

Figure 22.1. Non-food energy uses of biomass.

Conversion of (fresh) biological material into simple hydrocarbons or hydrogen can be achieved by a number of anaerobic fermentation processes, i.e. processes performed by suitable micro-organisms and taking place in the absence of oxygen. Such anaerobic "digestion" processes, the output of which is called *biogas*, work on the basis of most fresh biological material, wood excepted, provided that the proper conditions are maintained (temperature, population of micro-organisms, stirring, etc.). Thus, biological material in forms inconvenient for storage and use may be converted into liquid or (as considered in this chapter) gaseous fuels that can be utilised in a multitude of ways, like oil and natural gas products.

The "raw materials" that may be used are slurry or manure (e.g. from dairy farms or "industrial farming" involving large feedlots), city sewage and refuse, farming crop residues (e.g. straw or parts of cereal or fodder plants not normally harvested), or direct "fuel crops", such as ocean-grown algae or seaweeds, water hyacinths (in tropical climates) or fast-growing bushes or trees. The deep harvesting necessary in order to collect crop residues may not be generally advisable, owing to the role of these residues in preventing soil erosion. This objection may not be valid, if residues are returned to the fields (or the ocean) after energy extraction.

Among the fermentation processes, one set is particularly suited for producing gas from biomass in a wet process (cf. Fig. 22.1). It is the *anaerobic digestion*. It traditionally used animal manure as biomass feedstock, but other biomass sources can be used within limits that are briefly discussed in the following. The set of biochemical reactions making up the digestion process (a word indicating the close analogy to energy extraction from biomass by food digestion) is schematically illustrated in Fig. 22.2.

There are three discernible stages. In the first, complex biomass material is decomposed by a heterogeneous set of micro-organisms, not necessarily confined to anaerobic environments. These decompositions comprise hydrolysis of cellulosic material to simple glucose, using enzymes provided by the micro-organisms as catalysts. Similarly, proteins are decomposed to amino acids and lipids to long-chain acids. The significant result of this first phase is that most of the biomass is now water-soluble and in a simpler chemical form, suited for the next process step.

The second stage involves dehydrogenation (removing hydrogen atoms from the biomass material), such as changing glucose into acetic acid, carboxylation (removing carboxyl groups) of the amino acids, and breaking down the long-chain fatty acids into short-chain acids, again obtaining acetic acid as the final product. These reactions are fermentation reactions accomplished by a range of acidophilic (acid-forming) bacteria. Their optimum performance requires a pH environment in the range of 6–7 (slightly acid), but because the acids already formed will lower the pH of the solution, it is sometimes necessary to adjust the pH, for example, by adding lime.

Finally, the third phase is the production of biogas (a mixture of methane and carbon dioxide) from acetic acid by a second set of fermentation reactions performed by methanogenic bacteria. These bacteria require a strictly anaerobic (oxygen-free) environment. Often, all processes are made to take place in a single container, but separation of the processes into stages will allow greater efficiencies to be reached. The third phase takes of the order of weeks, the preceding phases on the order of hours or days, depending on the nature of the feedstock.

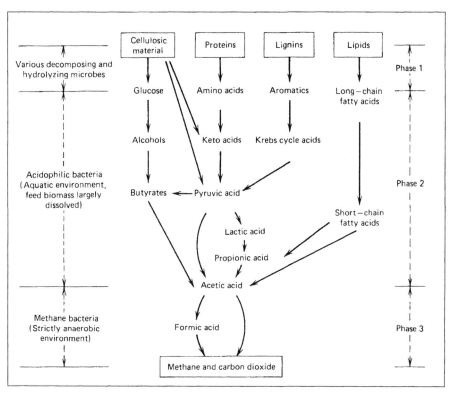

Figure 22.2. Simplified description of biochemical pathways in anaerobic digestion of biomass (based on Stafford *et al.*, 1981). Used with permission: from D. Stafford, D. Hawkes, and R. Horton, *Methane Production from Waste Organic Matter.* Copyright 1981 by The Chemical Rubber Co., CRC Press, Inc., Boza Raton, FL.

Starting from cellulose, the overall process may be summarised as

$$(C_6H_{10}O_5)_n + nH_2O \rightarrow 3nCO_2 + 3nCH_4 + 19n \text{ J mol}^{-1}. \quad (22.1)$$

The phase one reactions add up to

$$(C_6H_{10}O_5)_n + nH_2O \rightarrow nC_6H_{12}O_6. \quad (22.2)$$

The net result of the phase two reactions is

$$nC_6H_{12}O_6 \rightarrow 3nCH_3COOH, \tag{22.3}$$

with intermediate steps such as

$$C_6H_{12}O_6 \rightarrow 2C_2H_5OH + 2CO_2 \tag{22.4}$$

followed by dehydrogenation:

$$2C_2H_5OH + CO_2 \rightarrow 2CH_3COOH + CH_4. \tag{22.5}$$

The third-phase reactions then combine to

$$3nCH_3COOH \rightarrow 3nCO_2 + 3nCH_4. \tag{22.6}$$

In order for the digestion to proceed, a number of conditions must be fulfilled. The bacterial action is inhibited by the presence of metal salts, penicillin, soluble sulphides, or ammonia in high concentrations. Some source of nitrogen is essential for the growth of the micro-organisms. If there is too little nitrogen relative to the amount of carbon-containing material to be transformed, then bacterial growth will be insufficient and biogas production low. With too much nitrogen (a ratio of carbon to nitrogen atoms below around 15), "ammonia poisoning" of the bacterial cultures may occur. When the carbon–nitrogen ratio exceeds about 30, the gas production starts diminishing, but in some systems carbon–nitrogen values as high as 70 have prevailed without problems (Stafford *et al.*, 1981). Table 22.1 gives carbon–nitrogen values for a number of biomass feedstocks. It is seen that mixing feedstocks can often be advantageous. For instance, straw and sawdust would have to be mixed with some low C:N material, such as livestock urine or clover/lucerne (typical secondary crops that may be grown in temperate climates after the main harvest).

If digestion time is not a problem, almost any cellulosic material can be converted to biogas, even pure straw. One initially may have to wait for several months, until the optimum bacterial composition has been reached, but then continued production can take place, and despite low reaction rates an energy recovery similar to that of manure can be achieved with properly prolonged reaction times (Mardon, 1982).

Average manure production for fully bred cows and pigs (in Europe, Australia, and the Americas) is 40 and 2.3 kg wet weight d^{-1}, corresponding to 62 and 6.2 MJ d^{-1}, respectively. The equivalent biogas production may reach 1.2 and 0.18 m^3 d^{-1}. This amounts to 26 and 3.8 MJ d^{-1}, or 42 and 61% conversion efficiency, respectively (Taiganides, 1974). A discussion of the overall efficiency including transportation of biomass to the plant is given in Berglund and Börjesson (2002), finding maximum acceptable transport distances of 100–150 km.

The residue from the anaerobic digestion process has a higher value as a fertiliser than the feedstock. Insoluble organics in the original material are, to a large extent, made soluble, and nitrogen is fixed in the micro-organisms. Pathogen populations in the sludge are reduced. Stafford *et al.* (1981) found a 65–90% removal of *Salmonella* during anaerobic fermentation, and there is a significant reduction in streptococci, coliforms, and viruses, as well as an almost total elimination of disease-transmitting parasites such as *Ascaris*, hookworm, *Entamoeba*, and *Schistosoma*.

Material	Ratio	Material	Ratio
Sewage sludge	13:1	Bagasse	150:1
Cow dung	25:1	Seaweed	80:1
Cow urine	0.8:1	Alfalfa hay	18:1
Pig droppings	20:1	Grass clippings	12:1
Pig urine	6:1	Potato tops	25:1
Chicken manure	25:1	Silage liquor	11:1
Kitchen refuse	6–10:1	Slaughterhouse wastes	3–4:1
Sawdust	200–500:1	Clover	2.7:1
Straw	60–200:1	Lucerne	2:1

Table 22.1. Carbon–nitrogen ratios for various materials (based on Baader *et al.*, 1978; Rubins and Bear, 1942).

For this reason, anaerobic fermentation has been used fairly extensively as a cleaning operation in city sewage treatment plants, either directly on the sludge or after growing algae on the sludge to increase fermentation potential. Most newer plants make use of the biogas produced to fuel other parts of the treatment process, but with proper management, sewage plants may well be net energy producers (Oswald, 1973).

The other long-time experience with biogas and associated fertiliser production is in rural areas of a number of Asian countries, notably China and India. The raw materials here are mostly cow dung, pig slurry, and what in India is referred to as human night soil, plus in some cases grass and straw. The biogas is used for cooking, and the fertiliser residue is returned to the fields. The sanitary aspect of pathogen reduction lends strong support to the economic viability of these schemes.

The rural systems are usually based on simple one-compartment digesters with human labour for the filling and emptying of material. Operation is either in batches or with continuous new feed and removal of some 3–7% of the reactor content every day. Semi-industrialised plants have also been built during the last decade, for example, in connection with large pig-raising farms, where mechanised and highly automated collection of manure has been feasible. In some cases, these installations have utilised separate acid and methanation tanks.

Figure 22.3 shows an example of a town-scale digester plant, where the biogas is used in a combined electric power and district heat generating plant (Kraemer, 1981). Expected energy flows are indicated. Storage of biogas for the rural cooking systems is accomplished by variable-volume domes collecting the gas as it is produced (e.g., an inverted, water-locked can placed over the digester core). Biogas contains approximately 23 MJ m^{-3} and is therefore a medium-quality gas. CO_2 removal is necessary in order to achieve pipeline quality. An inexpensive way of achieving over 90% CO_2 removal is by water spraying. This way of producing compressed methane gas from biogas allows for automotive applications, such as the farmer producing tractor fuel on site. In New Zealand such uses have been developed since 1980 (see Fig. 22.4; Stewart and McLeod, 1980). Several demonstration experiences have recently been obtained in Sweden (Losciale, 2002).

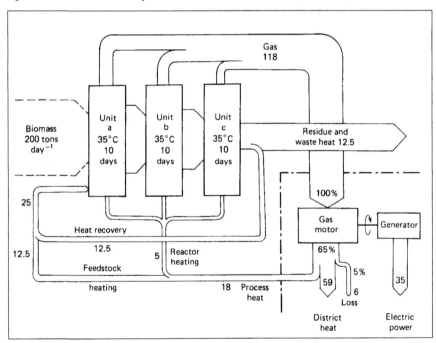

Figure 22.3. Calculated energy flows for a town biogas reactor plant, consisting of three successive units with ten days residence time in each. A biogas motor drives an electric generator, and the associated heat is, in part, recycled to the digestion process, while the rest is fed into the town district heating lines. Flows (all numbers without indicated unit) are in GJ d^{-1} (based on energy planning for Nysted commune, Denmark; Kraemer, 1981).

Storage of a certain amount of methane at ambient pressure requires over a thousand times more volume than the equivalent storage of oil. However, actual methane storage at industrial facilities uses pressures of about 140 times ambient (Biomass Energy Institute, 1978), so the volume penalty relative to oil storage would then be a factor of 9. Storage of methane in zeolitic material for later use in vehicles has been considered.

If residues are recycled, little environmental impact can be expected from anaerobic digestion. The net impact on agriculture may be positive, owing to nutrients being made more accessible and due to parasite depression. Undesirable contaminants, such as heavy metals, are returned to the soil in approximately the same concentrations as they existed before collection, unless urban pollution has access to the feedstock. The very fact that digestion depends on biological organisms may imply that warning signals in terms of poor digester performance may direct early attention to pollution of cropland or urban sewage systems. In any case, pollutant-containing waste, for example, from industry, should never be mixed with the energy-valuable biological material in urban refuse and sewage. The methane-forming bacteria are more sensitive to changes in environment, such as temperature and acidity, than the acid-forming ones.

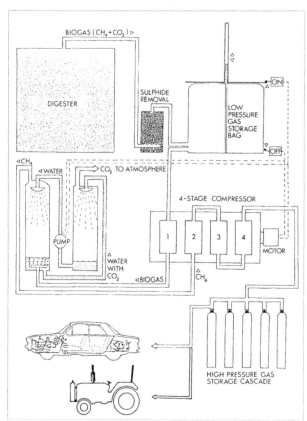

Figure 22.4. Schematic view of New Zealand scheme for methane production and vehicle use (from Stewart, D. and McLeod, R., *New Zealand Journal of Agriculture*, Sept. 1980, 9-24. With permission).

The digestion process itself does not emit pollutants if it operates correctly, but gas cleaning, such as H_2S removal, may lead to emissions. The methane gas itself shares many of the accident hazards of other gaseous fuels, being asphyxiating and explosive at certain concentrations in air (roughly 7–14% by volume). For rural cooking applications, the impacts may be compared with those of the fuels being replaced by biogas, notably wood burned in simple stoves. In these cases, as follows from the discussion in Chapter 21, the environment is dramatically improved by introducing biogas digesters.

An example of early biogas plants for use on a village scale in China, India and Pakistan is shown in Fig. 22.5. All the reactions take place in one compartment, which does not necessarily lead to optimum conversion efficiency. The time required for the acid-forming step is less than 24 h, whereas the duration of the second step should be 10–30 days. The heating of the fermentation tank required for application in many climatic regions may be derived from solar collectors, which could form the top cover of the tank. Alternatively, the top cover may be an inflatable dome serving as a store of gas, which can smooth out a certain degree of load variation. Some installations obtain the highest efficiency by batch operation, i.e. by leaving one batch of biological material in the tank for the entire fermentation period. The one shown in Fig. 22.5 allows continuous operation, i.e. a fraction of the slurry is removed every day and replaced by fresh biological material.

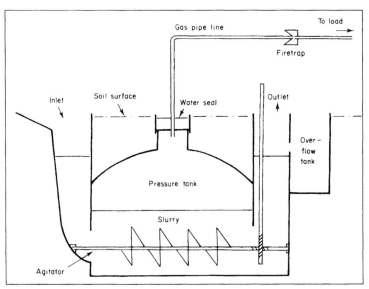

Figure 22.5. Single-chamber biogas plant (based on Chinese installations described by Appropriate Technology Development Organization, 1976).

Examples of predicted biogas production rates, for simple plants of the type shown in Fig. 22.5 and based on fluid manure from dairy cows, pigs or poultry, are shown in Table 22.2 and Fig. 22.6. Table 22.2 gives typical biogas production rates, per day and per animal, while Fig. 22.6 gives the conversion efficiencies measured, as functions of fermentation time (tank residence time), in a controlled experiment. The conversion efficiency is the ratio of the energy in the biogas (approximately 23 MJ m^{-3} of gas) and the energy in the manure, which would be released as heat by complete burning of the dry matter (some typical absolute values are given in Table 22.2). The highest efficiency is obtained with pigs' slurry, but the high bacteriological activity in this case occasionally has the side-effect of favouring bacteria other than those helping to produce biogas, e.g. ammonia-producing bacteria, the activity of which may destroy the possibility of further biogas production (Olsen, 1975).

Source	Manure per day		Biogas per day	
	kg wet weight	MJ	m³	MJ
Cows	40	62	1.2	26
Pigs	2.3	6.2	0.18	3.8
Hens	0.19	0.9	0.011	0.26

Table 2.2. Manure and potential biogas production for a typical animal per day (based on Taiganides, 1974).

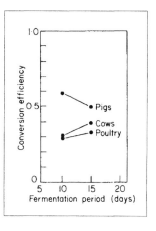

Figure 22.6. Measured conversion efficiencies (ratio between energy in biogas gas produced and energy in the manure) for simple biogas plants (like Fig. 22.5). Some 10–13 kg of fresh manure was added per day and per m³ of tank volume (based on Gramms et al., 1971).

As mentioned, the manure residue from biogas plants has a high value as fertiliser because the decomposition of organic material followed by hydrocarbon conversion leaves plant nutrients (e.g. nitrogen that was originally bound in proteins) in a form suitable for uptake. Malignant bacteria and parasites are not removed to as high a degree as by composting, owing to the lower process temperature.

Some city sewage plants produce biogas (by anaerobic fermentation of the sewage) as one step in the cleaning procedure, using the biogas as fuel for driving mechanical cleaning devices, etc. In this way it is in many cases possible to avoid any need for other energy inputs and in some cases to become a net energy producer (Danish Energy Agency, 1992). Figure 22.7 shows the system layout for a 300 t of biomass per day biogas plant accepting multiple types of feedstock and capable of delivering both power, process and district heat, and fertiliser. Figure 22.8 gives the measured performance data for nine large prototype biogas plants in Denmark.

The average production of the large-size Danish biogas plants was 35.1 m^3 per m^3 of biomass (at standard pressure, methane content in the biogas being on average 64%), or 806 MJ m^{-3} (the numbers are from 1992; Tafdrup, 1993). In-plant energy use amounted to 90 MJ m^{-3} distributed on 28 MJ electricity and 50 MJ heat, all produced by the biogas plant itself. Fuel used in transporting manure to the plant totalled 35 MJ, and the fertiliser value of the returned residue was estimated at 30 MJ. Thus, the net outside energy requirement is 5 MJ for a production of 716 MJ, or 0.7%, corresponding to an energy payback time of 3 days. If the in-plant biogas use is added, the energy consumption in the process is 13%. To this should be added the energy for construction of the plant, which has not been estimated. However, the best plants roughly break even economically, indicating that the overall energy balance is acceptable.

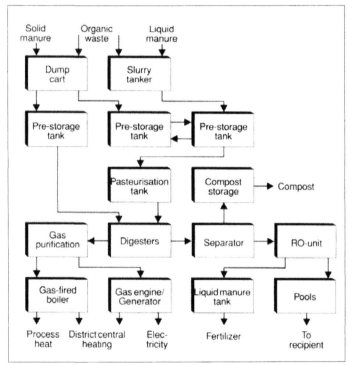

Figure 22.7. Layout of Lintrup biogas plant (Danish Energy Agency, 1992).

Greenhouse gas emissions

Using, as in the energy balance section above, the average of large Danish plants as an example, the avoided CO_2 emission from having the combined power and heat production use biomass instead of coal as a fuel is 68 kg per m^3 of biomass converted. Added should be emissions from transportation of biomass, estimated at 3 kg, and the avoided emissions from conventional production of fertiliser replaced by biogas residue, estimated at 3 kg. Reduced methane emissions, relative to the case of spreading manure directly on the fields, is of the order of 61 kg CO_2 equivalent (Tafdrup, 1993). As regards nitrous oxide, there is a possible gain by avoiding denitrification in the soil, but high uncertainty has made an actual estimate fortuitous at the present. The overall CO_2 reduction obtained by summing up the estimates given above is then 129 kg for each m^3 of biomass converted to biogas.

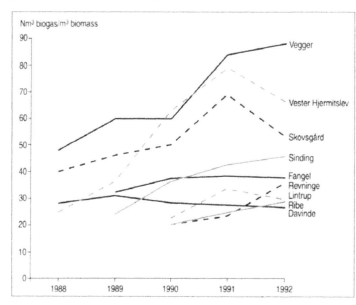

Figure 22.8. Annual average production efficiency (m^3 of biogas at standard pressure, denoted Nm^3, per m^3 of biomass feedstock) for nine community-size biogas plants in Denmark (Danish Energy Agency, 1992).

Other environmental effects
Compared with the current mix of coal, oil or natural gas plants, biogas plants have a 2–3 times lower SO_2 emission but a correspondingly higher NO_x emission. Higher ammonia content in the digested residue calls for greater care in using fertiliser from biogas plants, in order to avoid loss of ammonia. This is also true as regards avoiding loss of nutrients from fertiliser to waterways. Compared to spreading manure not refined by the biogas production process, a marked gain in fertiliser quality has been noted, including a much better defined composition, which will contribute to assisting correct dosage and avoiding losses to the environment. The dissemination of biogas plants removes the need for landfills, which is seen as an environmental improvement. Odour is moved from the fields (where manure and slurry would otherwise be spread) to the biogas plant, where it can be controlled by suitable measures (filters, etc.) (Tafdrup, 1993).

Hydrogen-producing cultures
Biochemical routes to fuel production include a number of schemes not presently developed to a technical or economic level of commercial interest.

Hydrogen is a fuel that may be produced directly by biological systems. Hydrogen enters in the process of photosynthesis, as it proceeds in green plants, where the net result of the process is

$$2H_2O + \text{solar radiation} \rightarrow 4e^- + 4H^+ + O_2.$$

However, the electrons and protons do not combine directly to form hydrogen,

$$4e^- + 4H^+ \rightarrow 2H_2,$$

but instead transfer their energy to a more complex molecule ($NADPH_2$; Sørensen, 2005), which is capable of driving the CO_2 assimilation process. By this mechanism, the plants avoid recombination of oxygen and hydrogen from the two processes mentioned above. Membrane systems keep the would-be reactants apart, and thus the energy-rich compound may be transported to other areas of the plant, where it takes part in plant growth and respiration.

Much thought has been given to modifications of plant material (e.g., by genetic engineering), in such a way that free hydrogen is produced on one side of a membrane and free oxygen on the other side (Berezin and Varfolomeev, 1976; Calvin, 1974; Hall *et al.*, 1979; Sørensen, 2005; 2006).

While dissociation of water (by light) into hydrogen and oxygen (photolysis; cf. Chapter 34) does not require a biological system, it is possible that utilisation of the process on a realistic scale can be more easily achieved if the critical components of the system, notably the membrane and the electron transport system, are of biological origin. Still, a breakthrough is required before any thought can be given to practical application of direct biological hydrogen production cultures.

CHAPTER

BIOLOGICAL CONVERSION INTO LIQUID FUELS

Anaerobic fermentation processes may be used not only to produce gases but also to produce liquid fuels from biological raw materials. An example is the ethanol production (22.4) from glucose, known as standard yeast fermentation in the beer, wine and liquor industries. It has to take place in steps, such that the ethanol is removed (by distillation or dehydrator application) whenever its concentration approaches a value (around 12%), which would impede reproduction of the yeast culture.

In order to reduce the primary biological material (e.g. molasses, cellulose pulp or citrus fruit wastes) to glucose, the hydrolysis process (22.6) may be used. Some decomposition takes place in any acid solution, but in order to obtain complete hydrolysis, specific enzymes must usually be provided, either directly or by adding micro-organisms capable of forming such enzymes. The yeast fungi themselves contain enzymes capable of decomposing polysaccharides into glucose. The theoretical maximum efficiency of glucose-to-ethanol conversion (described in more detail below) is 97%, and, according to Calvin (1977), the Brazilian alcohol industry already in 1974 obtained 14% of the energy in the raw sugar input, in the form of ethanol produced by fermentation of just the molasses residues from sugar refining, i.e. in addition to the crystallised sugar produced. A more recent figure is 25% (see Fig. 23.2) for an optimised plant design but including some indirect energy inputs (EC, 1994; energy for transportation of biomass from the place of production to the ethanol plant is still not included).

Mechanical energy input, e.g. for stirring, could be covered by the fermentation wastes if they were burned in a steam power plant. In the European example (EC, 1994), these inputs amount to about a third of the energy inputs through the sugar itself.

Alternative fermentation processes based on molasses or other sugar-containing materials produce acetone–butanol, acetone–ethanol or butanol–isopropanol mixtures, when the proper bacteria are added. In addition, carbon dioxide and small amounts of hydrogen are formed (see e.g. Beesch, 1952; Keenan, 1977).

Conversion of biomass into liquid fuels appears in the overview of Fig. 22.1 showing the conversion routes open for biofuel generation. Also fossil biomass may sustain such conversion (e.g. from coal to oil; cf. Chapter 24). Among the non-food energy uses of biomass, there are several options leading to liquid fuels, which may serve as a substitute for oil products. As indicated in Fig. 22.1, liquid end products appear as the result of either biochemical conversion using fermentation bacteria or a thermochemical conversion process involving gasification and, for example, further methanol synthesis. These processes, which convert biomass into liquid fuels that are easy to store, are discussed below, but first the possibility of direct production of liquid fuels by photosynthesis is presented.

Direct photosynthetic production of hydrocarbons

Oil from the seeds of many plants, such as rape, olive, groundnut, corn, palm, soy bean, and sunflower, is used as food or in the preparation of food. Many of these oils will burn readily in diesel engines and can be used directly or mixed with diesel oil of fossil origin, as they are indeed in several pilot projects around the world. However, in most of these cases the oil does not constitute a major fraction of the total harvest yield, and any expansion of these crops to provide an excess of oil for fuel use would interfere with food production. A possible exception is palm oil, because inter-cropping of palm trees with other food crops may provide advantages such as retaining moisture and reducing wind erosion.

Much interest is therefore concentrated on plants that yield hydrocarbons and that, at the same time, are capable of growing on land unsuited for food crops. Calvin (1977) first identified the *Euphorbia* family as an interesting possibility. The rubber tree, *Hevea brasiliensis*, is of this family, and its rubber is a hydrocarbon–water emulsion, the hydrocarbon of which (polyisoprene) has a large molecular weight, about a million, making it an elastomer. However, other plants of the genus *Euphorbia* yield latex of much lower molecular weight, which could be refined in much the same way as crude oil. In the tropical rainforests of Brazil, Calvin found a tree, *Cobaifera langsdorfii*, which annually yields some 30 litres of virtually pure diesel fuel (Maugh, 1979). Still, the interest centres on species that are likely to grow in arid areas such as the deserts of the southern United States, Mexico, Australia, and so on.

Possibilities include *Euphorbia lathyris* (gopher plant*)*, *Simmondsia chinensis* (jojoba), *Cucurdia foetidissima* (buffalo gourd) and *Parthenium argentatum* (guayule). The gopher plant has about 50% sterols (on a dry weight basis) in its latex, 3% polyisoprene (rubber), and a number of terpenes. The sterols are suited as feedstocks for replacing petroleum in chemical applications. Yields of first-generation plantation experiments in California are 15–

25 barrels of crude oil equivalent or some 144 GJ ha^{-1} (i.e., per 10^4 m^2). In the case of *Hevea*, genetic and agronomic optimisation has increased yields by a factor of 2000 relative to those of wild plants, so quite high hydrocarbon production rates should be achievable after proper development (Calvin, 1977; Johnson and Hinman, 1980; Tideman and Hawker, 1981). Other researchers are less optimistic (Stewart *et al.*, 1982; Ward, 1982). Currently, rapeseed-based bio-diesel oil is produced in several European countries, by simply squeezing the oil out of the seeds. The consumer cost of plant oils used for food (olive oil, sunflower oil, etc.) is currently some ten times higher than the cost of diesel oil, so the reason that rapeseed oil production is seen as profitable by European farmers is a subsidy given by the EU Commission for reducing food-production acreage.

Alcohol fermentation

The ability of yeast and bacteria such as *Zymomonas mobilis* to ferment sugar-containing material to form alcohol is well known from beer, wine, and liquor manufacture. If the initial material is cane sugar, the fermentation reaction may be summarised as

$$C_6H_{12}O_6 \rightarrow 2C_2H_5OH + 2CO_2. \tag{23.1}$$

The energy content of ethanol is 30 MJ kg^{-1}, and its octane rating is 89–100. With alternative fermentation bacteria, the sugar may be converted into butanol, $C_2H_5(CH_2)_2OH$. In Brazil, the cost of ethanol has recently come down to that of gasoline (Johansson, 2002), but with possible environmental externalities not accounted for.

In most sugar-containing plant material, the glucose molecules exist in polymerised form such as starch or cellulose, of the general structure $(C_6H_{10}O_5)_n$. Starch or hemicellulose is degraded to glucose by hydrolysis (cf. Fig. 22.1), while lignocellulose resists degradation owing to its lignin content. Lignin glues the cellulosic material together to keep its structure rigid, whether it be crystalline or amorphous. Wood has high lignin content (about 25%), and straw also has considerable amounts of lignin (13%), while potato or beet-starch contains very little lignin.

Some of the lignin seals may be broken by pre-treatment, ranging from mechanical crushing to the introduction of swelling agents causing rupture (Ladisch *et al.*, 1979).

The hydrolysis process is given by (22.2). In earlier times, hydrolysis was always achieved by adding an acid to the cellulosic material. During both world wars, Germany produced ethanol from cellulosic material by acid hydrolysis, but at very high cost. Acid recycling is incomplete; with low acid concentration the lignocelluloses is not degraded, and with high acid concentration the sugar already formed from hemicellulose is destroyed.

Consequently, alternative methods of hydrolysis have been developed, based on enzymatic intervention. Bacterial (e.g., of the genus *Trichoderma*) or fungal (such as *Sporotrichum pulverulentum*) enzymes have proved capable of converting cellulosic material, at near ambient temperatures, to some 80% glucose and a remainder of cellodextrins (which could eventually be fermented, but in a separate step with fermentation micro-organisms other than those responsible for the glucose fermentation) (Ladisch *et al.*, 1979).

The residue left behind after the fermentation process (23.1) can be washed and dried to give a solid product suitable as fertiliser or as animal feed. The composition depends on the original material, in particular with respect to lignin content (small for residues of molasses, beets, etc., high for straws and woody material, but with fibre structure broken as a result of the processes described above). If the lignin content is high, direct combustion of the residue is feasible, and it is often used to furnish process heat to the final distillation.

The outcome of the fermentation process is a water–ethanol mixture. When the alcohol fraction exceeds about 10%, the fermentation process slows down and finally halts. Therefore, an essential step in obtaining fuel alcohol is to separate the ethanol from the water. Usually, this is done by distillation, a step that may make the overall energy balance of the ethanol production negative. The sum of agricultural energy inputs (fertiliser, vehicles, machinery) and all process inputs (cutting, crushing, pre-treatment, enzyme recycling, heating for different process steps from hydrolysis to distillation), as well as energy for transport, is, in existing operations such as those of the Brazilian alcohol programme (Trinidade, 1980), around 1.5 times the energy outputs (alcohol and fertiliser if it is utilised). However, if the inputs are domestic fuels, for example, combustion of residues from agriculture, and if the alcohol produced is used to displace imported oil products, the balance might still be quite acceptable from a national economic point of view.

If, further, the lignin-containing materials of the process are recovered and used for process heat generation (e.g. for distillation), then such energy should be counted not only as input but also as output, making the total input and output energy roughly balance. Furthermore, more sophisticated process design, with cascading heat usage and parallel distillation columns operating with a time displacement such that heat can be reused from column to column (Hartline, 1979), could reduce the overall energy inputs to 55–65% of the outputs.

Radically improved energy balances would emerge if distillation could be replaced by a less energy intensive separation method. Several such methods for separating water and ethanol have been demonstrated on a laboratory scale, including: drying with desiccants such as calcium hydroxide, cellulose, or starch (Ladisch and Dyck, 1979); gas chromatography using rayon to retard water, while organic vapours pass through; solvent extraction using dibutyl phthalate, a water-immiscible solvent of alcohols; and passage through semipermeable membranes or selective zeolite absorbers (Hartline, 1979) and phase separation (APACE, 1982). The use of dry cellulose or starch appears particularly attractive, because over 99% pure alcohol can be obtained with less than 10% energy input, relative to the combustion energy of the alcohol. Furthermore, the cellulosic material may be cost-free, if it can be taken from the input stream to the fermentation unit and returned to it after having absorbed water (the fermentation reaction being "wet" anyway). The energy input of this scheme is for an initial distillation, bringing the ethanol fraction of the aqueous mixture from the initial 5–12% up to about 75%, at which point the desiccation process is started. As can be seen from Fig. 23.1, the distillation energy is modest up to an alcohol content of 90% and then starts to rise rapidly. The drying process thus substitutes for the most energy-expensive part of the distillation process.

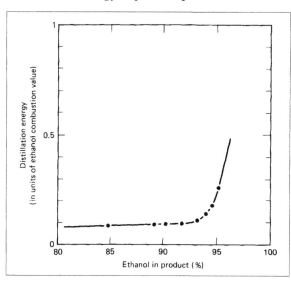

Figure 23.1. Distillation energy for ethanol–water mixture, as a function of ethanol content (assumed initial ethanol fraction 12%) (based on Ladisch and Dyck, 1979).

The ethanol fuel can be stored and used in the transportation sector much the same way as gasoline. It can be mixed with gasoline or can fully replace gasoline in spark ignition engines with high compression ratios (around 11). The knock resistance and high octane number of ethanol make this possible, and with pre-heating of the alcohol (using combustion heat that is recycled), the conversion efficiency can be improved. Several countries presently use alcohol–gasoline blends with up to 10% ethanol. This does not require any engine modification. Altering the gasoline Otto engines may be inconvenient in a transition period, but if alcohol distribution networks are implemented and existing gas stations modified, then the car engines could be optimised for alcohol fuels without regard to present requirements. A possible alternative to spark ignition engines is compression ignition engines, where auto-ignition of the fuel under high compression (a ratio of 25) replaces spark or glow plug ignition. With additives or chemical transformation into acetal, alcohol fuels could be used in this way (Bandel, 1981). Ethanol does not blend with diesel oil, so mixtures do require the use of special emulsifiers (Reeves *et al.*, 1982). However, diesel oil can be mixed with other biofuels without problems, e.g. the plant oils (rapeseed oil, etc.) presently in use in Germany.

A number of concerns with regard to the environmental impacts of the ethanol fermentation energy conversion chain must be considered. First of all, the biomass being used may have direct uses as food or may be grown in competition with production of food. The reason is, of course, that the easiest ethanol fermentation is obtained by starting with a raw material with as high a content of elementary sugar as possible, i.e. starting with sugar cane or cereal grain. Since sugar cane is likely to occupy prime agricultural land, and cereal production must increase with increasing world population, neither of these biomass resources should be used as fermentation inputs. However, residues from cereal production and from necessary sugar production (present sugar consumption is in many regions of the world too high from a health and nutrition point of view) could be used for ethanol fermentation, together with urban refuse, extra crops on otherwise committed land, perhaps aquatic crops and forest renewable resources. Concerns about proper soil management, recycling nutrients, and covering topsoil to prevent erosion are very appropriate in connection with the enhanced tillage utilisation that characterises combined food and ethanol production schemes.

The issues of competition with food and other (industry feedstock) uses of the biomass used for ethanol production could be reduced if only biomass residues were employed, and if nutrients were returned to the fields after energy extraction. This requires further enzyme additions for degradation of ligno-cellulosic components and may lead to slightly higher ethanol production costs. On the other hand, the biomass resource potential is greatly enlarged, as some 90% of harvested biomass does not end up as food (Sørensen, 2002b). The enzymes and catalysts needed for producing ethanol from agricultural residues and household waste have already been developed (Ögren *et al.*, 2007; Li *et al.*, 2007).

The hydrolysis process involves several potential environmental impacts. If acids are used, corrosion and accidents may occur, and substantial amounts of water would be required to clean the residues for reuse. Most acid would be recovered, but some would follow the sewage stream. Enzymatic hydrolysis would seem less cumbersome. Most of the enzymes would be recycled, but some might escape with wastewater or residues. Efforts should be made to ensure that they are made inactive before any release.

This is particularly important when, as envisaged, the fermentation residues are to be brought back to the fields or used as animal feed. A positive impact is the reduction of pathogenic organisms in residues after fermentation. Transport of biomass could involve dust emissions, and transport of ethanol might lead to spills (in insignificant amounts, as far as energy is concerned, but with possible local environmental effects), but overall the impacts from transport would be very small.

Finally, the combustion of ethanol in engines or elsewhere leads to pollutant emissions. Compared with gasoline combustion, emissions of carbon monoxide and hydrocarbons diminish, while those of nitrous oxides, aromatics, and aldehydes increase (Hespanhol, 1979), assuming that modified ignition engines are used. With special ethanol engines and exhaust controls, critical emissions may be controlled. In any case, the lead pollution still associated with gasoline engines in some countries would be eliminated.

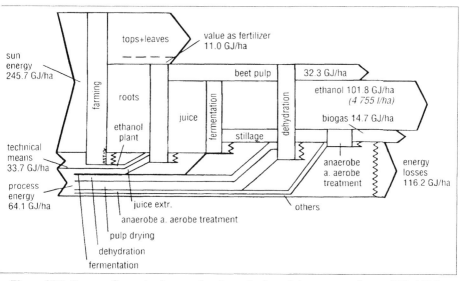

Figure 23.2. Energy flows in the production of ethanol from sugar beets (EC, 1994). Energy inputs to biomass growth, harvesting and transport are not indicated.

The energy balance of current ethanol production from biomass is not very favourable. A European study has estimated the energy flows for a number of feedstocks (EC, 1994). The highest yield of about 100 GJ ha^{-1} is found for sugar beets, shown in Fig. 23.2, but the process energy inputs and allotted life-cycle inputs into technical equipment are as large as the energy of the ethanol produced. A part of this may be supplied from biogas co-produced with the ethanol, but the overall energy efficiency remains low.

CHAPTER 24

THERMOCHEMICAL CONVERSION TO GASE- OUS AND OTHER FUELS

Before discussing the conversion of fresh biomass, the gasification of coal is briefly discussed because of its potential importance for continued use of fossil biomass (coal being the largest such source) and also because of its similarity to processes relevant for other solid biomass.

Inefficient conversion of coal to oil has historically been used by isolated coal-rich but oil-deficient nations (Germany during World War II, South Africa). Coal is gasified to carbon monoxide and hydrogen, which is then, by the Fischer–Tropsch process (passage through a reactor, e.g. a fluidised bed, with a nickel, cobalt, or iron catalyst), partially converted into hydrocarbons. Sulphur compounds have to be removed as they would impede the function of the catalyst. The reactions involved are of the form

$$(2n + 1)H_2 + nCO \rightarrow C_nH_{2n+2} + nH_2O,$$
$$(n + 1)H_2 + 2nCO \rightarrow C_nH_{2n+2} + nCO_2$$

and conversion efficiencies range from 21 to 55% (Robinson, 1980). Further separation of the hydrocarbons generated may then be performed, for instance, gasoline corresponding to the range $4 \leq n \leq 10$ in the above formulae.

Alternative coal liquefaction processes involve direct hydrogenation of coal under suitable pressures and temperatures. Pilot plants have been operating in the United States, producing up to 600 t a day (slightly different processes are named "solvent refining", "H-coal process", and "donor solvent process"; cf. Hammond, 1976). From an energy storage point of view, either coal or the converted product may be stockpiled.

For use in the transportation sector, production of liquid hydrocarbons such as methanol from natural gas could also be advantageous. In the long term, methanol is likely to be produced from renewable biomass sources as described in Section 24.2.

Conversion of fossil biomass such as coal into a gas is considered a way of reducing the negative environmental impacts of coal utilisation. However, in some cases the impacts have only been moved but not eliminated. Consider, for example, a coal-fired power plant with 99% particle removal from flue gases. If it were to be replaced by a synthetic gas-fired power plant with gas produced from coal, then sulphur could be removed at the gasification plant using dolomite-based scrubbers. This would practically eliminate the sulphur oxide emissions, but on the other hand, dust emissions from the dolomite processing would represent particle emissions twice as large as those avoided at the power plant by using gas instead of coal (Pigford, 1974). Of course, the dust is emitted at a different location.

This example, as well as the health impacts associated with coal mining, whether on the surface or in mines (although not identical), has sparked interest in methods of gasifying coal *in situ*. Two or more holes are drilled. Oxygen (or air) is injected through one, and a mixture of gases, including hydrogen and carbon oxides, emerges at the other hole. The establishment of proper communication between holes, and suitable underground contact surfaces, has proved difficult, and recovery rates are modest.

The processes involved would include

$$2C + O_2 \rightarrow 2CO, \tag{24.1}$$

$$CO + H_2O \leftrightarrow CO_2 + H_2. \tag{24.2}$$

The stoichiometric relation between CO and H_2 can then be adjusted using the shift reaction (24.2), which may proceed in both directions, depending on steam temperature and catalysts. This opens the way for methane synthesis through the reaction

$$CO + 3H_2 \rightarrow CH_4 + H_2O. \tag{24.3}$$

At present, the emphasis is on improving gasifiers using coal already extracted. Traditional methods include the Lurgi fixed-bed gasifier (providing gas under pressure from non-caking coal at a conversion efficiency often as low as 55%) and the Koppers–Totzek gasifier (oxygen input, the produced gas unpressurised, also of low efficiency).

Improved process schemes include the hy-gas process, requiring a hydrogen input; the bi-gas concept of brute force gasification at extremely high temperatures; and the slagging Lurgi process, capable of handling powdered coal (Hammond, 1976).

Promising, but still at an early stage of development, is catalytic gasification (e.g. potassium catalyst), where all processes take place at a common, relatively low temperature, so that they can be combined in a single reactor (Fig. 24.1). The primary reaction here is

$$C + H_2O \rightarrow H_2 + CO \tag{24.4}$$

(H_2O being in the form of steam above 550°C), to be followed by (24.2) and (24.3). The catalyst allows all processes to take place at 700°C. Without catalyst, the gasification would have to take place at 925°C and the shift reaction and methanation at 425°C, that is, in a separate reactor where excess hydrogen or carbon monoxide would be lost (Hirsch et al., 1982).

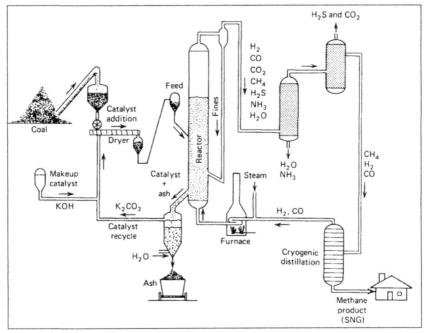

Figure 24.1. Schematic diagram of catalytic gasification process (SNG is synthetic natural gas). (From Hirsch *et al.*, 1982. Reprinted from *Science*, **215**, 121-127, 8, January 1982, with permission. Copyright 1982 American Association for the Advancement of Science.)

In a coal gasification scheme, storage would be (of coal) before conversion. Peat can be gasified in much the same way as coal, as can wood (with optional subsequent methanol conversion as described in Section 24.2).

24.1 Thermochemical gasification of fresh biomass

Gasification of biomass, and particularly wood and other lignin-containing cellulosic material, has a long history. The processes may be viewed as "combustion-like" conversion, but with less oxygen available than needed for burning. The ratio of oxygen available and the amount of oxygen that would allow complete burning is called the "equivalence ratio". For equivalence ratios below 0.1 the process is called "pyrolysis", and only a modest fraction of the biomass energy is found in the gaseous product – the rest being in char and oily residues. If the equivalence ratio is between 0.2 and 0.4, the process is called a proper "gasification". This is the region of maximum energy transfer to the gas (Desrosiers, 1981).

Chemical reaction	Energy consumed (kJ g^{-1})^a	Products / process
$C_6H_{10}O_5 \rightarrow 6C + 5H_2 + 2.5\,O_2$	5.94[b]	Elements, dissociation
$C_6H_{10}O_5 \rightarrow 6C + 5H_2O(g)$	–2.86	Charcoal, charring
$C_6H_{10}O_5 \rightarrow 0.8\,C_6H_8O + 1.8\,H_2O(g) + 1.2\,CO_2$	–2.07[c]	oily residues, pyrolysis
$C_6H_{10}O_5 \rightarrow 2C_2H_4 + 2CO_2 + H_2O(g)$	0.16	Ethylene, fast pyrolysis
$C_6H_{10}O_5 + \frac{1}{2}O_2 \rightarrow 6CO + 5H_2$	1.85	Synthesis gas, gasification
$C_6H_{10}O_5 + 6H_2 \rightarrow 6"CH_2" + 5\,H_2O(g)$	–4.86[d]	Hydrocarbons, –generation
$C_6H_{10}O_5 + 6O_2 \rightarrow 6CO_2 + 5\,H_2O(g)$	–17.48	heat, combustion

Table 24.1. Energy change for idealised cellulose thermal conversion reactions. (Source: T. Reed (1981), in *Biomass Gasification* (T. Reed, ed.), reproduced with permission. Copyright 1981, Noyes Data Corporation, Park Ridge, NJ)
[a] Specific reaction heat.
[b] The negative of the conventional heat of formation calculated for cellulose from the heat of combustion of starch.
[c] Calculated from the data for the idealised pyrolysis oil C_6H_8O (ΔH_c = – 745.9 kcal mol^{-1}, ΔH_f = 149.6 kcal g^{-1}, where H_c = heat of combustion and H_f = heat of fusion).
[d] Calculated for an idealised hydrocarbon with ΔH_c as above. H_2 is consumed.

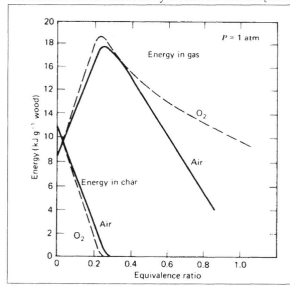

Figure 24.2. Calculated energy content in gas and char produced by equilibrium processes between air (or oxygen) and biomass, as a function of equivalence ratio. (From Reed, 1981. Reprinted from *Biomass Gasification* (T. Reed, ed.), with permission. Copyright 1981, Noyes Data Corporation, Park Ridge, NJ).

The chemical processes involved in biomass gasification are similar to the reactions (24.1)–(24.4) for coal gasification. Table 24.1 lists a number of reactions involving polysaccharidic material, including pyrolysis and gasification. In addition to the chemical reaction formulae, the table gives enthalpy changes for idealised reactions (i.e., neglecting the heat required to bring the reactants to the appropriate reaction temperature). Figure 24.2 gives the energy of the final products, gas and char, as a function of the equivalence ratio, still based on an idealised thermodynamical calculation. The specific heat of the material is 3 kJ g^{-1} of wood at the peak of energy in the gas, increasing to 21 kJ g^{-1} of wood for combustion at equivalence ratio equal to unity. Much of this sensible heat can be recovered from the gas, so that process heat inputs for gasification can be kept low.

Figure 24.3 gives the equilibrium composition calculated as a function of the equivalence ratio. By equilibrium composition is understood the composition of reaction products occurring after the reaction rates and reaction temperature have stabilised adiabatically. The actual processes are not necessarily adiabatic; in particular the low-temperature pyrolysis reactions are not. Still, the theoretical evaluations assuming equilibrium conditions serve as a useful guideline for evaluating the performance of actual gasifiers.

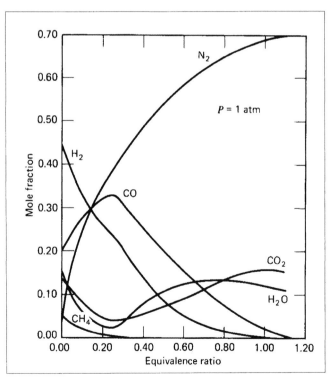

Figure 24.3. Calculated gas composition resulting from equilibrium processes between air and biomass, as a function of equivalence ratio. (From Reed, 1981. Reprinted from *Biomass Gasification* (T. Reed, ed.), with permission. Copyright 1981, Noyes Data Corporation, Park Ridge, NJ).

The idealised energy change calculation of Table 24.1 assumes a cellulosic composition such as the one given in (22.2). For wood, the average ratios of carbon, hydrogen and oxygen are 1:1.4:0.6 (Reed, 1981).

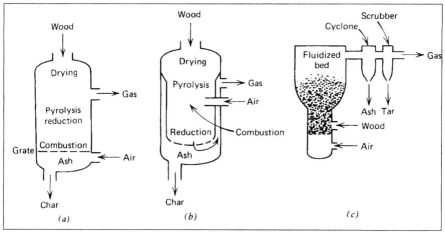

Figure 24.4. Gasifier types: (a) updraft, (b) downdraft, and (c) fluidised bed.

Figure 24.4 shows three examples of wood gasifiers: the updraft, the downdraft, and the fluidised bed types. The drawback of the updraft type is a high rate of oil, tar, and corrosive chemical formation in the pyrolysis zone. This problem is solved by the downdraft version, where such oils and other matter pass through a hot charcoal bed in the lower zone of the reactor and become cracked to simpler gases or char. The fluidised bed reactor may prove superior for large-scale operations, because passage time is smaller. The drawback of this is that ash and tars are carried along with the gas and have to be removed later in cyclones and scrubbers. Several variations on these gasifier types have been suggested (Drift, 2002; Gøbel et al., 2002).

The gas produced by gasification of biomass is a "medium-quality gas", meaning a gas with burning value in the range 10–18 MJ m^{-3}. This gas may be used directly in Otto or diesel engines, it may be used to drive heat pump compressors, or alternatively, it may by upgraded to pipeline-quality gas (about 30 MJ m^{-3}) or converted to methanol, as discussed in Section 24.2.

Environmental impacts derive from biomass production, collection (e.g. forestry work) and transport to gasification site, from the gasification and related processes, and finally from the use made of the gas. The gasification residues – ash, char, liquid waste water, and tar – have to be disposed of. Char may be recycled to the gasifier, while ash and tars could conceivably be utilised in the road or building construction industry. The alternative of landfill disposal would represent a recognised impact. Investigations of emissions from combustion of producer gas indicate low emissions of nitrous oxides and hydrocarbons, as compared with emissions from combustion of natural gas. In one case, carbon monoxide emissions were found to be higher than for natural gas burning, but it is believed that this problem can be rectified as more experience in adjusting air-to-fuel ratios is gained (Wang et al., 1982).

24.2 Methanol from biomass

There are various ways of producing methanol from biomass sources, as indicated in Fig. 22.1. Starting from wood or isolated lignin, the most direct routes are by liquefaction or by gasification. The pyrolysis alternative gives only a fraction of the energy in the form of a producer gas.

By high-pressure hydrogenation, biomass may be transformed into a mixture of liquid hydrocarbons suitable for further refining or synthesis of methanol (Chartier and Meriaux, 1980), but all methanol production schemes so far have used a synthesis gas, which may be derived from wood gasification or coal gasification. The low-quality "producer gas" resulting directly from the wood gasification (used in cars throughout Europe during World War II) is a mixture of carbon monoxide, hydrogen gas, carbon dioxide, and nitrogen gas. If air is used for gasification, the energy conversion efficiency is about 50%, and if pure oxygen is used instead, some 60% efficiency is possible, and the gas produced has less nitrogen content (Robinson, 1980). Gasification or pyrolysis could conceivably be performed with heat from (concentrating) solar collectors, for example, in a fluidised bed gasifier maintained at 500°C.

The producer gas is cleaned, CO_2 and N_2 as well as impurities are removed (the nitrogen by cryogenic separation), and methanol is generated at elevated pressure by the reaction

$$2H_2 + CO \rightarrow CH_3OH. \tag{24.5}$$

The carbon monoxide and hydrogen gas (possibly with additional CO_2) is called the "synthesis gas", and it is usually necessary to use a catalyst in order to maintain the proper stoichiometric ratio between the reactants of (24.5) (Cheremisinoff et al., 1980). A schematic process diagram is shown in Fig. 24.5.

An alternative is biogas production from the biomass (Chapter 22), followed by the methane to methanol reaction,

$$2CH_4 + O_2 \rightarrow 2CH_3OH, \tag{24.6}$$

also used in current methanol production from natural gas (Wise, 1981). Change of the H_2/CO stoichiometric ratio for (24.5) is obtained by the "shift reaction" (24.2). Steam is added or removed in the presence of a catalyst (iron oxide, chromium oxide).

The conversion efficiency of the synthesis gas to methanol step is about 85%, implying an overall wood to methanol energy efficiency of 40–45%. Improved catalytic gasification techniques raise the overall conversion efficiency to some 55% (Faaij and Hamelinck, 2002). The currently achieved efficiency is about 50%, but not all life-cycle estimates of energy inputs have been included or performed (EC, 1994).

The octane number of methanol is similar to that of ethanol, but the heat of combustion is less, amounting to 18 MJ kg^{-1}. However, the engine efficiency of methanol is higher than that of gasoline, by at least 20% for current motor car engines, so an "effective energy content" of 22.5 MJ kg^{-1} is sometimes quoted (EC, 1994). Methanol can be mixed with gasoline in standard engines, or used in specially designed Otto or diesel engines, such as a spark ignition engine run on vaporised methanol, with the vaporisation energy being recovered from the coolant flow (Perrin, 1981). Uses are similar to those of ethanol, but several differences exist in the assessment of environmental impacts, from production to use (e.g. toxicity of fumes at filling stations).

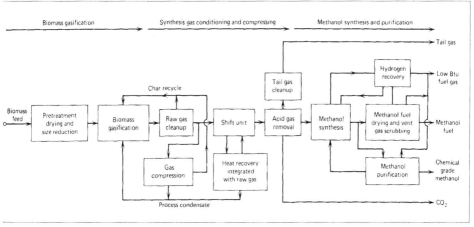

Figure 24.5. Schematic flow diagram for biomass to methanol conversion process (From Wan, Simmins, and Nguyen, 1981. Reprinted from *Biomass Gasification* (T. Reed, ed.), with permission. Copyright 1981, Noyes Data Corp., Park Ridge, NJ).

The future cost of methanol fuel is expected to reach US$ 8/GJ (Faaij and Hamelinck, 2002). The gasification can be made in closed environments, where all emissions are collected, as well as ash and slurry. Cleaning processes in the methanol formation steps will recover most catalysts in reusable form, but other impurities would have to be disposed of along with the gasification products. Precise schemes for waste disposal have not been formulated, but it seems unlikely that all nutrients could be recycled to agri- or silviculture as in the case of ethanol fermentation (SMAB, 1978). However, the production of ammonia by a process similar to the one yielding methanol is an alternative use of the synthesis gas. Production of methanol from *eucalyptus* rather than from woody biomass has been studied in Brazil (Damen *et al.*, 2002). More fundamental studies aiming to better understand the way in which methanol production relies on degradation of lignin are ongoing (Minami *et al.*, 2002).

BENT SØRENSEN 2007

VII. ENERGY TRANSMISSION

CHAPTER 25

HEAT TRANSMISSION

25.1 General remarks on transmission

In the past, transport of energy has often been in the form of carrying fuels to the site needed. With regard to renewable energy resources, such transport may be useful in connection with biomass-derived energy, either by transporting the biological materials themselves or by conversion into biofuels, which may be more convenient to move. For most other types of renewable energy, the resource itself cannot be "moved" (possible exceptions may exist, such as diverting a river flow to the place of hydropower utilisation). Instead, an initial conversion process may be performed, and the emerging energy form may be transmitted to the load areas, where it may be used directly or subjected to a second conversion process before delivery to the actual users.

Like fuels (which represent chemical energy), heat, mechanical and possibly electrical energy may be stored, and the storage "containers" may be transported. Alternatively, energy may be transmitted through a suitable transmission system, which may involve pipeline transmission (for heat, fuels and certain types of mechanical energy, e.g. pressure or kinetic energy of a gas or a fluid), electric transmission lines (for electricity) or radiant transmission (for heat or electricity).

Energy transmission is used not only to deliver energy from convenient sites of generation (such as where the renewable resources are) to the dominant sites of energy use, but also to deal with mismatch between the time distribution of (renewable) energy generation and time variations in demand. As such, energy transmission and energy storage may supplement each other. Some demands may be displaceable, while others are time urgent. The latter ones often have a systematic variation over the hours of the day and over the seasons. This may be taken advantage of by long-distance transmission of energy across time zones (east–west).

25.2 District heating lines

Most heating systems involve the transport of sensible heat in a fluid or gas (such as water or air) through pipes or channels. Examples are the solar heating system illustrated in Fig. 16.9, the heat pump illustrated in Fig. 6.1 and the geothermal heating plants illustrated in Fig. 7.1. Solar heating systems and heat pumps may be used in a decentralised manner, with an individual system providing heat for a single building, but they may also be used on a community scale, with one installation providing heat for a building block, a factory complex, a whole village or a city of a certain size. Many combined heat and power (CHP) systems involve heat transmission through distances of 10–50 km (decentralised and centralised CHP plants). In some regions, pure heating plants are attached to a district heating grid.

Assuming that a hot fluid or gas is produced at the central conversion plant, the transmission may be accomplished by pumping the fluid or gas through a pipeline to the load points. The pipeline may be placed underground and the tubing insulated in order to reduce conduction and convection of heat away from the pipe. If the temperature of the surrounding medium can be regarded as approximately constant, equal to T_{ref}, the variation in fluid (or gas) temperature $T^{fluid}(x)$ along the transmission line (with the path-length co-ordinate denoted x) can be evaluated by the same expression (6.1) used in the simple description of a heat exchanger in Chapter 6. Temperature variations across the inside area of the pipe perpendicular to the stream-wise direction are not considered.

The rate of heat loss from the pipe section between the distances x and $(x+dx)$ is then of the form

$$dE/dx = J_m\, C_p^{fluid}\, dT^{fluid}(x)/dx = h'\, (T_{ref} - T^{fluid}(x)), \tag{25.1}$$

where h' for a cylindrical insulated pipe of inner and outer radii r_1 and r_2 is related to the heat transfer coefficient λ^{pipe} (describing conductivity plus convection terms) by

$$h' = 2\pi\, \lambda^{pipe} / \log(r_2/r_1).$$

Upon integration from the beginning of the transmission pipe, $x = x_1$, (25.1) gives, in analogy to (6.1),

$$T^{fluid}(x) = T_{ref} + (T^{fluid}(x_1) - T_{ref})\exp(-h'(x-x_1)/J_m\, C_p^{fluid}), \tag{25.2}$$

where J_m is the mass flow rate and C_p^{fluid} is the fluid (or gas) heat capacity. The total rate of heat loss due to the temperature gradient between the fluid in the pipe and the surroundings is obtained by integrating (25.1) from x_1 to the end of the line, x_2,

$$\Delta E = \int_{x_1}^{x_2} \frac{dE}{dx} \, dx = -J_m \, C_p^{fluid} \, (T^{fluid}(x_1) - T_{ref}) \left(1 - \exp\left(-\frac{h'(x_2 - x_1)}{J_m \, C_p^{fluid}} \right) \right).$$

The relative loss of heat supplied to the pipeline, along the entire transmission line, is then

$$\frac{\Delta E}{E} = -\left(1 - \exp\left(-\frac{h'(x_2 - x_1)}{J_m \, C_p^{fluid}} \right) \right), \qquad (25.3)$$

provided that the heat entering the pipe, E, is measured relative to a reservoir of temperature T_{ref}.

The total transmission loss will be larger than (25.3) because of friction in the pipe. The distribution of flow velocity, v, over the pipe inside cross section depends on the roughness of the inner walls (v being zero at the wall), and the velocity profile generally depends on the magnitude of the velocity (say at the centreline). Extensive literature exists on the flow in pipes (see e.g. Grimson, 1971). The net result is that additional energy must be provided in order to make up for the pipe losses. For an incompressible fluid moving in a horizontal pipe of constant dimensions, flow continuity demands that the velocity field is constant along the pipeline, and the friction can in this case be described by a uniformly decreasing pressure in the stream-wise direction. A pump must provide the energy flux necessary to compensate for the pressure loss,

$$\Delta E^{pump} = -Av \, \Delta P,$$

where A is the area of the inner pipe cross section and ΔP is the total pressure drop along the pipeline.

Transmission lines presently in use for city district heating have average heat losses of 10–15%, depending on insulation thickness (see e.g. WEC, 1991). The pump energy added is in the form of high-quality mechanical work, but with proper dimensioning of the tubes it may be kept to a small fraction of the heat energy transmitted. This energy is converted into heat by the frictional dissipation, and some of this heat may actually be credited to the heat transmission. Maximum heat transmission distances currently considered economic are around 30 km, with the possible exception of some geothermal installations. Figure 25.1 shows an integrated design that allows much easier installation than earlier pipes with separate insulation. In countries with a high penetration of combined heat and power production, such as Denmark, a heat storage facility holding some 10 h of heat load is sometimes added to each plant, in order that the (fuel-based) plant does not have to generate more electric power than needed, say, at night-time when the heating load is large (Danish Energy Agency, 1993).

Heat pipes

A special heat transmission problem exists in cases where very large amounts of heat have to be delivered to or removed from a small region, for example, in connection with highly concentrating solar energy collectors. Such heat transfer may be accomplished by a heat pipe, a device described in Section 4.2 in relation to thermionic generators.

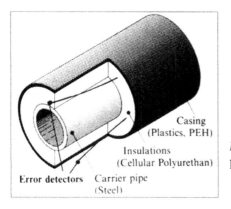

Casing
(Plastics, PEH)

Insulations
(Cellular Polyurethan)

Error detectors   Carrier pipe
(Steel)

Figure 25.1. Pre-insulated district heating pipe (Danish Energy Agency, 1993).

CHAPTER 26

POWER TRANSMISSION

26.1 Normal conducting lines

At present, electric current is transmitted in major utility grids, as well as distributed locally to each load site by means of conducting wires. Electricity use is dominated by alternating current (AC), as far as utility networks are concerned, and most transmission over distances up to a few hundred kilometres is by AC. For transmission over longer distances (e.g. by ocean cables), conversion to direct current (DC) before transmission and back to AC after transmission is common. Cables are either buried in the ground (with appropriate electric insulation) or are overhead lines suspended in the air between masts, without electrical insulation around the wires. Insulating connections are provided at the tower fastening points, but otherwise the low electric conductivity of air is counted on. This implies that the losses will comprise conduction losses depending on the instantaneous state of the air (the "weather situation"), in addition to the ohmic losses connected with the resistance R of the wire itself, $E^{heat} = RI^2$, with I being the current. The leak current between the elevated wire and the ground depends on the potential difference as well as on the integrated resistivity, such that the larger the voltage, the further the wires must be placed from the ground.

Averaged over different meteorological conditions, the losses in a standard AC overhead transmission line (138–400 kV, at an elevation of some 15–40 m) are currently a little under 1% per 100 km of transmission (Hammond *et al.*, 1973), but the overall transmission losses of utility networks, including the finely branched distribution networks in the load areas, may for many older, existing grids amount to some 12–15% of the power production, for a grid extending over a land area of about 10^4 km^2 (Blegaa *et al.*, 1976). Losses are down to 5–6% for the best systems installed at present and are expected to decrease further to the level of 2–3% in the future, when the currently best technologies penetrate further (Kuemmel *et al.*,

1997). This loss is calculated relative to the total production of electricity at the power plants attached to the common grid, and thus includes certain in-plant and transformer losses. The numbers also represent annual averages for a power utility system occasionally exchanging power with other utility systems through interconnecting transmission lines, which may involve transmission distances much longer than the linear extent of the load area being serviced by the single utility system in question.

The trend is to replace overhead lines by underground cables, primarily for visual and environmental reasons. This has already happened for the distribution lines in most of Europe and is increasingly also being required for transmission lines. In Japan and the United States, overhead lines are still common.

Underground transmission and distribution lines range from simple co-axial cables to more sophisticated constructions insulated by a compressed gas. Several trans-ocean cables (up to 1000 km) have been installed in the Scandinavian region in order to bring the potentially large surpluses of hydropower production to the European continent. The losses through these high-voltage (up to 1000 kV) DC lines are under 0.4% per 100 km, to which should be added the 1–2% transmission loss occurring at the thyristor converters on shore that transform AC into DC and vice versa (Ch. 19 in IPCC, 1996b). The cost of these low-loss lines is currently approaching that of conventional AC underwater cables (about 2 euro kW^{-1} km^{-1}; Meibom et al., 1999; Wizelius, 1998).

One factor influencing the performance of underground transmission lines is the slowness of heat transport in most soils. In order to maintain the temperature within the limits required by the materials used, active cooling of the cable could be introduced, particularly if large amounts of power have to be transmitted. For example, the cable may be cooled to 77 K (liquid nitrogen temperature) by means of refrigerators spaced at intervals of about 10 km (cf. Hammond et al., 1973). This allows increased amounts of power to be transmitted in a given cable, but the overall losses are hardly reduced, since the reduced resistance in the conductors is probably outweighed by the energy spent on cooling. According to (3.1), the cooling efficiency is limited by a Carnot value of around 0.35, i.e. more than three units of work have to be supplied in order to remove one unit of heat at 77 K.

Off-shore issues
The power from an off-shore wind farm is transmitted to an on-shore distribution hub by means of one or more undersea cables, the latter providing redundancy that in the case of large farms adds security against cable disruption or similar failures. Current off-shore wind farms use AC cables of up to 150 kV (Eltra, 2003). New installations use cables carrying all three leads plus control wiring. In the interest of loss minimisation for larger in-

stallations, it is expected that future systems may accept the higher cost of DC–AC conversion (on shore, the need for AC–DC conversion at sea depends on the generator type used), similar to the technology presently in use for many undersea cable connections between national grids (e.g. between Denmark and Norway or Sweden). Recent development of voltage source-based high voltage direct current control systems to replace the earlier thyristor-based technology promises better means of regulation of the interface of the DC link to the on-shore AC system (Ackermann, 2002).

26.2 Superconducting lines

For DC transmission, the ohmic losses may be completely eliminated by use of superconducting lines. A number of elements, alloys and compounds become superconducting when cooled to a sufficiently low temperature. Physically, the onset of superconductivity is associated with the sudden appearance of an energy gap between the "ground state", i.e. the overall state of the electrons, and any excited electron state (similar to the situation illustrated in Fig. 14.3, but for the entire system rather than for individual electrons). A current, i.e. a collective displacement (flow) of electrons, will not be able to excite the system away from the "ground state" unless the interaction is strong enough to overcome the energy gap. This implies that no mechanism is available for the transfer of energy from the current to other degrees of freedom, and thus the current will not lose any energy, which is equivalent to stating that the resistance is zero. In order that the electron system remains in the ground state, the thermal energy spread must be smaller than the energy needed to cross the energy gap. This is the reason why superconductivity occurs only below a certain temperature, which may be quite low (e.g. 9 K for niobium, 18 K for niobium–tin, Nb_3Sn). However, there are other mechanisms that in more complex compounds can prevent instability, thereby explaining the findings in recent years of materials that exhibit superconductivity at temperatures approaching ambient (Pines, 1994; Demler and Zhang, 1998).

For AC transmission, a superconducting line will not be loss-free, owing to excitations caused by the time variations of the electromagnetic field (cf. Hein, 1974), but the losses will be much smaller than for normal lines. It is estimated that the amount of power that can be transmitted through a single cable is in the gigawatt range. This figure is based on suggested designs, including the required refrigeration and thermal insulation components within overall dimensions of about 0.5 m (cable diameter). The power required for cooling, i.e. to compensate for heat flow into the cable, must be considered in order to calculate the total power losses in transmission.

For transmission over longer distances it may, in any case, be an advantage to use direct current, despite the losses in the AC–DC and DC–AC conversions (a few per cent as discussed above). Future intercontinental transmission using superconducting lines has been discussed, notably by Nielsen and Sørensen (1996), Sørensen and Meibom (2000), and Sørensen (2004). Motivation for such thoughts is of course the location of some very promising renewable energy sites far from the areas of load. Examples would be solar installations in the Sahara or other desert areas, or wind power installations at isolated rocky coastlines of northern Norway or in Siberian highlands.

Finally, radiant transmission of electrical energy may be mentioned. The technique for transmitting radiation and re-collecting the energy (or some part of it) is well developed for wavelengths near or above visible light. Examples are laser beams (stimulated atomic emission) and microwave beams (produced by accelerating charges in suitable antennas), ranging from the infrared to the wavelengths used for radio and other data transmission (e.g. between satellites and ground-based receivers). Large-scale transmission of energy in the form of microwave radiation has been proposed in connection with satellite solar power generation, but is not currently considered practical. Short distance transmission of signals, e.g. between computers and peripheral units, do involve only minute transfers of energy and is already in widespread use.

FUEL TRANSMISSION

Fuels such as natural gas, biogas, hydrogen and other energy-carrier gases may be transmitted through pipelines, at the expense of a fairly modest amount of pumping energy (at least for horizontal transfer). Pipeline oil transmission is also in abundant use. Alternatives used for sea transportation between continents are containers onboard ships for solid fuels, oil, compressed gases or liquefied gases. Similar containers are in use for shorter distance transport by rail or road. Higher energy densities may be obtained by some of the energy storage devices discussed in Part IX below, such as metal hydrides or carbon nanotubes, e.g. for hydrogen transport. Light container materials are, of course, preferable in order to reduce the cost of moving fuels by vessel, whether on land or at sea.

Current natural gas networks consist of plastic distribution lines operated at pressures of 0.103 to about 0.4 MPa and steel transmission lines operated at pressures of 5-8 MPa. With some upgrading of valves, some modern natural gas pipelines could be used for the transmission of hydrogen (Sørensen et al., 2001; Sørensen, 2005). Certain steel types may become brittle with time, as a result of hydrogen penetration into the material, and cracks may develop. It is believed that H_2S impurities in the hydrogen stream increases the problem, but investigations of the steel types currently used for new pipelines indicate little probability of damage by hydrogen (Pöpperling et al., 1982; Kussmaul and Deimel, 1995).

Mechanical devices have been used to transfer mechanical energy over short distances, but mechanical connections with moving parts are not practical for distances of transfer, which may be considered relevant for transmission lines. However, mechanical energy in such forms as hydraulic pulses can be transmitted over longer distances in feasible ways, as, for example, mentioned in connection with wave energy conversion devices placed in open oceans (Chapter 13).

VILLA VISION - BENT SØRENSEN 1995

VIII. HEAT STORAGE

CHAPTER 28

HEAT CAPACITY STORAGE

Storage of energy provides a way of adjusting to variations in the energy demand, i.e. a way of meeting a load with a time-dependence different from that of generation. For fuel-type energy, storage can help burn the fuel more efficiently, by avoiding those situations where demand variations would otherwise require regulation of combustion rates beyond what is technically feasible or economic. For renewable energy sources of a fluctuating nature, storage can help make energy systems including such sources as dependable as the conventional systems.

Ideal requirements of energy storage include rapid access and versatility of the energy form in which energy from the store is delivered. Conversion of one type of stored energy into another form of stored energy could be advantageous. For example, the production of electricity in large fossil or nuclear power plants may involve long start-up times including additional costs when used for load levelling, while the use of pumped water storage allows delivery upon demand in less than a minute. The economic feasibility of such energy storage depends on the relative fixed and variable costs of the different converter types and on the cost and availability of different fuels. Another example is a fuel-driven automobile, which operates at or near peak power only during short intervals of acceleration. If short-term storage is provided, e.g. batteries, which can accumulate energy produced off-peak by the automobile engine and deliver the power for starting and acceleration, then the capacity of the primary engine may be greatly reduced.

In connection with renewable energy resources, of which many are intermittent and of a fluctuating power level, supplementing conversion by energy storage is essential if the actual demand is to be met at all times. The only alternative would seem to be fuel-based back-up conversion equipment (e.g. using renewable biofuels), but this, of course, is just a kind of energy storage, which in the long run may require fuels provided by conversion processes based on renewable primary energy sources.

28.1 Storage of heat

Heat capacity, or "sensible heat" storage, is accomplished by changing the temperature of a material without changing its phase or chemical composition. The amount of energy stored by heating a piece of material of mass m from temperature T_0 to temperature T_1 at constant pressure is

$$E = m \int_{T_0}^{T_1} c_P \, dT, \qquad (28.1)$$

where c_P is the specific heat capacity at constant pressure.

Energy storage at low temperatures is needed in renewable systems such as solar absorbers delivering space heating, hot water, and eventually heat for cooking (up to 100°C). The actual heat storage devices may be of modest size, aiming at delivering heat during the night after a sunny day, or they may be somewhat larger, capable of meeting the demand during a number of consecutive overcast days. Finally, the storage system may provide seasonal storage of heat, as required at high latitudes where seasonal variations in solar radiation are large, and, furthermore, heat loads are to some extent inversely correlated with the length of the day.

Another aspect of low-temperature heat storage (as well as of some other energy forms) is the amount of decentralisation. Many solar absorption systems are conveniently placed on existing rooftops, that is, in a highly decentralised fashion. A sensible-energy heat store, however, typically loses heat from its container, insulated or not, in proportion to the surface area. The relative heat loss is smaller, the larger the store dimensions, and thus more centralised storage facilities, for example, of communal size, may prove superior to individual installations. This depends on an economic balance between the size advantage and the cost of additional heat transmission lines for connecting individual buildings to a central storage facility. One should also consider other factors, such as the possible advantage in supply security offered by the common storage facility (which would be able to deliver heat, for instance, to a building with malfunctioning solar collectors).

Water storage

Heat energy intended for later use at temperatures below 100°C may conveniently be stored as hot water, owing to the high heat capacity of water (4180 J kg^{-1} K^{-1} or 4.18 × 10^6 J m^{-3} K^{-1} at standard temperature and pressure), combined with the fairly low thermal conductivity of water (0.56 J m^{-1} s^{-1} K^{-1} at 0°C, rising to 0.68 J m^{-1} s^{-1} K^{-1} at 100°C).

Most space heating and hot water systems of individual buildings include a water storage tank, usually in the form of an insulated steel container with a capacity corresponding to less than a day's hot water usage and often only a small fraction of a cold winter day's space heating load. For a one-family dwelling, a 0.1-m^3 tank is typical in Europe and the United States.

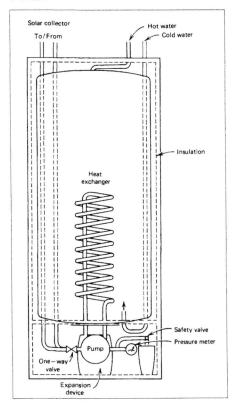

Figure 28.1. Water container for heat storage with possibility of temperature stratification (e.g. for use in connection with solar collectors) (Ellehauge, 1981).

A steel hot water container may look like the one sketched in Fig. 28.1. It is cylindrical with a height greater than the diameter in order to make good temperature stratification possible, an important feature if the container is part of a solar heating system. A temperature difference of up to 50°C between the top and bottom water can be maintained, with substantial improvements (over 15%) in the performance of the solar collector heating system, because the conversion efficiency of the collector declines (see Fig. 28.2) with the temperature difference between the water coming into the collector and the ambient outdoor temperature (Koppen *et al.*, 1979). Thus, the water from the cold lower part of the storage tank would be used as input to the solar collector circuit, and the heated water leaving the solar absorber would be delivered to a higher-temperature region of the storage tank, normally the top layer. The take-out from the storage tank to load (directly or through a heat exchanger) is also from the top of the tank, because the solar system will, in this case, be able to cover load over a longer period of the year (and possibly for the entire year). There is typically a minimum temperature required for the fluid carrying heat to the load areas, and during winter, the solar collector system may not always be able to raise the entire storage tank volume to a temperature above this threshold. Thus, temperature stratification in storage containers is often a helpful feature. The minimum load-input temperatures are around 45–50°C for space heating through water-filled "radiators" and "convectors," but only 25–30°C for water-filled floor heating systems and air-based heating and ventilation duct systems.

For hot water provision for a single family, using solar collector systems with a few square meters of collectors (1 m^2 in sunny climates, 3–5 m^2 at high latitudes), a storage tank of around 0.3 m^3 is sufficient for diurnal storage, while a larger tank is needed if consecutive days of zero solar heat absorption can be expected. For complete hot water and space heating solar systems, storage requirements can be quite substantial if load is to be met at all times and the solar radiation has a pronounced seasonal variation.

Most solar collector systems aiming at provision of both hot water and space heating for a single-family dwelling have a fairly small volume of storage and rely on auxiliary heat sources. This is the result of an economic trade-off due to the rapid reduction in solar collector gains with increasing coverage, that is, the energy supplied by the last square meter of collector added to the system becomes smaller and smaller as the total collector area increases. Of course, the gain is higher with increased storage for fixed collector area over some range of system sizes, but this gain is very modest (cf. Sørensen, 2004).

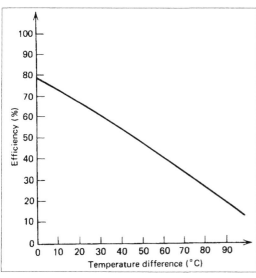

Figure 28.2. Efficiency curve for flat-plate solar collector as a function of the temperature difference between the average temperature of the fluid, which is removing heat from the collector, and the ambient, outside air temperature. The efficiency is the percentage of the absorbed solar radiation, which is transferred to the working fluid. The curve is based on measurements for a selective-surface collector, tilted 45° and receiving 800 W m^{-2} of solar radiation at incident angles below 30°. Wind speed along the front of the collector is 5 m s^{-1} (Svendsen, 1980).

For this reason, many solar space-heating systems only have diurnal storage, say, a hot water storage. In order to avoid boiling, when the solar radiation level is high for a given day, the circulation from storage tank to collector is disconnected whenever the storage temperature is above some specified value (e.g. 80°C), and the collector becomes stagnant. This is usually no problem for simple collectors with one layer of glazing and black paint absorber, but with multilayered cover or selective-surface coating of the absorber, the stagnation temperatures are often too high and materials would become damaged if these situations were allowed to occur. Instead, the storage size may be increased to such a value that continuous circulation of water from storage through collectors can be maintained during such sunny days, without violating maximum storage temperature requirements at any time during the year. If the solar collector is so efficient that it still has a net gain above 100°C (such as the one shown in Fig. 28.2), this heat gain must be balanced by heat losses from piping and from the storage itself, or the store must be large enough for the accumulated temperature rise during the most sunny periods of the year to be acceptable. In high-latitude climatic regions, this can be achieved with a few square metres of water storage, for collector areas up to about 50 m^2.

Larger amounts of storage may be useful if the winter is generally sunny, but a few consecutive days of poor solar radiation do occur from time to time. This, for example, is the situation for an experimental house in Regina, Saskatchewan (Besant et al., 1979). The house is super-insulated and is designed to derive 33% of its heating load from passive gains of south-facing windows, 55% from activities in the house (body heat, electric appliances), and the remaining 12% from high-efficiency (evacuated tube) solar collectors. The collector area is 18 m^2 and there is a 13-m^3 water-storage tank. On a January day with outdoor temperatures between –25°C (afternoon) and –30°C (early morning), about half the building heat loss is provided by indirect gains. The other half (about 160 MJ d^{-1}) must be drawn from the store, in order to maintain indoor temperatures of 21°C, allowed to drop to 17°C between midnight and 0700 h. On sunny January days, the amount of energy drawn from the storage reduces to about 70 MJ d^{-1}. Since overcast periods of much over a week occur very rarely, the storage size is such that 100% coverage can be expected, from indirect and direct solar sources, in most years.

The situation is very different in, for instance, Denmark. Although the heating season has only 2700 Celsius degree days (due to Gulf Stream warming), as compared with 6000 degree days for Regina, there is very little solar gain (through windows or to a solar collector) during the months of November through February. A modest storage volume, even with a large solar collector area, is therefore unable to maintain full coverage during the winter period, as indicated by the variations in storage temperatures in a concrete case, shown in Fig. 28.3 (the water store is situated in the attic, losing heat to about ambient air temperatures; this explains why the storage temperatures approach freezing in January).

In order to achieve near 100% coverage under conditions such as the Danish ones, very large water tanks would be required, featuring facilities to maintain stable temperature stratification and containing so much insulation (over 1 m) that truly seasonal mismatch between heat production and use can be handled. This is still extremely difficult to achieve for a single house, but for a communal system for a number of reasonably close buildings and with the storage facility (and maybe also collectors) placed centrally, 100% coverage should be possible (cf. simulations made in Sørensen, 2004).

Community-size storage facilities
With increasing storage container size, the heat loss through the surface – for a given thickness of insulation – will decrease per unit of heat stored. There are two cases, depending on whether the medium surrounding the container (air, soil, etc.) is rapidly mixing. Consider first the case of a storage container surrounded by air.

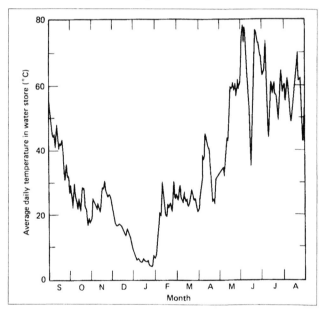

Figure 28.3. Measured average daily temperature in a 5.5-m³ water storage tank (having 0.2 m of rock wool insulation) fed by 50 m² of solar flat-plate collector (two layers of glass cover, non-selective absorber, tilt angle 38°). The solar system covers 12% of the building heat load (based on Jørgensen *et al.*, 1980).

The container may be cylindrical such as the one illustrated in Fig. 28.1. The rate of heat loss is assumed proportional to the surface area and to the temperature difference between inside and outside, with the proportionality constant being denoted U. It is sometimes approximated in the form

$$U = 1/(x/\lambda + \mu),$$

where x is the thickness of insulation, λ is the thermal conductivity of the insulating material (about 0.04 W m^{-1} per °C for mineral wool) and μ (around 0.1 m^2 W^{-1} per °C) is a parameter describing the heat transfer in the boundary layer air [may be seen as a crude approximation to (16.12)].

The total heat loss rate is

$$P^{loss} = 2\pi R \, (R + L) \, U \, (T_s - T_a), \tag{28.2}$$

where R and L are radius and height of the cylinder, T_s is the average temperature of the water in the store, and T_a is the outside ambient air temperature. The fraction of the stored heat energy lost per unit time is

$$\frac{P^{loss}}{E^{sens}} = \frac{2U(1 + R/L)}{R \, c_p^{water} \, \rho^{water}}, \tag{28.3}$$

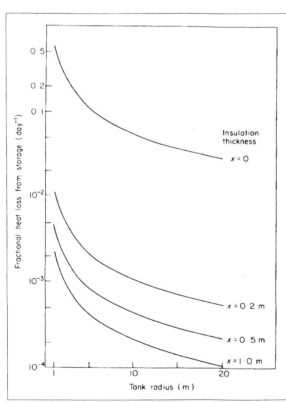

Figure 28.4. Fractional heat loss from a cylindrical storage tank with different degrees of insulation, assuming well-mixed conditions inside as well as outside the tank.

that is, the loss is independent of the temperatures and inversely proportional to a linear system dimension (c_p is heat capacity and ρ is density).

Figure 28.4 shows, for $L = 3R$, the ratio of heat loss rate P^{loss} and stored energy E^{sens}, according to (28.3), as a function of R and for different values of the insulation thickness. This ratio is independent of the temperature difference, owing to the linear assumption for the heat loss rate. If the storage times required imply maximum fractional losses of 10–20%, uninsulated tanks in the size range 5 m $\leq R \leq$ 20 m are adequate for a few days of storage. If storage times around a month are required, some insulation must be added to tanks in the size range under consideration. If seasonal storage is needed, the smallest tank sizes will not work even with a metre of mineral or glass wool wrapping. Community-size tanks, however, may serve for seasonal storage, with a moderate amount of insulation (0.2 m in the specific example).

A hot water tank with R = 11.5 m and L = 32 m has been used since 1978 by a utility company in Odense, Denmark, in connection with combined production of electricity and heat for district heating (Jensen, 1981). The hot water store is capable of providing all necessary heating during winter electricity peak hours, during which the utility company wants a maximum electricity production. With a hot water store two or three times larger, the cogenerating power units could be allowed to follow the electricity demand, which is small relative to the heat demand during winter nights.

A hot water store of similar magnitude, around 13 000 m³, may serve a solar-heated community system for 50–100 one-family houses, connected to the common storage facility by district heating lines. The solar collectors may still be on individual rooftops, or they may be placed centrally, for example, in connection with the store. In the first case, more piping and labour are required for installation, but in the second case, land area normally has to be dedicated to the collectors. Performance is also different for the two systems, as long as the coverage by solar energy is substantially less than 100%, and the auxiliary heat source feeds into the district heating lines. The reason is that when storage temperature is below the minimum required, the central solar collector will perform at high efficiency (Fig. 28.2), whereas individual solar collectors will receive input temperatures already raised by the ancillary heat source and thus not perform as well. Alternatively, auxiliary heat should be added by individual installations on the load side of the system, but, unless the auxiliary heat is electrically generated, this is inconvenient if the houses do not already possess a fuel-based heating system.

Most cost estimates speak against storage containers placed in air. If the container is buried underground (possibly with its top facing the atmosphere), the heat escaping the container surface will not be rapidly mixed into the surrounding soil or rock. Instead, the region closest to the container will reach a higher temperature, and a temperature gradient through the soil or rock will slowly be built up. An exception is soil with ground water infiltration. Here the moving water helps to mix the heat from the container into the surroundings. However, if a site can be found with no ground water (or at least no ground water in motion), then the heat loss from the store will be greatly reduced, and the surrounding soil or rock can be said to function greatly reduced, and the surrounding soil or rock can be said to function as an extension of the storage volume.

As an example, let us consider a spherical water tank embedded in homogeneous soil. The tank radius is denoted R, the water temperature is T_s and the soil temperature far away from the water tank is T_0. If the transport of heat can be described by a diffusion equation, then the temperature distribution as a function of distance from the centre of the storage container may be written (Shelton, 1975)

$$T(r) = T_0 + (T_s - T_0)\, R/r, \tag{28.4}$$

where the distance r from the centre must be larger than the tank radius R in order for the expression to be valid. The corresponding heat loss is

$$P^{sens} = \int_{sphere} \lambda\, \partial\, T(r)/\partial\, r\; dA = -\lambda\, (T_s - T_0)\, 4\pi R, \tag{28.5}$$

where λ is the heat conductivity of the soil and (28.5) gives the heat flux out of any sphere around the store, of radius $r \geq R$. The flux is independent of r. The loss relative to the heat stored in the tank itself is

$$P^{loss} / E^{sens} = -3\lambda / (R^2\, c_p^{water}\, \rho^{water}). \tag{28.6}$$

Compared to (28.3), it is seen that the relative loss from the earth-buried store is declining more rapidly with increasing storage size than the loss from a water store in air or other well-mixed surroundings. The fractional loss goes as R^{-2} rather than as R^{-1}.

Practical considerations in building an underground or partly underground water store suggest an upside-down obelisk shape and a depth around 10 m for a 20 000-m^2 storage volume. The obelisk is characterised by tilting sides, with a slope as steep as feasible for the soil type encountered. The top of the obelisk (the largest area end) would be at ground level or slightly above it, and the sides and bottom would be lined with plastic foil not penetrable by water. Insulation between lining and ground can be made with mineral wool foundation elements or similar materials. As a top cover, a sailcloth held in bubble shape by slight overpressure is believed to be the least expensive solution. Top insulation of the water in the store can be floating foam material. If the bubble cloth is impermeable to light, algae growth in the water can be avoided (Danish Department of Energy, 1979).

Two community-size seasonal hot water stores placed underground are operating in Sweden. They are both shaped as cut cones. One is in Studsvik. Its volume is 610 m^3, and 120 m^2 of concentrating solar collectors float on the top insulation, which can be turned to face the sun. Heat is provided for an office building of 500-m^2 floor area. The other system is in the Lambohov district of Linköping. It serves 55 semidetached one-family houses having a total of 2600-m^2 flat-plate solar collectors on their roofs. The storage is 10 000 m^3 and situated in solid rock (excavated by blasting). Both installations have operated since 1979, and they furnish a large part of the heat loads of the respective buildings. Figure 28.5 gives the temperature changes by season for the Studsvik project (Andreen and Schedin, 1980; Margen, 1980; Roseen, 1978).

Another possibility is to use existing ponds or lake sections for hot water storage. Top insulation would normally be required, and, in the case of lakes used only in part, an insulating curtain should separate the hot and the cold water.

A numerical integration of the equations for the spread of temperature anomalies in geological formations [cf. Chapter 3 of Sørensen (2004)] will add information about the time required for reaching the steady-state situation. According to calculations such as Shelton's (1975), this may be about one year for typical large heat store cases.

The assumption of a constant temperature throughout the storage tank may not be valid, particularly not for large tanks, owing to natural stratification that leaves the colder water at the bottom and the warmer water at the top. Artificial mixing by mechanical stirring is, of course, possible, but for many applications temperature stratification would be preferable. This could be used to advantage, for example, with storage tanks in a solar heating system (see Fig. 16.9), by feeding the collector with the colder water from the lower part of the storage and thus improving the efficiency (16.23) of the solar collector.

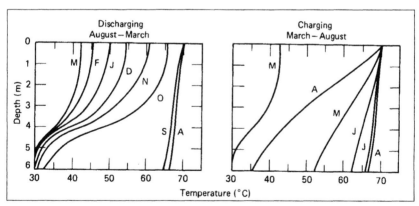

Figure 28.5. Temperature profiles in the Studsvik hot water store, monthly during (right) charging period March (M) to August (A) and during (left) discharging period from August (A) to March (M) (based on Roseen, 1978).

An alternative would be to divide the tank into physically separated sub-units (cf. Duffie and Beckman, 1974), but this requires a more elaborate control system to introduce and remove heat from the different units in an optimised way, and if the storage has to be insulated the sub-units should at least be placed with common boundaries (tolerating some heat transfer among the units) in order to keep the total insulation requirement down.

Multi-unit storage systems may be most appropriate for uninsulated storage in the ground, owing to the participation of the soil between the storage units. Figure 28.6 shows a possible arrangement (Brüel *et al.*, 1976) based on cylindrical heat exchangers placed in one or more circles around a central heat exchanger cylinder in the soil. If no active aquifers are traversing the soil region, a steady-state temperature distribution with a high central temperature can be built up. Used in connection with, for example, a flat-plate solar collector, it is possible to choose to take the fluid going to the collector circuit from a low temperature and to deliver the return fluid from the solar collector to successive heat exchangers of diminishing temperatures in order to maximise the collector efficiency as well as the amount of heat transferred to the storage. Dynamic simulations of storage systems of this type, with definite boundary conditions and source terms, have been performed by Zlatev and Thomsen (1976).

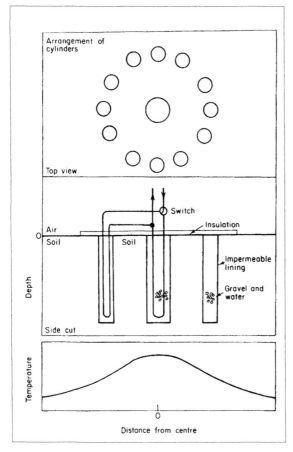

Figure 28.6. Uninsulated heat storage in soil based on arrangements of cylindrical holes (e.g. drilled with use of water flushing-type drills) (based on Brüel *et al.*, 1976).

Use of other materials, such as gravel, rock or soil, has been considered in connection with heating systems at temperatures similar to those relevant for water. Despite volume penalties of a factor 2–3 (depending on void volume), these materials may be more convenient than water for some applications. The main problem is to establish a suitable heat transfer surface for the transfer of heat to and from the storage. For this reason, gravel and rock stores have been mostly used with air, large volumes of which can be blown through the porous material, as a transfer fluid.

For applications in other temperature ranges, other materials may be preferred. Iron (e.g. scrap material) is suitable for temperatures extending to several hundred degrees Celsius. In addition, for rock-based materials the temperatures would not be limited to the boiling point of water.

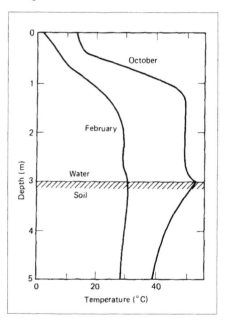

Figure 28.7. Temperature profiles for solar salt gradient pond in Miamisburg, Ohio, for the months of October and February (maximum and minimum temperature in convective layer). The continuation of the profiles in the soil beneath the pond is also included. (From Wittenberg and Harris, 1979. Reprinted with permission from *Proceedings of 14th Intersociety Energy Conversion Engineering Conference.* Copyright 1979 American Chemical Society.)

Solar ponds and aquifer storage

A solar pond is a natural or artificial hot water storage system much like the ones described above, but with the top water surface exposed to solar radiation and operating like a solar collector. In order to achieve both collection and storage in the same medium, layers from top to bottom have to be "inversely stratified", that is, stratified with the hottest zone at the bottom and the coldest one at the top. This implies that thermal lift must be opposed, either by physical means such as placing horizontal plastic barriers to separate the layers or by creating a density gradient in the pond, which provides gravitational forces to overcome the buoyancy forces. This can be done by adding certain salts to the pond, taking advantage of the higher density of the more salty water (Rabl and Nielsen, 1975).

An example of a solar pond of obelisk shape is the 5200-m³ pond installed at Miamisburg, Ohio. Its depth is 3 m, and the upper half is a salt gradient layer of NaCl, varying from 0% at the top to 18.5% at a 1.5-m depth. This gradient layer opposes upward heat transport and thus functions as a top insulation without impeding the penetration of solar radiation. The bottom layer has a fixed salt concentration (18.5%) and contains heat exchangers for withdrawing energy. In this layer, convection may take place without problems. Most of the absorption of solar radiation takes place at the bottom surface (this is why the pond should be shallow), and the heat is subsequently released to the convective layer. At the very top, however, some absorption of infrared solar radiation may destroy the gradient of temperature.

Figure 28.7 shows temperature gradients for the Miamisburg pond during its initial loading period (no load connected). From the start of operation in late August the first temperature maximum occurred in October, and the subsequent minimum occurred in February. The two situations are shown in Fig. 28.7. The temperature in the ground just below the pond was also measured. In October, the top layer disturbance can be seen, but in February, it is absent, due to ice cover on the top of the pond.

Numerical treatment of seasonal storage in large top-insulated or solar ponds may be by time simulation, or by a simple approximation, in which solar radiation and pond temperature are taken as sine functions of time, with only the amplitude and phase as parameters to be determined. This is a fairly good approximation because of the slow response of a large seasonal store, which tends to be insensitive to rapid fluctuations in radiation or air temperature. However, when heat is extracted from the store, it must be checked that disturbance of the pond's temperature gradient will not occur, say, on a particularly cold winter day, where the heat extraction is large. Still, heat extraction can, in many cases, also be modelled by sine functions, and if the gradient structure of the pond remains stable, such a calculation gives largely realistic results.

In Israel, solar ponds are being operated for electricity generation by use of Rankine cycle engines with organic working fluids in order to be able to accept the small temperature difference available. A correspondingly low thermodynamic efficiency must be accepted (Winsberg, 1981).

Truly underground storage of heat may also take place in geological formations capable of accepting and storing water, such as rock caverns and aquifers. In the aquifer case, it is important that water transport be modest, that is, that hot water injected at a given location stay approximately there and exchange heat with the surroundings only by conduction and diffusion processes. In such cases, it is estimated that high cycle efficiencies (85% at a temperature of the hot water some 200°C above the undisturbed aquifer temperature – the water being under high pressure) can be attained after breaking the system in, that is, after having established stable temperature gradients in the surroundings of the main storage region (Tsang *et al.*, 1979).

Medium- and high-temperature storage
In relation to industrial processes, temperature regimes are often defined as medium in the interval from 100 to 500°C and high above 500°C. These definitions may also be used in relation to thermal storage of energy, but it may be useful to single out also the lower medium temperature range from 100°C to about 300°C, as the discussion below indicates.

Materials suitable for heat storage should have a large heat capacity, they must be stable in the temperature interval of interest, and it should be convenient to add heat to or withdraw heat from them.

Material	Temperature interval (°C)	Mass spec. heat (kJ kg^{-1} °C^{-1})	Volume spec. heat (MJ m^{-3} °C^{-1})	Heat conductivity (W m^{-1} °C^{-1})
Solids				
Sodium chloride	< 800	0.92	2.0	9[a, b]
Iron (cast)	< 1500	0.46	3.6	70[b]–34[c]
Rock (granite)	< 1700	0.79	2.2	2.7[b]
Bricks		0.84	1.4	0.6
Earth (dry)		0.79	1.0	1.0
Liquids				
Water	0–100	4.2	4.2	0.6
Oil ("thermal")	–50 to 330	2.4	1.9	0.1
Sodium	98 to 880	1.3	1.3	85[b]–60[c]
Diethylene glycol	–10 to 240	2.8	2.9	

Table 28.1 Heat capacities of various materials (Kaye and Laby, 1959; Kreider, 1979; Meinel and Meinel, 1976). All quantities have some temperature dependence. Standard atmospheric pressure has been assumed; that is, all heat capacities are c_p's.
[a] Less for granulates with air-filled voids.
[b] At 1000°C.
[c] At 700°C.

The last of these requirements can be fulfilled in different ways. Either the material itself should possess good heat conductivity, as do metals, for example, or it should be easy to establish heat transfer surfaces between the material and some other suitable medium. If the transfer medium is a liquid or a gas, it could be passed along the transfer surface at a velocity sufficient for the desired heat transfer, even if the conductivities of the transfer fluid and the receiving or delivering material are small. If the storage material is arranged in a finite geometry, such that insufficient transfer is obtained by a single pass, then the transfer fluid may be passed along the surface several times. This is particularly relevant for transfer media such as air, which has very low heat conductivity, and when air is used as a transfer fluid, it is important that the effective transfer surface be large. This may be achieved for storage materials of granular form such as pebble or rock beds, where the nodule size and packing arrangement can be such that air can be forced through and reach most of the internal surfaces with as small an expenditure of compression energy as possible.

These considerations lie behind the approaches to sensible heat storage, which are exemplified by the range of potential storage materials listed in Table 28.1. Some are solid metals, where transfer has to be by conduction through the material. Others are solids that may exist in granular form for blowing by air or another gas through the packed material. They exhibit more modest heat conductivity. The third group comprises liquids, which may serve as both heat storage materials and transfer fluids. The dominating path of heat transfer may be conduction, advection (moving the entire fluid), or convection (turbulent transport). For highly conducting materials such as liquid sodium, little transfer surface is required, but for the other materials listed substantial heat exchanger surfaces may be necessary.

Solid metals, such as cast iron, have been used for high-temperature storage in industry. Heat delivery and extraction may be by passing a fluid through channels drilled into the metal. For the medium- to high-temperature interval the properties of liquid sodium (cf. Table 28.1) make this a widely used material for heat storage and transport, despite the serious safety problems (sodium reacts explosively with water). It is used in nuclear breeder reactors and in concentrating solar collector systems, for storage at temperatures between 275 and 530°C in connection with generation of steam for industrial processes or electricity generation. The physics of heat transfer to and from metal blocks and of fluid behaviour in pipes is a standard subject covered in several textbooks (see e.g. Grimson, 1971).

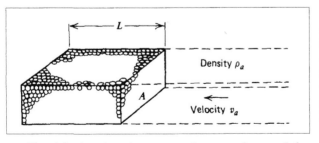

Figure 28.8. Rock bed sensible heat store. Air of density ρ_a and velocity v_a may be blown through the bed cross-section A, travelling the length L of the bed.

Fixed beds of rock or granulate can be used for energy storage at both low and high temperatures, normally using air blown through the bed to transfer heat to and from the store. The pressure drop ΔP across a rock bed of length L, such as the one illustrated in Fig. 28.8 where air is blown through the entire cross-sectional area A, may be estimated as (Handley and Heggs, 1968)

$$\Delta P \approx \rho_a v_a^2 \, L \, d_s^{-1} \, m_s^2 \, (368 + 1.24 \, Re/m_s)/(Re(1 - m_s)^3), \qquad (28.7)$$

where ρ_a and v_a are density and velocity of the air passing through the bed in a steady-state situation, d_s is the equivalent spherical diameter of the rock particles, and m, their mixing ratio, is one minus the air fraction in the volume $L \times A$. Re is the Reynolds number describing the ratio between "inertial" and "viscous" forces on the air passing between the rock particles. Re may be estimated as $\rho_a v_a d_a/\mu$, where μ is the dynamic viscosity of air. If the rock particles are not spherical, the equivalent diameter may be taken as

$$d_s = (6ALm_s/n)^{1/3},$$

where n is the number of particles in the entire bed volume. The estimate (28.7) assumes the bed to be uniform and the situation stationary. The total surface area of the particles in the bed is given by

$$A_s = 6m_s \, AL/d_s = n \pi d_s^2.$$

Optimal storage requires that the temperature gradient between the particle surfaces and their interior be small and that the pressure drop (28.7) also be small, leading to optimum particle diameters of a few centimetres and a void fraction of about 0.5 (implying that m_s is also about 0.5).

Organic materials such as diethylene glycol or special oil products (Table 28.1) are suitable for heat storage between 200 and 300°C and have been used in concentrating solar collector test facilities (Grasse, 1981). Above 300°C, the oil decomposes.

Despite low volume heat capacities, gaseous heat storage materials could also be considered, such as steam (water vapour), which is often stored under pressure, in cases where steam is the form of heat energy to be used later (in industrial processes, power plants, etc.).

LATENT HEAT AND CHEMICAL TRANS-FORMATION STORAGE

Energy associated with structural or phase change

The energy associated with a change of phase for a given material can be used to store energy. The phase change may be one of melting or evaporating, or it may be associated with a structural change, e.g. in lattice form or content of crystal-bound water. When heat is added or removed from a given material, a number of changes may take place successively, or in some cases simultaneously, involving phase changes as well as energy storage in thermal motion of molecules (i.e. both latent and sensible heat). The total energy change, which can serve as energy storage, is given by the change in enthalpy.

Solid–solid phase transitions are observed in one-component, binary, and ternary systems, as well as in single elements. An example of the latter is solid sulphur, which occurs in two different crystalline forms, a low-temperature orthorhombic form and a high-temperature monoclinic form (cf. Moore, 1972). However, the elementary sulphur system has been studied merely out of academic interest in contrast to the one-component systems listed in Table 29.1. Of these systems, which have been studied in relation to practical energy storage, Li_2SO_4 exhibits both the highest transition temperature T_t and the highest latent heat for the solid–solid phase change ΔH_{ss}. Pure Li_2SO_4 has a transition from a monoclinic to a face-centred cubic structure with a latent heat of 214 KJ kg^{-1} at 578°C. This is much higher than the heat of melting (–67 KJ kg^{-1} at 860°C). Another one-component material listed in Table 29.1 is Na_2SO_4, which has two transitions at 201 and 247°C, with the total latent heat of both transitions being ~ 80 KJ kg^{-1}.

Recently, mixtures of Li_2SO_4 with Na_2SO_4, K_2SO_4 and ZnSO have been studied. Also, some ternary mixtures containing these and other sulphates were included in a Swedish investigation (cf. Sjoblom, 1981). Two binary systems (Li_2SO_4–Na_2SO_4, 50 mol% each, T_t = 518°C; and 60% Li_2SO_4–40% $ZnSO_4$, T_t = 459°C) have high values of latent heat, ~190 KJ kg^{-1}, but they exhibit a strong tendency for deformation during thermal cycling. A number of ternary salt mixtures based on the most successful binary compositions have been studied experimentally, but there is a lack of knowledge of both phase diagrams, structures, and re-crystallisation processes that lead to deformation in these systems.

Material	Transition temperature T_t (°C)	Latent heat ΔH_{ss} (kJ kg^{-1})
V_2O_2	72	50
FeS	138	50
KHF_2	196	135
Na_2SO_4	210, 247	80
Li_2SO_4	578	214

Table 29.1. Solid–solid transition enthalpies ΔH_{ss} (Fittipaldi, 1981).

Salt hydrates
The possibility of energy storage by use of incongruently melting salt hydrates has been intensely investigated, starting with the work of Telkes (Telkes, 1952, 1976). The molten salt consists of a saturated solution and additionally some undissolved anhydrous salt because of its insufficient solubility at the melting temperature, considering the amount of released crystal water available. Sedimentation will develop, and a solid crust may form at the interface between layers. In response to this, stirring is applied, for example, by keeping the material in rotating cylinders (Herrick, 1982), and additives are administered in order to control agglomeration and crystal size (Marks, 1983).

An alternative is to add extra water to prevent phase separation. This has led to a number of stable heat of fusion storage systems (Biswas, 1977; Furbo, 1982). Some melting points and latent heats of salt hydrates are listed in Table 29.2. Here we use as an example Glauber salt ($Na_2SO_4 \cdot 10H_2O$), the storage capacity of which is illustrated in Fig. 29.1, both for the pure hydrate and for a 33% water mixture used in actual experiments. Long-term verification of this and other systems in connection with solar collector installations has been carried out by the Commission of the European Union. For hot water systems the advantage over sensible heat water stores is minimal, but this may change when space heating is included, because of the seasonal storage need (Furbo, 1982).

Salt hydrates release water when heated and release heat when they are formed. The temperatures at which the reaction occurs vary for different compounds, ranging from 30 to 80°C; this makes possible a choice of storage systems for a variety of water-based heating systems such as solar, central, and district heating. Table 29.2 shows the temperatures T_m of incongruent melting and the associated quasi-latent heat Q (or ΔH) for some hydrates that have been studied extensively in heat storage system operation. The practical use of salt hydrates faces physicochemical and thermal problems such as super-cooling, non-congruent melting, and heat transfer difficulties imposed by locally low heat conductivities (cf., e.g. Achard *et al.*, 1981).

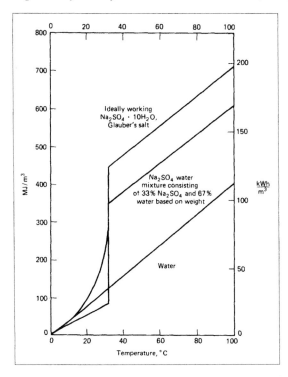

Figure 29.1. Heat storage capacity as a function of temperature for ideally melting Glauber salt, for Glauber salt plus extra water, and for pure water (Furbo, 1982).

Generally, storage in a chemical system with two separated components, of which one draws low-temperature heat from the environment and the other absorbs or delivers heat at fairly low (30–100°C) or medium (100–200°C) temperature, is referred to as a *chemical heat pump* (McBride, 1981). Simply stated, the chemical heat pump principle consists in keeping a substance in one of two containers, although it would prefer to be in the other one. In a classical example, the substances are water vapour and sulphuric acid. Because the pressure over sulphuric acid is much lower than over liquid water (see Fig. 29.2), the water vapour will prefer to move from the water surface to the H_2SO_4 surface and become absorbed there, with a heat gain deriving in part from the mixing process and in part from the heat of evaporation. The temperature of the mixture is then above what is needed at the water source. Heat is stored when the temperature of the sulphuric acid/water container is made still higher, so that the equilibrium pressure of vapour above the acid surface at this temperature becomes higher than that above the water surface at its temperature. The pressure gradient will in this situation move water vapour back to the water surface for condensation.

Hydrate	Incongruent melting point, T_m (°C)	Specific latent heat ΔH (MJ m^{-3})
CaCl$_2$ · 6H$_2$O	29	281
Na$_2$SO$_4$ · 10H$_2$O	32	342
Na$_2$CO$_3$ · 10H$_2$O	33	360
Na$_2$HPO$_4$ · 12H$_2$O	35	205
Na$_2$HPO$_4$ · 7H$_2$O	48	302
Na$_2$S$_2$O$_3$ · 5H$_2$O	48	346
Ba(OH)$_2$ · 8H$_2$O	78	655

Table 29.2. Characteristics of salt hydrates.

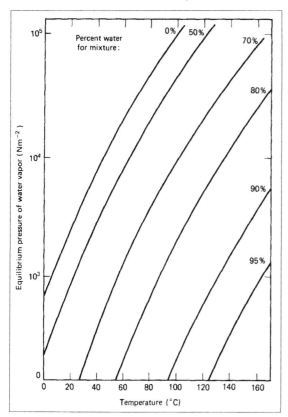

Figure 29.2. Equilibrium pressure of water vapour over sulphuric acid/water mixtures as a function of temperature and percentage of water in the mixture (Christensen, 1981).

A similar system, but with the water being attached as crystal water in a salt (i.e., salt hydration), is the Na$_2$S/water system developed and tested in Sweden but not really having made it in the marketplace ("System Tepidus"; Bakken, 1981). Figure 29.3 shows this chemical heat pump, which is charged by the reaction

$$Na_2S + 5H_2O \text{ (vapour)} \rightarrow Na_2S \cdot 5H_2O + 312 \text{ kJ mol}^{-1}. \qquad (29.1)$$

The heat for the evaporation is taken from a reservoir of about 5°C, that is, a pipe extending through the soil at a depth of a few metres (as in commercial electric heat pumps with the evaporator buried in the lawn), corresponding to roughly 10°C in the water container (B in Fig. 29.3) owing to heat exchanger losses. The water vapour flows to the Na_2S container (A in Fig. 29.3) through a connecting pipe that has been evacuated of all gases other than water vapour and where the water vapour pressure is of the order of 1% of atmospheric pressure. During charging, the temperature in the sodium sulphide rises to 65–70°C, owing to the heat formed in the process (29.1). When the temperatures in containers A and B and the equilibrium pressures of the water vapour are such that they correspond to each other by a horizontal line in the pressure–temperature diagram shown in Fig. 29.4, the flow stops and the container A has been charged.

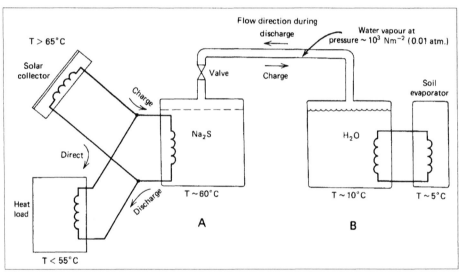

Figure 29.3. Schematic picture of a chemical heat pump operating between a sodium sulphide and a water container and based on the formation of the salt hydrate $Na_2S \cdot 5H_2O$. Typical temperatures are indicated. There is a low-temperature heat source connected to the water container and a high-temperature heat source (a solar collector) connected to the salt container, along with the heat demand (load). A switch allows either load or solar collector to be connected.

To release the energy, a load area of temperature lower than the container A is connected to it, and heat is transferred through a heat exchanger. Lowering the temperature in A causes a pressure gradient to form in the connecting pipe, and new energy is drawn from B to A. In order to prevent the heat reservoir (the soil) from cooling significantly, new heat must be added to compensate for the heat withdrawn. This takes place continuously by transfer processes in the soil (solar radiation absorbed at the soil surface is conducted to the subsurface location of the evaporator pipes). However, in the long term a lower temperature would develop in the soil environment if no active makeup heat were supplied. This is done by leading surplus heat from a solar collector to the sodium sulphide container when the solar heat is not directly required in the load area. When the temperature of container A is raised in this way, the pressure gradient above the salt will be in the direction of driving water vapour to container B, thereby removing some of the crystallisation water from the salt.

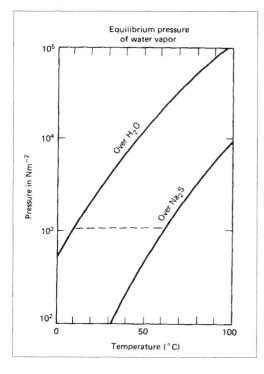

Figure 29.4. Equilibrium pressure of water vapour over water and over sodium sulphide as a function of temperature. For a given pressure (dashed line), the equilibrium temperatures in the water and the salt containers will differ by roughly 55°C.

The two actual installations of this type of chemical heat pump are a one-family dwelling with a storage capacity of 7000 kWh (starting operation in 1979) and an industrial building (Swedish Telecommunications Administration) with 30 000 kWh worth of storage, operated from 1980. Future applications may comprise transportable heat stores, since container A may be detached (after closing the valve indicated in Fig. 29.3) and carried to another site. Once the container is detached, the sensible heat is lost, as the container cools from its 60°C to ambient temperatures, but this only amounts to 3–4% of the energy stored in the $Na_2S \cdot 5H_2O$ (Bakken, 1981).

It should be mentioned that a similar loss will occur during use of the storage unit in connection with a solar heating system. This is because of the intermittent call upon the store. Every time it is needed, its temperature increases to 60°C (i.e. every time the valve is opened), using energy to supply the sensible heat, and every time the need starts to decrease, there is a heat loss associated either with making up for the heat transfer to the surroundings (depending on the insulation of the container) to keep the container temperature at 60°C or, if the valve has been closed, with the heat required to re-heat the container to its working temperature. These losses could be minimised by using a modular system, where only one module at a time is kept at operating temperature, ready to supply heat if the solar collector is not providing enough. The other modules would then be at ambient temperatures except when they are called upon to become re-charged or to replace the unit at standby.

The prototype systems were not economically viable, but the developers estimate that the system cost in regular production may reach 4–5 euro per kWh of heat supplied for a 15-m^3 storage system for a detached house, with half the cost taken up by the solar collector system and the other half by the store. For transport, container sizes equivalent to 4500 kWh of 60°C heat are envisaged. However, the charge-rate capacity of roughly 1 W kg^{-1} may be insufficient for most applications.

Although the application of the chemical heat pump considered above is for heating, the concept is equally useful for cooling applications. Here the load would simply be connected to the cold container. Several projects are underway to study various chemical reactions of the gas/liquid or gas/solid type based on pressure differences, either for cooling alone or for both heating and cooling. The materials are chosen on the basis of temperature requirements and long-term stability while allowing many storage cycles. For example, $NaI–NH_3$ systems have been considered for air conditioning purposes (Fujiwara *et al.*, 1981). A number of ammoniated salts that react on heat exchange could be contemplated.

Chemical reactions
The use of high-temperature chemical heat reactions in thermal storage systems is fairly new and to some extent related to attempts to utilise high-temperature waste heat and to improve the performance of steam power plants (cf. Golibersuch *et al.*, 1976). The chemical reactions that are used to store the heat allow, in addition, upgrading of heat from a lower temperature level to a higher temperature level, a property that is not associated with phase transition or heat capacity methods.

Conventional combustion of fuel is a chemical reaction in which the fuel is combined with an oxidant to form reaction products and surplus heat. This type of chemical reaction is normally irreversible, and there is no easy

29. LATENT HEAT AND CHEMICAL STORAGE

way that the reverse reaction can be used to store thermal energy. The process of burning fuel is a chemical reaction whereby energy in one form (chemical energy) is transformed into another form (heat) accompanied by an increase in entropy. In order to use such a chemical reaction for storage of heat, it would, for example, in the case of hydrocarbon, require a reverse process whereby the fuel (hydrocarbon) could be obtained by adding heat to the reaction products carbon dioxide and water. So, use of chemical heat reactions for thermal energy storage requires suitable reversible reactions.

The change in bond energy for a reversible chemical reaction may be used to store heat, but although a great variety of reversible reactions are known, only a few have so far been identified as being technically and economically acceptable candidates. The technical constraints include temperature, pressure, energy densities, power densities, and thermal efficiency. In general, a chemical heat reaction is a process whereby a chemical compound is dissociated by heat absorption, and later, when the reaction products are recombined, the absorbed heat is again released. Reversible chemical heat reactions can be divided into two groups: thermal dissociation reactions and catalytic reactions. The thermal dissociation reaction may be described as

$$AB \; (+\Delta H \text{ at } T_1 \text{ and } p_1) \leftrightarrow A + B \; (-\Delta H \text{ at } T_2 \text{ and } p_2), \qquad (29.2)$$

indicating that the dissociation takes place by addition of heat ΔH to AB at temperature T_1 and pressure p_1, whereas the heat is released $(-\Delta H)$ at the reverse reaction at temperature T_2 and pressure p_2. The reciprocal reaction (from right to left) occurs spontaneously if the equilibrium is disturbed, that is, if $T_2 < T_1$ and $p_2 > p_1$. To avoid uncontrolled reverse reaction, the reaction products must therefore be separated and stored in different containers. This separation of the reaction products is not necessary in catalytic reaction systems where both reactions (left to right and right to left) require a catalyst in order to obtain acceptable high reaction velocities. If the catalyst is removed, neither of the reactions will take place even when considerable changes in temperature and pressure occur. This fact leads to an important advantage, namely, that the intrinsic storage time is, in practice, very large and, in principle, infinite. Another advantage of closed-loop heat storage systems employing chemical reactions is that the compounds involved are not consumed, and because of the high energy densities (in the order of magnitude: 1 MWh m^{-3} compared to that of the sensible heat of water at ΔT = 50 K, which is 0.06 MWh m^{-3}), a variety of chemical compounds are economically acceptable.

The interest in high-temperature chemical reactions is derived from the work of German investigators on the methane reaction

$$Q + CH_4 + H_2O \; \leftrightarrow \; CO + 3H_2, \qquad (29.3)$$

(Q being heat added) which was studied in relation to long-distance transmission of high-temperature heat from nuclear gas reactors (cf. Schulten *et al.*, 1974). The transmission system called "EVA-ADAM", an abbreviation of the German *Einzelrohrhrversuchsanlage und Anlage zur dreistufigen adiabatischen Methanisierung*, is being further developed at the nuclear research centre at Jülich, Germany. It consists of steam reforming at the nuclear reactor site, transport over long distances of the reformed gas ($CO + 3H_2$), and methanation at the consumer site where heat for electricity and district heating is provided (cf. e.g. Harth *et al.*, 1981).

Closed loop system	Enthalpy[a] ΔH^0 (kJ mol^{-1})	Temperature range (K)
$CH_4 + H_2O \leftrightarrow CO + 3H_2$	206 (250)[b]	700–1200
$CH_4 + CO_2 \leftrightarrow 2CO + 2H_2$	247	700–1200
$CH_4 + 2H_2O \leftrightarrow CO_2 + 4H_2$	165	500–700
$C_6H_{12} \leftrightarrow C_6H_6 + 3H_2$	207	500–750
$C_7H_{14} \leftrightarrow C_7H_8 + 3H_2$	213	450–700
$C_{10}H_{18} \leftrightarrow C_{10}H_8 + 5H_2$	314	450–700

Table 29.3. High-temperature closed-loop chemical C-H-O reactions (Hanneman *et al.*, 1974; Harth *et al.*, 1981).
[a] Standard enthalpy for complete reaction.
[b] Including heat of evaporation of water.

The reaction in (29.3) is a suitable candidate for energy storage that can be accomplished as follows: heat is absorbed in the endothermic reformer where the previously stored low-enthalpy reactants (methane and water) are converted into high-enthalpy products (carbon monoxide and hydrogen). After heat exchange with the incoming reactants, the products are then stored in a separate vessel at ambient temperature conditions, and although the reverse reaction is thermodynamically favoured, it will not occur at these low temperatures and in the absence of a catalyst. When the heat is needed, the products are recovered from storage and the reverse, exothermic reaction (methanation) is run (cf. Golibersuch *et al.*, 1976). Enthalpies and temperature ranges for some high-temperature closed-loop C-H-O systems, including the reaction (29.3), are given in Table 29.3. The performance of the cyclohexane to benzene and hydrogen system (listed fourth in Table 29.3) has been studied in detail by Italian workers, and an assessment of a design storage plant has been made (cf. Cacciola *et al.*, 1981). The complete design storage plant consists of hydrogenation and dehydrogenation multistage adiabatic reactors, storage tanks, separators, heat exchangers, and multistage compressors. Thermodynamic requirements are assured by independent closed-loop systems circulating nitrogen in the dehydrogenation and hydrogen in the hydrogenation units.

A number of ammoniated salts are known to dissociate and release ammonia at different temperatures, including some in the high-temperature range (cf. Yoneda *et al.*, 1980). The advantages of solid–gas reactions in general are high heats of reaction and short reaction times. This implies, in principle, high energy and power densities. However, poor heat and mass transfer characteristics in many practical systems, together with problems of sagging and swelling of the solid materials, lead to reduced densities of the total storage system.

Metal hydride systems are primarily considered as stores of hydrogen, as will be discussed later. However, they have also been contemplated for heat storage, and in any case the heat-related process is integral in getting the hydrogen into and out of the hydride. The formation of a hydride MeH_x (metal plus hydrogen) is usually a spontaneous exothermic reaction:

$$Me + \tfrac{1}{2} x\, H_2 \rightarrow MeH_x + Q, \tag{29.4}$$

which can be reversed easily by applying the amount of heat Q,

$$MeH_x + Q \rightarrow Me + \tfrac{1}{2} x\, H_2. \tag{29.5}$$

Thus, a closed-loop system, where hydrogen is not consumed but is pumped between separate hydride units, may be used as a heat store. High-temperature hydrides such as MgH_2, Mg_2NiH_2 and TiH_2 have, owing to their high formation enthalpies (e.g. for MgH_2, $\Delta H \geq 80$ kJ per mol of H_2; for TiH_2, $\Delta H > 160$ KJ per mol of H_2), heat densities of up to 3 MJ kg^{-1} or 6 GJ m^{-3} in a temperature range extending from 100 to 600°C (cf. Buchner, 1980).

TASMANIA - BENT SØRENSEN, 1998

IX. HIGH-QUALITY ENERGY STORAGE

CHAPTER

PUMPED HYDRO STORAGE

30.1. Overview of the storage of high-quality energy

A number of storage systems may be particularly suitable for the storage of "high-quality" energy, such as mechanical energy or electric energy. If the energy to be stored is derived from a primary conversion delivering electricity, for example, then one needs an energy storage system, which will allow the regeneration of electricity with a high cycle efficiency, i.e. with a large fraction of the electricity input recovered in the form of electricity. Thermal stores, such as the options mentioned in the previous section, may even at T = 800–1500°C (metals, etc.) not achieve this, because of the Carnot limit to the efficiency of electricity regeneration, combined with energy losses from storage walls. Thermal stores fed via a heat pump drawing from a suitable reference reservoir may reach tolerable cycle efficiencies more easily.

Table 30.1 gives an indication of the actual or estimated energy densities and cycle efficiencies of various storage systems. The theoretical maximum of energy density is, in some cases, considerably higher than the values quoted. For comparison, energy densities of thermal stores and a number of fuels are also given. Some of the fuels (methanol, wood and hydrogen) may be produced by conversion based on renewable energy sources (without having to wait for fossilisation processes to occur). The cycle efficiency is defined with the assumption that the initial energy form is electricity or another high-quality energy form, and the value quoted for hydrogen is based on electrolysis of water as the first step in the storage cycle. Methanol may also be reversibly transformed into hydrogen and carbon oxide to play the role of a closed storage cycle (Prengle and Sun, 1976). The most striking feature is the low-volume energy density of nearly all the reversible storage concepts considered, relative to that of solid or liquid fossil fuels.

The magnitudes of stores that in practice may be associated with renewable energy use are discussed in Sørensen (2004) for various future energy systems. For comparison, reservoirs of fossil fuels may be found in Chapter 21, Table 21.1.

Storage form	Energy density		Cycle efficiency
	kJ kg^{-1}	MJ m^{-3}	
Conventional fuels			
Crude oil	42 000	37 000	
Coal	32 000	42 000	
Dry wood	12 500[a]	10 000	
Synthetic fuels			
Hydrogen, gas	120 000	10	0.4–0.6
Hydrogen, liquid	120 000	8 700	
Hydrogen, metal hydride	2 000–9 000	5 000–15 000	
Methanol	21 000	17 000	
Ethanol	28 000	22 000	
Thermal – low quality			
Water, 100°C → 40°C	250	250	
Rocks, 100°C → 40°C	40–50	100–140	
Iron, 100°C → 40°C	~30	~230	
Thermal – high quality			
Rocks, e.g. 400°C → 200°C	~160	~430	
Iron, e.g. 400°C → 200°C	~100	~800	
Inorganic salts, heat of fusion > 300°C	> 300	> 300	
Mechanical			
Pumped hydro, 100 m head	1	1	0.65–0.8
Compressed air		~15	0.4–0.5
Flywheels, steel	30–120	240–950	
Flywheels, advanced	> 200	> 100	~0.95
Electrochemical			
Lead–acid	40–140	100–900	0.7–0.8
Nickel–cadmium	~350	~350	varying
Lithium ion (other advanced batteries)	700 (> 400)	1400 (> 300)	0.7 (> 0.8)
Superconducting		~100	~0.85

Table 30.1. Energy density by weight and volume for various storage forms, based on measured data or expectations for practical applications. For the storage forms aimed at storing and regenerating high-quality energy (electricity), cycle efficiencies are also indicated. Hydrogen gas density is quoted at ambient pressure and temperature. For compressed air energy storage, both electricity and heat inputs are included on equal terms in estimating the cycle efficiency (with use of Jensen and Sørensen, 1984).
[a] Oven-dry wood may reach values up to 20 000 kJ kg^{-1}.

30.2 Pumped hydro storage

The total exploitable hydro potential is of the order of 10^{12} W on average over the year (Sørensen, 2004), and only the fraction of this associated with reservoirs can be considered relevant for energy storage. Those river flows that have to be tapped as they come may be interesting as energy sources, but not as energy storage options.

30. PUMPED HYDRO STORAGE

The hydro reservoirs feeding into turbine power plants may be utilised for storage of electric energy generated by non-hydropower plants (e.g. wind or photovoltaic energy converters), provided that all the power plants are connected by a common grid, and provided that transmission capacity is sufficient to accommodate the extra burden of load-levelling storage-type operation of the system. The storage function in a system of this kind is primarily obtained by displacement of load. This means that the hydropower units are serving as backup for the non-hydro generators by providing power when non-hydropower production falls short of load. The small start-up time for hydro turbines (½–3 minutes) makes this mode of operation convenient. When there is a surplus power generation from the non-hydro units, then the hydro generation is decreased, and non-hydro produced power is transmitted to the load areas otherwise served by hydropower (Sørensen, 1981; Meibom et al., 1999). In this way, there is no need to pump water up into the hydro reservoirs, as long as the non-hydropower generation stays below the combined load of hydro and non-hydro load areas. To fulfil this condition, the relative sizes of the different types of generating units must be chosen carefully.

When the surplus energy to be stored exceeds the amounts that can be handled in the displacement mode described above, then upward pumping of water into the hydro reservoirs may be considered by use of two-way turbines, so that the energy can be stored and recovered by the same installation. Alternatively, pumped storage may utilise natural or artificially constructed reservoirs not associated with any exploitable hydropower.

Figure 30.1 shows an example of the layout of a pumped storage facility. Installations where reservoirs are not part of a hydro flow system are typically intended for short-term storage. They may be used for load-levelling purposes, providing a few hours of peak load electric power per day, based on nighttime pumping. In terms of average load covered, the storage capacities of these installations are below 24 h. On the other hand, some of the natural reservoirs associated with hydro schemes have storage capacities corresponding to one or more years of average load (e.g. the Norwegian hydro system; cf. Sørensen, 1981; Meibom et al., 1999). Pumping schemes for such reservoirs could serve for long-term storage of energy.

If no natural elevated reservoirs are present, pumped storage schemes may be based on underground lower reservoirs and surface level upper reservoirs. The upper reservoirs may be lakes or oceans. The lower ones should be excavated or should make use of natural cavities in the underground. If excavation is necessary, a network of horizontal mine shafts, such as the one illustrated in Fig. 30.2, may be employed in order to maintain structural stability against collapse (Blomquist et al., 1979; Hambraeus, 1975).

The choice of equipment is determined primarily by the size of head, that is, the vertical drop available between the upper and the lower reservoir.

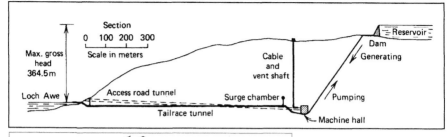

Figure 30.1 (above). Layout of pumped hydro storage system at Cruachan in Scotland.

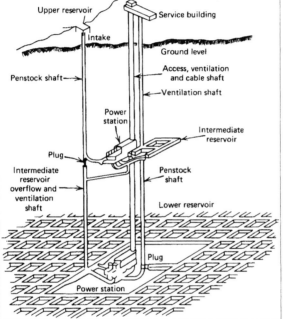

Figure 30.2 (left). Schematic layout of underground pumped hydro storage system. From Blomquist *et al.*, 1979. Reprinted with permission from *the Proceedings of the 14th Intersociety Energy Conversion Engineering Conference.* Copyright 1979 American Chemical Society.

Figure 11.1 shows in a schematic form the three most common types of hydro turbines. The Kaplan turbine (called a Nagler turbine if the position of the rotor blades cannot be varied) is most suited for low heads, down to a few metres. Its rotor has the shape of a propeller, and the efficiency of converting gravitational energy into shaft power is high (over 0.9) for the design water velocity, but lower for other water speeds. The efficiency drop away from the design water velocity is rapid for the Nagler turbine, less so for the Kaplan version. These turbines are inefficient for upward pumping, although they can be made to accept water flow from either side (André, 1976). A displacement pump may be used in a "tandem arrangement" (i.e. separate turbine and pump). The electric generator is easily made reversible, so that it may serve either as generator or as motor.

For larger heads, the Francis and Pelton turbines may be used. Francis turbines have a set of fixed guiding blades leading the water onto the rotating blades (the "runner") at optimum incident angle. It can be used with water heads up to about 600 m in the simple version illustrated in Fig. 11.1b, but multistage versions have been considered, guiding the water through a number of runners (five for an actual French installation; cf. Blomquist *et al.*, 1979). In this way, heads above 1000 m can be accepted, and the arrangement may be completely reversible, with very modest losses. For either pumping or generating, the turbine efficiency at design water flow may be over 0.95, but for decreasing flow the efficiency drops. Typical overall efficiencies of the storage cycle (pumping water up by use of surplus electric power, regenerating electric power based on downward flow through turbines) are around 0.8 for existing one-stage Francis turbine installations. Shifting from pumping to generating takes about 1 min (Hambraeus, 1975). The total cycle efficiency of the multistage Francis turbines for heads of 1000–1500 m is about 0.7 (Blomquist *et al.*, 1979).

If the head is larger than the limit for single-stage, reversible Francis turbines, an alternative to the multistage Francis turbines is offered by the tandem units consisting of separate impulse turbines and pumps. The pump units for pumping upward over height differences exceeding 1000 m are usually multistage pumps (six stages for an actual installation in Italy), with efficiency over 0.9 being achieved. The impulse turbine part is of Pelton type (see Fig. 11.1a), consisting of a bucket-wheel being driven by the impulse of one or more water jets created by passing the water through nozzles. The power for this process is the pressure force created by the column of water, from the turbine placed at the lower reservoir level to the upper reservoir level. The pressure energy can be converted partially or fully into linear kinetic energy according to the requirements of the different turbine types,

$$mg\,\Delta z = W^{pot}_{initial} = m'\,\tfrac{1}{2}u^2 + (m - m')\,P\,\rho^{-1} = W^{kin} + H.$$

Here the initial potential energy associated with the head Δz is transformed into a kinetic energy part associated with partial mass m' moving with velocity u and a pressure energy part with the enthalpy H given by the pressure P over the density of water ρ, times the remaining mass $m-m'$. The conversion efficiency of Pelton turbines is about 0.9 over a wide range of power levels, and the tandem arrangement of separate turbine and pump (but generator/motor, turbine, and pump all mounted on a common shaft) allows quick shifts between generating and pumping or vice versa.

The losses in conversion are associated in part with "leakage", that is, with water that passes round the turbine without contributing to power, and in part with energy dissipation in the form of heat, for example, due to friction (cf. Angrist, 1976). Further losses are associated with evaporation of water, especially from solar-exposed upper reservoirs.

Excavation for underground storage limits the application to short-term storage (up to about 24 h of average load) because the cost scales approximately linearly with storage capacity. For large natural reservoirs, seasonal energy storage can be considered, since the cost has a large component determined by the maximum load requirement and therefore becomes fairly independent of storage capacity beyond a certain point, as long as the reservoir is available.

CHAPTER

FLYWHEELS

Mechanical energy may be stored in the form of rotational motion under conditions of low frictional losses. A flywheel is such a rotating structure, capable of receiving and delivering power through its shaft of rotation. Friction is diminished by use of high-performance bearings, and the entire rotating structure may be enclosed in a space with partial vacuum or filled with an inert gas.

The amount of energy stored in a body of mass distribution $\rho(x)$ rotating about a fixed axis with angular velocity ω is

$$W = \tfrac{1}{2} I \, \omega^2, \qquad (31.1)$$

with the moment of inertia I given by

$$I = \int \rho(x) \, r^2 \, dx.$$

It would appear from these expressions that high angular velocity and a majority of the mass situated at large distance r from the axis of rotation would lead to high amounts of energy stored. The relevant question to ask, however, is how to obtain the highest energy density, given material of a certain strength.

The strength of materials is quantified as the tensile strength, defined as the highest stress not leading to a permanent deformation or breaking of the material. If the material is inhomogeneous, the breaking stresses, and hence the tensile strengths, are different in different directions. For fibrous materials, there are characteristic tensile strengths in the direction of the fibres and perpendicular to the direction of the fibres, the former in some cases being orders of magnitude larger than the latter ones.

The components in x-, y-, and z-directions of the force per unit volume, f, are related to the stress tensor τ_{ij} by

$$f_i = \Sigma_j \, \partial \, \tau_{ij} / \partial x_j, \qquad (31.2)$$

and the tensile strength σ_i in a direction specified by i is

$$\sigma_i = \max (\Sigma_j \, \tau_{ij} \, n_j \,), \qquad (31.3)$$

where n is a unit vector normal to the "cut" in the material (see Fig. 31.1) and the maximum sustainable stress in the direction i is to be found by varying the direction of n, that is, varying the angle of slicing. In other words, the angle of the cut is varied until the stress in the direction i is maximum, and the highest value of this maximum stress not leading to irreversible damage defines the tensile strength. If the material is isotropic, the tensile strength is independent of direction and may be denoted σ.

Figure 31.1. Definition of internal stress force (Jensen and Sørensen, 1984).

Consider now a flywheel such as the one illustrated in Fig. 31.2, rotating with angular velocity ω about a fixed axis. The mass distribution is symmetric around the axis of rotation, that is, invariant with respect to rotations through any angle θ about the rotational axis. It is further assumed that the material is homogeneous, so that the mass distribution is fully determined by the mass density ρ and the variation of disk width $b(r)$ as a function of radial distance r from the axis, for a shape symmetric about the midway plane normal to the axis of rotation.

The internal stress forces (31.2) plus any external forces f_{ext} determine the acceleration of a small volume of the flywheel situated at the position x:

$$\rho \, \frac{d^2 x_i}{dt^2} = f_{ext,i} + \sum_j \frac{\partial \tau_{ij}}{\tau \, x_j}, \qquad (31.4)$$

which is the Newtonian equation of motion. Integrated over some volume V, the force becomes

$$F_i = \int_V f_{ext,i} \, dx + \int_V \sum_j \frac{\partial \tau_{ij}}{\partial x_j} \, dx = F_{ext,i} + \int_V \sum_j \tau_{ij} \, n_j \, da, \qquad (31.5)$$

where in the last line the volume integral over V has been transformed into a surface integral over the surface A enclosing the volume V, with n being a unit vector normal to the surface.

The constant-stress disk

For uniform rotation with constant angular velocity ω, the acceleration on the left-hand side of (31.4) is radial and given by $r\omega^2$ at the distance r from the axis. Disregarding gravitational forces, the centrifugal force alone must be balanced by the internal stresses, and one may proceed to find the conditions under which all parts of the material experience the same stress τ. If τ equals the tensile strength σ or a specified fraction of it (in order to have a safety margin, as one always would in practice), then the material is utilised optimally, and the energy stored is the maximum that can be achieved using the given material properties.

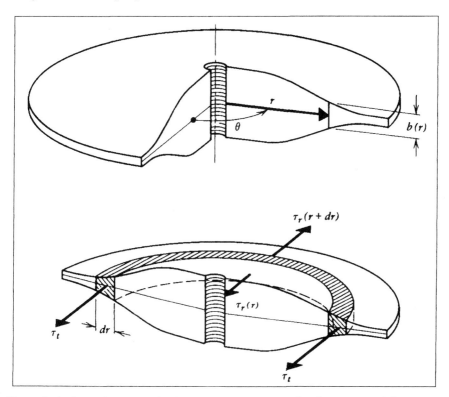

Figure 31.2. Co-ordinates and other quantities used in the description of flywheels. The lower part of the figure illustrates the half-torus shape, confined between radii r and $r + dr$, used in the evaluation of stresses in the direction perpendicular to the cut (Jensen and Sørensen, 1984).

Taking the volume V as that enclosed between the radius r and $r + dr$ and between the centre angle $\theta = -x/2$ and $\theta = x/2$, with the full widths being $b(r)$ and $b(r+dr)$, the balance between the centrifugal force and the internal stresses is obtained from (31.4) and (31.5),

$$2\rho\, r\omega^2 b(r)\, r\, dr = 2\tau((r + dr)\, b(r + dr) - r\, b(r) - b(r)\, dr). \tag{31.6}$$

The factors of 2 come from the angular integrals over $\cos\theta$. The first two terms on the right-hand side of (31.6) are derived from the radial stresses, while the last term represents the tangential stresses on the cuts halving the torus shape considered (cf. Fig. 31.1). To first order in dr, (31.6) may be rewritten as

$$\rho\, r^2 \omega^2 b(r) = \tau\, r\, db(r)/dr, \tag{31.7}$$

from which the disc thickness leading to constant stress is found as

$$b(r) = b_0 \exp\left(-\tfrac{1}{2}\,\rho\, r^2 \omega^2/\tau\right). \tag{31.8}$$

The optimum shape is seen to be an infinitely extending disc of exponentially declining thickness.

Other flywheel shapes
The approach used above may be generalised. Instead of assuming constant stress, one may assume the shape of the flywheel to be known [i.e., $b(r)$ is known, the material still being homogeneous and the shape symmetrical around the axis of rotation as well as upon reflection in the midway plane perpendicular to the axis]. Then the stresses will have to be calculated as a function of rotational speed ω. Owing to the assumptions made, there are only two radially varying stress functions to consider, the radial stress $\tau_r(r)$ and the tangential stress $\tau_t(r)$, both depending only on the distance r from the axis of rotation. Stress components parallel to the axis of rotations are considered absent. Considering again a half-torus shape (see Fig. 31.2), the forces perpendicular to the cut plane may be written in a generalisation of (31.6):

$$2\rho\, r^2 \omega^2 b(r)\, dr = 2(\tau_r\,((r + dr)\, b(r + dr)\,(r + dr) - r\,\tau_r(r)\, b(r)) - \tau_t(r)\, b(r)\, dr. \tag{31.9}$$

To first order in dr, this gives, after rearrangement,

$$\tau_t(r) = \rho\, r^2\,\omega^2 + \tau_r(r) + d\,\tau_r(r)/dr + r\,\tau_r(r)\, db(r)/(b(r)\, dr). \tag{31.10}$$

This is one equation relating radial and tangential stresses. In order to determine the stresses, a second relation must be established. This is the relation between stresses and strains, corresponding to Hooke's law in the theory of elasticity. Introduce deformation parameters ε_t and ε_r for tangential and radial stretching by the relations

$$2\pi\Delta r = 2\pi\varepsilon_t, \tag{31.11}$$

$$d(\Delta r)/dr = \varepsilon_r = \varepsilon_t + r\, d\varepsilon_t/dr. \tag{31.12}$$

where the first equation gives the tangential elongation of the half-torus confined between r and $r + dr$ (see Fig. 31.2), and (31.12) gives the implied radial stretching. Then the stress–strain relations may be written

$$\varepsilon_t(r) = Y^{-1}(\tau_t(r) - \mu\, \tau_r(r)), \tag{31.13}$$

$$\varepsilon_r(r) = Y^{-1}(\tau_r(r) - \mu\, \tau_t(r)), \tag{31.14}$$

where the strength of compression Y is called Young's module and μ is Poisson's ratio, being the ratio between Y and the corresponding quantity Z measuring the strength of expansion in directions perpendicular to the compression (Shigley, 1972). Eliminating the deformations from (31.12)–(31.14), one obtains a new relation between the stresses:

$$(1 + \mu)\,(\tau_r(r) - \tau_t(r)) = r\,d\tau_t(r)\,/\,dr - r\mu\,d\tau_r(r)\,/\,dr. \tag{31.15}$$

Inserting (31.10) into (31.15) results in a second-order differential equation for the determination of $\tau_r(r)$. The solution depends on materials properties through ρ, Y and μ and on the state of rotation through ω. Once the radial stress is determined, the tangential one can be evaluated from (31.10).

As an example, consider a plane disc of radius r_{max}, with a centre hole of radius r_{min}. In this case, the derivatives of $b(r)$ vanish, and the solution to (31.15) and (31.10) is

$$\tau_r(r) = (3 + \mu)\rho\,\omega^2\,(r^2_{min} + r^2_{max} - r^2_{min}\,r^2_{max}\,/\,r^2 - r^2)\,/8 \tag{31.16}$$

$$\tau_t(r) = (3 + \mu)\rho\,\omega^2\,(r^2_{min} + r^2_{max} + r^2_{min}\,r^2_{max}\,/\,r^2 - (1 + 3\mu)\,r^2\,/\,(3 + \mu))\,/8.$$

The radial stress rises from zero at the inner rim, reaches a maximum at $r = (r_{min}r_{max})^{1/2}$, and then declines to zero again at the outer rim. The tangential stress is maximum at the inner rim and declines outwards. Its maximum value exceeds the maximum value of the radial stress for most relevant values of the parameters (μ is typically around 0.3).

Comparing (31.10) with (31.1) and the expression for I, it is seen that the energy density W in (31.1) can be obtained by isolating the term proportional to ω^2 in (31.10), multiplying it by $\frac{1}{2}r$, and integrating over r. The integral of the remaining terms is over a stress component times a shape-dependent expression, and it is customary to use an expression of the form

$$W\,/\,M = \sigma K_m\,/\,\rho, \tag{31.17}$$

where $M = \int \rho b(r)r\,d\theta\,dr$ is the total flywheel mass and σ is the maximum stress [cf. (31.3)]. K_m is called the shape factor. It depends only on geometry, if all stresses are equal as in the "constant stress disc", but as the example of a flat disc has indicated [see (31.16)], the material properties and the geometry cannot generally be factorised. Still, the maximum stress occurring in the flywheel may be taken out as in (31.17), in order to leave a dimensionless quantity K_m to describe details of the flywheel construction (also, the factor ρ has to be there to balance ρ in the mass M, in order to make K_m dimensionless). The expression (31.17) may now be read in the following way: given a maximum acceptable stress σ, there is a maximum energy storage density given by (31.17). It does not depend on ω, and it is largest for light materials and for large design stresses σ. The design stress is typically chosen as a given fraction ("safety factor") of the tensile strength. If the tensile strength itself is used in (31.17), the physical upper limit for energy storage is obtained, and using (31.1), the expression gives the maximum value of ω for which the flywheel will not fail by deforming permanently or by disintegrating.

Flywheel performance
Some examples of flywheel shapes and the corresponding calculated shape factors K_m are given in Table 31.1. The infinitely extending disc of constant stress has a theoretical shape factor of unity, but for a practical version with finite truncation, K_m of about 0.8 can be expected. A flat, solid disc has a shape factor of 0.6, but if a hole is pierced in the middle, the value reduces to about 0.3. An infinitely thin rim has a shape factor of 0.5 and a finite one of about 0.4, and a radial rod or a circular brush (cf. Fig. 31.3) has K_m equal to one-third.

Shape	K_m
Constant stress disc	1
Flat, solid disc ($\mu = 0.3$)	0.606
Flat disc with centre hole	~0.3
Thin rim	0.5
Radial rod	1/3
Circular brush	1/3

Table 31.1. Flywheel shape factors.

According to (31.17), the other factors determining the maximum energy density are the maximum stress and the inverse density, in the case of a homogeneous material. Table 31.2 gives tensile strengths and/or design stresses with a safety factor included and gives densities for some materials contemplated for flywheel design.

For automotive purposes, the materials with the highest σ/ρ values may be contemplated, although they are also generally the most expensive. For stationary applications, weight and volume are less decisive, and low material cost becomes a key factor. This is the reason for considering cellulosic materials (Hagen *et al.*, 1979). One example is plywood discs, where the disc is assembled from layers of unidirectional plies, each with different orientation. Using (31.17) with the unidirectional strengths, the shape factor should be reduced by almost a factor of 3. Another example in this category is paper roll flywheels, that is, hollow, cylindrically wound shapes, for which the shape factor is $K_m = (1 + (r_{min}/r_{max})^2)/4$ (Hagen et al., 1979). The specific energy density would be about 15 kJ kg^{-1} for the plywood-construction and some 27 kJ kg^{-1} for "super-paper" hollow torus shapes.

Material	Density (kg m^{-3})	Tensile strength (10^6 N m^{-2})	Design stress (10^6 N m^{-2})
Birch plywood	700	125	30
"Super-paper"	1100	335	
Aluminium alloy	2700	500	
Mild steel	7800		300
Maraging steel	8000	2700	900
Titanium alloy	4500		650
Carbon fibre (40% epoxy)	1550	1500	750
E-glass fibre (40% epoxy)	1900	1100	250
S-glass fibre (40% epoxy)	1900	1750	350
Kevlar fibre (40% epoxy)	1400	1800	1000

Table 31.2. Properties of materials considered for flywheels (Davidson et al., 1980; Hagen et al., 1979).

Unidirectional materials may be used in configurations such as the flywheel illustrated in Fig. 31.3b, where tangential (or "hoop") stresses are absent. Volume efficiency is low (Rabenhorst, 1976). Generally, flywheels made from filament have an advantage in terms of high safety, because disintegration into a large number of individual threads makes a failure easily contained. Solid flywheels may fail by expelling large fragments, and for safety such flywheels are not proper in vehicles, but may be placed underground for stationary uses.

Approximately constant stress shapes (cf. Fig. 31.3a) are not as volume efficient as flat discs. Therefore, composite flywheels of the kind shown in Fig. 31.3c have been contemplated (Post and Post, 1973). Concentric flat rings (e.g. made of Kevlar) are separated by elastomers that can eliminate breaking stresses when the rotation creates differential expansion of adjacent rings. Each ring must be made of a different material in order to keep the variations in stress within a small interval. The stress distribution inside each ring can be derived from the expressions in (31.16), assuming that the elastomers fully take care of any interaction between rings. Alternatively, the elastomers can be treated as additional rings, and the proper boundary conditions can be applied (see e.g. Toland, 1975).

Flywheels of the types described above may attain energy densities up to 200 kJ kg^{-1}. The problem is to protect this energy against frictional losses. Rotational speeds would typically be 3–5 revolutions per second. The commonly chosen solution is to operate the flywheel in near vacuum and to avoid any kind of mechanical bearings. Magnetic suspension has recently become feasible for units of up to about 200 t, using permanent magnets made from rare-earth cobalt compounds and electromagnetic stabilisers (Millner, 1979). In order to achieve power input and output, a motor generator is inserted between the magnetic bearing suspension and the flywheel rotor. If the motor is placed inside the vacuum, a brush-less type is preferable.

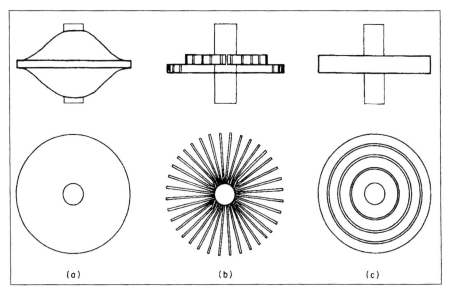

(a)   (b)   (c)

Figure 31.3. Different flywheel concepts. The upper line gives side views, and the lower line gives top views.

For stationary applications, the weight limitations may be circumvented. The flywheel could consist of a horizontally rotating rim wheel of large dimensions and weight, supported by rollers along the rim or by magnetic suspension (Russell and Chew, 1981; Schlieben, 1975). Energy densities about 3000 kJ kg^{-1} could, in principle, be achieved by using fused silica composites (cf. fibres of Table 31.2), if the installations were placed underground in order to allow reduced safety margins. Unit sizes would be up to 10^5 kg.

32

COMPRESSED GAS STORAGE

Gases tend to be much more compressible than solids or fluids, and investigations of energy storage applications of elastic energy on a larger scale have therefore concentrated on the use of gaseous storage media.

Storage on a smaller scale may make use of steel containers, such as the ones common for compressed air used in mobile construction work. In this case the volume is fixed and the amount of energy stored in the volume is determined by temperature and pressure. If air is treated as an ideal gas, the (thermodynamic) pressure P and temperature T are related by the equation of state

$$PV = v \mathscr{R} T, \tag{32.1}$$

where V is the volume occupied by the air, v is the number of moles in the volume, and $\mathscr{R} = 8.315$ J K^{-1} mol^{-1}. The pressure P corresponds to the stress in the direction of compression for an elastic cube, except that the sign is reversed (in general, the stress equals $-P$ plus viscosity-dependent terms). The container may be thought of as a cylinder with a piston, enclosing a given number of moles of gas, say, air, and the compressed air is formed by compressing the enclosed air from standard pressure at the temperature of the surroundings, that is, increasing the force f_x applied to the piston, while the volume decreases from V_0 to V. The amount of energy stored is

$$W = A \int_{x_0}^{x} f_x \, dx = - \int_{V_0}^{V} P \, dV, \tag{32.2}$$

where A is the cylinder cross-sectional area, x and x_0 are the piston positions corresponding to V and V_0, and P is the pressure of the enclosed air.

For large-scale storage applications, underground cavities have been considered. The three possibilities investigated until now are salt domes, cavities in solid rock formations, and aquifers.

Cavities in salt deposits may be formed by flushing water through the salt. The process has, in practical cases, been extended over a few years, in which case the energy spent (and cost) has been very low (Weber, 1975). Salt domes are salt deposits extruding upward toward the surface, therefore allowing cavities to be formed at modest depths.

Rock cavities may be either natural or excavated, and the walls are properly sealed to ensure air-tightness. If excavated, they are much more expensive to make than salt caverns.

Aquifers are layers of high permeability, permitting underground water flows along the layer. In order to confine the water stream to the aquifer, there have to be encapsulating layers of little or no permeability above and below the water-carrying layer. The aquifers usually do not stay at a fixed depth, and thus, there will be slightly elevated regions where a certain amount of air can become trapped without impeding the flow of water. This possibility of air storage (under the elevated pressure corresponding to the depth involved) is illustrated in Fig. 32.1c.

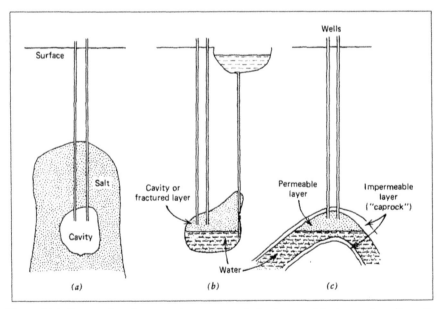

Figure 32.1. Types of underground compressed air storage: (a) storage in salt cavity, (b) rock storage with compensating surface reservoir, and (c) aquifer storage.

Figure 32.1 illustrates the forms of underground air storage mentioned: salt, rock, and aquifer storage. In all cases, the site selection and preparation process is fairly delicate. Although the general geology of the area considered is known, the detailed properties of the cavity will not become fully disclosed until the installation is complete. The ability of the salt cavern to keep an elevated pressure may not live up to expectations based on sample analysis and pressure tests at partial excavation. The stability of a natural rock cave, or of a fractured zone created by explosion or hydraulic methods, is also uncertain until actual full-scale pressure tests have been conducted. For the aquifers, the decisive measurements of permeability can only be made at a finite number of places, so that surprises are possible due to rapid permeability change over small distances of displacement (cf. Adolfson *et al.*, 1979).

The stability of a given cavern during operation of a compressed air storage system is influenced by two design features: the temperature variations and the pressure variations. It is possible to keep the cavern wall temperature nearly constant, either by cooling the compressed air before letting it down into the cavern or by performing the compression so slowly that the temperature only rises to the level prevailing on the cavern walls. The latter possibility (isothermal compression) is impractical for most applications, because excess power must be converted at the rate at which it comes. Most systems therefore include one or more cooling steps. With respect to the problem of pressure variations, when different amounts of energy are stored, the solution may be to store the compressed air at constant pressure but variable volume. In this case either the storage volume itself should be variable, as it is by aquifer storage (when variable amounts of water are displaced), or the underground cavern should be connected to an open reservoir (Fig. 32.1b), so that a variable water column may take care of the variable amounts of air stored at the constant equilibrium pressure prevailing at the depth of the cavern. This kind of compressed energy storage system may alternatively be viewed as a pumped hydro storage system, with extraction taking place through air-driven turbines rather than through water-driven turbines.

Adiabatic storage

Consider now the operation of a variable-pressure type of system. The compression of ambient air takes place approximately as an adiabatic process, that is, without heat exchange with the surroundings. Denoting by γ the ratio between the partial derivatives of pressure with respect to volume at constant entropy and at constant temperature,

$$(\partial P / \partial V)_S = \gamma (\partial P / \partial V)_T, \tag{32.3}$$

the ideal gas law (5.33) gives $(\partial P / \partial V)_T = -P/V$, so that for constant γ,

$$PV^\gamma = P_0 V_0^\gamma. \tag{32.4}$$

The constant on the right-hand side is here expressed in terms of the pressure P_0 and volume V_0 at a given time. For air at ambient pressure and temperature, $\gamma = 1.40$. The value decreases with increasing temperature and increases with increasing pressure, so (32.4) is not entirely valid for air. However, in the temperature and pressure intervals relevant for practical application of compressed air storage, the value of γ varies less than $\pm10\%$ from its average value.

Inserting (32.4) into (32.2) we get the amount of energy stored,

$$W = -\int_{V_0}^{V} P_0 \left(\frac{V_0}{V}\right)^{\gamma} dV = \frac{P_0 V_0}{\gamma - 1}\left(\left(\frac{V_0}{V}\right)^{\gamma-1} - 1\right),$$ (32.5)

or, alternatively,

$$W = \frac{P_0 V_0}{\gamma - 1}\left(\left(\frac{P}{P_0}\right)^{(\gamma-1)/\gamma} - 1\right).$$ (32.6)

More precisely, this is the work required for the adiabatic compression of the initial volume of air. This process heats the air from its initial temperature T_0 to a temperature T, which can be found by rewriting (32.1) in the form

$$T / T_0 = PV / (P_0 V_0)$$

and combining it with the adiabatic condition (32.4),

$$T = T_0 (P / P_0)^{(\gamma-1)/\gamma}.$$ (32.7)

Since desirable pressure ratios in practical applications may be up to about $P/P_0 = 70$, maximum temperatures exceeding 1000 K can be expected. Such temperature changes would be unacceptable for most types of cavities considered, and the air is therefore cooled before transmission to the cavity. Surrounding temperatures for underground storage are typically about 300 K for salt domes and somewhat higher for storage in deeper geological formations. Denoting this temperature T_s, the heat removed if the air is cooled to T_s at constant pressure amounts to

$$H = c_P(T - T_s),$$ (32.8)

where c_P is the heat capacity at constant pressure. Ideally, the heat removed would be kept in a well-insulated thermal energy store, so that it can be used to re-heat the air when it is taken up from the cavity to perform work by expansion in a turbine, with the associated pressure drop back to ambient pressure P_0. Viewed as a thermodynamic process in a temperature–entropy (T, S)-diagram, the storage and retrieval processes in the ideal case look as indicated in Fig. 32.2. The process leads back to its point of departure, indicating that the storage cycle is loss-free under the idealised conditions assumed so far.

In practice, the compressor has a loss (of maybe 5–10%), meaning that not all the energy input (electricity, mechanical energy) is used to perform compression work on the air. Some is lost as friction heat and so on. Further, not all the heat removed by the cooling process can be delivered to re-heat the air. Heat exchangers have finite temperature gradients, and there may be losses from the thermal energy store during the time interval between cooling and re-heating. Finally, the exhaust air from actual turbines has temperatures and pressures above ambient. Typical loss fractions in the turbine may be around 20% of the energy input, at the pressure considered above (70 times ambient) (Davidson *et al.*, 1980). If under 10% thermal losses could be achieved, the overall storage cycle efficiency would be about 65%.

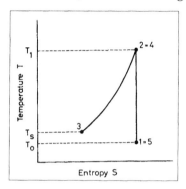

Figure 32.2. Idealised operation of adiabatic compressed air storage system. The charging steps are 1-2 adiabatic compression and 2-3 isobaric cooling to cavern temperature. The unloading steps are 3-4 isobaric heating and 4-5 adiabatic expansion through turbine. The diagram follows a given amount of air, whereas an eventual thermal energy store is external to the "system" considered. T_0 is surface ambient temperature, T_1 is temperature after compression, and T_s is the cavern temperature.

Figure 32.3. Operation of compressed air storage systems with finite losses. The solid path corresponds to a scheme with two compression and two cooling stages, temperature adjustment of the stored air, reheating, and a single turbine stage with reject air injected in the atmosphere (open cycle 1-8). As an alternative, the path from step 7 to step 12 exhibits two heating and expansion steps, corresponding to the operating plant at Huntorf, Germany. See text for further details.

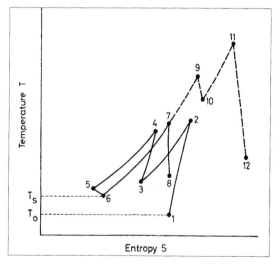

The real process may look as indicated in Fig. 32.3 in terms of temperature and entropy changes. The compressor loss in the initial process 1-2 modifies the vertical line to include an entropy increase. Further, the compression has been divided into two steps (1-2 and 3-4) in order to reduce the maximum temperatures. Correspondingly, there are two cooling steps (2-3 and 4-5), followed by a slight final cooling performed by the cavity surroundings (5-6). The work retrieval process involves in this case a single step 6-7 of re-heating by use of heat stored from the cooling processes (in some cases more than one re-heating step is employed). Finally, 7-8 is the turbine stage, which leaves the cycle open by not having the air reach the initial temperature (and pressure) before it leaves the turbine and mixes into the ambient atmosphere. Also, this expansion step shows deviations from adiabaticity, seen in Fig. 32.3 as an entropy increase.

There are currently a number of operating installations. The earliest full-scale compressed storage facility has operated since 1978 at Huntorf in Germany. It is rated at 290 MW and has about 3×10^5 m^3 storage volume (Lehmann, 1981). It does not have any heat recuperation, but it has two fuel-based turbine stages, implying that the final expansion takes place from a temperature higher than any of those involved in the compression stages (and also at higher pressure). This is indicated in Fig. 32.3 as 7-9-10-11-12, where steps 7-9 and 10-11 represent additional heating based on fuel, while steps 9-10 and 11-12 indicate expansion through turbines. If heat recuperation is added, as it has been in the 110-MW plant operated by the Alabama Electric Corp. (USA) since 1991, this will move point 7 upward towards point 9, and point 8 will move in the direction of 12, altogether representing an increased turbine output (Linden, 2003).

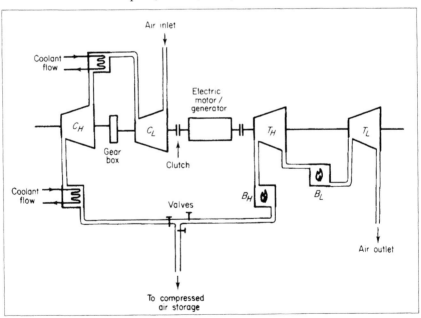

Figure 32.4. Layout of the Huntorf compressed air storage facility. Compressors are denoted *C*, turbines are *T*, and burners are *B*. The subscripts *H/L* stand for high/low pressure.

32. COMPRESSED GAS STORAGE

The efficiency calculation is changed in the case of additional fuel combustion. The additional heat input may be described by (32.8) with appropriate temperatures substituted, and the primary enthalpy input H_0 is obtained by dividing H by the fuel to heat conversion efficiency. The input work W_{in} to the compressor changes in the case of a finite compressor efficiency η_c from (32.6) to

$$W_{in} = \frac{P_0 V_0}{\gamma - 1}\left(\left(\frac{P}{P_0}\right)^{(\gamma-1)/(\eta_c)} - 1\right). \tag{32.9}$$

The work delivered by the turbine receiving air of pressure P_1 and volume V_1, and exhausting it at P_2 and V_2, with a finite turbine efficiency η_t, is

$$W_{out} = \frac{P_1 V_1}{\gamma - 1}\left(1 - \left(\frac{P_2}{P_1}\right)^{\eta_t(\gamma-1)/\gamma}\right), \tag{32.10}$$

which except for the appearance of η_t is just (32.6) rewritten for the appropriate pressures and volume.

Now, in case there is only a single compressor and a single turbine stage, the overall cycle efficiency is given by

$$\eta = W_{out}/(W_{in} + H_0). \tag{32.11}$$

For the German compressed air storage installation mentioned above, η is 0.41. Of course, if the work input to the compressor is derived from fuel (directly or through electricity), W_{in} may be replaced by the fuel input W_0 and a fuel efficiency defined as

$$\eta_{fuel} = W_{out} / (W_0 + H_0) \tag{32.12}$$

If W_{in}/W_0 is taken as 0.36, η_{fuel} for the example becomes 0.25, which is 71% of the conversion efficiency for electric power production without going through the store. The German installation is used for providing peak power on weekdays, charging during nights and weekends.

Figure 32.4 shows the physical layout of the German plant, and Fig. 32.5 shows the layout of a more advanced installation with no fuel input, corresponding to the two paths illustrated in Fig. 32.3.

Aquifer Storage

The aquifer storage system shown in Fig. 32.1c would have an approximately constant working pressure, corresponding to the average hydraulic pressure at the depth of the air-filled part of the aquifer. According to (32.2) the stored energy in this case simply equals the pressure P times the volume of air displacing water in the aquifer. This volume equals the physical volume V times the porosity p, that is, the fractional void volume accessible to intruding air (there may be additional voids that the incoming air cannot reach), so the energy stored may be written

$$W = pVP. \tag{32.13}$$

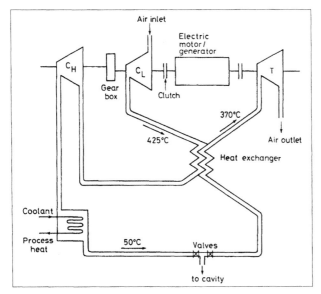

Figure 32.5. Layout of an "advanced" compressed air storage facility, with heat recuperation and no fuel input (symbols are explained in legend to Fig. 32.4).

Typical values are $p = 0.2$ and P around 6×10^6 N m^{-2} at depths of some 600 m, with useful volumes of 10^9 to 10^{10} m^3 for each site. Several such sites have been investigated with the idea of storing natural gas.

An important feature of an energy storage aquifer is the time required for charging and emptying. This time is determined by the permeability of the aquifer. The permeability is basically the proportionality factor between the flow velocity of a fluid or gas through the sediment and the pressure gradient causing the flow. The linear relationship assumed may be written

$$v = - K \, (\eta\rho)^{-1}\partial P/\partial s, \tag{32.14}$$

where v is the flow velocity, η is the viscosity of the fluid or gas, ρ is its density, P is the pressure, and s is the path length in the downward direction. K is the permeability, being defined by (32.14). In metric (SI) units, the permeability has the dimension of m^2. The unit of viscosity is m^2 s^{-1}. Another commonly used unit of permeability is the *darcy*. One darcy equals 1.013×10^{12} m^2. If filling and emptying of the aquifer storage are to take place in a matter of hours rather than days, the permeability has to exceed 10^{11} m^2. Sediments such as sandstone are found with permeabilities ranging from 10^{10} to 3×10^{12} m^2, often with considerable variation over short distances.

In actual installations, losses occur. The cap-rock bordering the aquifer region may not have negligible permeability, implying a possible leakage loss. Friction in the pipes leading to and from the aquifer may cause a loss of pressure, as may losses in the compressor and turbine. Typically, losses of about 15% are expected in addition to those of the power machinery. Large aquifer stores for natural gas are in operation, e.g. at Stenlille, Denmark: 10^9 m³ total volume, from which 3.5×10^8 m³ of gas (at atmospheric pressure 0.1 MPa) can be extracted. The actual storage pressure is 17 MPa (Energinet, 2007). The Danish grid-company further operates an excavated salt dome gas storage at Lille Thorup of slightly smaller volume, but allowing 4.2×10^8 m³ of natural gas to be extracted at at 0.1 MPa (storage pressure 23 MPa).

Hydrogen storage

Hydrogen can be stored like other gases, compressed in suitable containers capable of taking care of the high diffusivity of hydrogen, as well as sustaining the pressures required to bring the energy density up to useful levels. However, the low energy density of gaseous hydrogen has brought alternative storage forms into the focus of investigation, such as liquid hydrogen and hydrogen trapped inside metal hydride structures or inside carbon-based or other types of nanotubes (cf. Sørensen, 2005).

Hydrogen is an energy carrier, not a primary energy form. The storage cycle therefore involves both the production of hydrogen from primary energy sources and the retrieval of the energy form demanded by a second conversion process.

Hydrogen production

Conventional hydrogen production is by catalytic steam reforming of methane (natural gas) or gasoline with water vapour. The process, which typically takes place at 850°C and 2.5 MPa, is

$$C_nH_m + n\ H_2O \rightarrow n\ CO + (n+m/2)\ H_2, \tag{32.15}$$

followed by the catalytic shift reaction

$$CO + H_2O \rightarrow CO_2 + H_2. \tag{32.16}$$

Finally, CO_2 is removed by absorption or membrane separation. The heat produced by (32.16) often cannot be directly used for (32.15). For heavy hydrocarbons, including coal dust, a partial oxidation process is currently in use (Zittel and Wurster, 1996). An emerging technology is high-temperature plasma-arc gasification, on the basis of which a pilot plant operates on natural gas at 1600°C in Norway (at Kvaerner Engineering). The advantage of this process is the pureness of the resulting products (in energy terms, 48% hydrogen, 40% carbon and 10% water vapour) and therefore low environmental impacts. Since all three main products are useful energy carriers, the conversion efficiency may be said to be 98% minus the energy needed for the process. However, conversion of natural gas to carbon is not normally desirable, and the steam can be used only locally, so a 48% efficiency is more meaningful.

Production of hydrogen from biomass may be achieved by biological fermentation or by high-temperature gasification similar to that of coal. These processes were described in more detail in Chapter 22. Production of hydrogen from (wind- or solar-produced) electricity may be achieved by conventional electrolysis (as first demonstrated by Faraday in 1820 and widely used since about 1890) or by reversible fuel cells, with current efficiencies of about 70% and over 90% (in the laboratory), respectively.

Electrolysis conventionally uses an aqueous alkaline electrolyte, with the anode and cathode areas separated by a microporous polymer diaphragm (replacing earlier asbestos diaphragms),

$$H_2O \rightarrow H_2 + \frac{1}{2} O_2, \tag{32.17}$$

$$\Delta H = \Delta G + T\,\Delta S. \tag{32.18}$$

At 25°C, the change in free energy, ΔG, is 236 kJ mol^{-1}, and the electrolysis would require a minimum amount of electric energy of 236 kJ mol^{-1}, while the difference between enthalpy and free energy changes, $\Delta H - \Delta G$, in theory could be heat from the surroundings. The energy content of hydrogen (equal to ΔH) is 242 kJ mol^{-1} (lower heating value), so the $T\,\Delta S$ could exceed 100%. However, if heat at 25°C is used, the process is exceedingly slow. Temperatures used in actual installations are so high that the heat balance is positive and cooling has to be applied. This is largely a consequence of electrode overvoltage mainly stemming from polarisation effects. The cell potential V for water electrolysis may be expressed by

$$V = V_r + V_a + V_c + Rj, \tag{32.19}$$

where V_r is the reversible cell potential. The overvoltage has been divided into the anodic and cathodic parts V_a and V_c. The current is j, and R is the internal resistance of the cell. The three last terms in (32.19) represent the electrical losses, and the voltage efficiency η_V of an electrolyser operating at a current j is given by

$$\eta_V = V_r / V, \tag{32.20}$$

while the thermal efficiency is

$$\eta_t = \Delta H / \Delta G = |\Delta H / (n \mathscr{F} V)|, \tag{32.21}$$

with the Faraday constant being $\mathscr{F} = 96493$ coulombs mol^{-1} and n being the number of moles transferred in the course of the overall electrochemical reaction to which ΔG relates.

Efforts are being made to increase the efficiency above the current 50 to 80% (for small to large electrolysers) by increasing operating temperature and optimising electrode materials and cell design; in that case the additional costs should be less than for the emerging solid-state electrolysers, which are essentially fuel cells operated in reverse fashion, i.e. using electric power to produce hydrogen and oxygen from water in an arrangement and with reaction schemes formally the same as those of fuel cells described in Chapter 19. If the same fuel cell allows operation in both directions, it is called a *reversible fuel cell*.

A third route contemplated for hydrogen production from water is thermal decomposition of water. As the direct thermal decomposition of the water molecule requires temperatures exceeding 3000 K, which is not possible with presently available materials, attempts have been made to achieve decomposition below 800°C by an indirect route using cyclic chemical processes. Such thermochemical or water-splitting cycles were originally designed to reduce the required temperature to the low values attained in nuclear reactors, but could, of course, be used with other technologies generating heat at around 400°C. An example of the processes studied (Marchetti, 1973) is the three-stage reaction

$$6FeCl_2 + 8H_2O \rightarrow 2Fe_3O_4 + 12HCl + 2H_2 \ (850°C)$$
$$2Fe_3O_4 + 3Cl_2 + 12HCl \rightarrow 6FeCl_3 + 6H_2O + O_2 \ (200°C)$$
$$6FeCl_3 \rightarrow 6FeCl_2 + 3Cl_2 \ (420°C).$$

The first of these reactions still requires a high temperature, implying a need for energy to be supplied, in addition to the problem of the corrosive substances involved. The research is still a long way from having created a practical technology.

The process of water photodissociation by light has received some attention (Sørensen, 2005). There have been attempts to imitate the natural photosynthetic process, using semiconductor materials and membranes to separate the hydrogen and oxygen formed by application of light (Calvin, 1974; Wrighton *et al.*, 1977). So far no viable reaction scheme has been found, either for artificial photodissociation or for hybrid processes using heat and chemical reactions in side processes (Hagenmuller, 1977).

Processing of the hydrogen produced involves removal of dust and sulphur, plus other impurities depending on the source material (e.g. CO_2 if biogas is the source).

Hydrogen storage forms
The storage forms relevant for hydrogen are connected with the physical properties of hydrogen, such as the energy density shown in Table 30.1 for hydrogen in various forms. Combustion and safety-related properties are shown in Table 32.1 and compared with those of methane, propane and gasoline. The high diffusivity has implications for container structure, together with the large range of flammability/explosivity for all applications.

Property	Unit	Hydrogen	Methane	Propane	Gasoline
Minimum energy for ignition	10^{-3} J	0.02	0.29		0.24
Flame temperature	°C	2045	1875		2200
Auto-ignition temperature in air	°C	585	540	510	230–500
Maximum flame velocity	m s^{-1}	3.46	0.43	0.47	
Range of flammability in air	vol.%	4–75	5–15	2.5–9.3	1.0–7.6
Range of explosivity in air	vol.%	13–65	6.3–13.5		1.1–3.3
Diffusion coefficient in air	10^{-4} m^2 s^{-1}	0.61	0.16		0.05

Table 32.1. Safety-related properties of hydrogen and other fuels (with use of Zittel and Wurster, 1996).

Compressed storage in gaseous form. The low-volume density of hydrogen at ambient pressure (Table 30.1) makes compression necessary for energy storage applications. Commercial hydrogen containers presently use pressures of 20–30×10^6 Pa, with corresponding energy densities of 1900–2700×10^6 J m^{-3}, which is still less than 10% of that of oil. Research is in progress for increasing pressures to about 70 MPa, using high-strength composite materials such as Kevlar fibres. Inside liners of carbon fibres (earlier glass/aluminium) are required to reduce permeability. Compression energy requirements influence storage cycle efficiencies and involve long transfer times. The work required for isothermal compression from pressure p_1 to p_2 is of the form

$$W = A\,T \log\,(p_2/p_1),$$

where A is the hydrogen gas constant 4124 J K^{-1} kg^{-1} times an empirical, pressure-dependent correction (Zittel and Wurster, 1996). To achieve the compression, a motor rated at Bm must be used, where m is the maximum power throughput and B, depending on engine efficiency, is around 2.

Liquid hydrogen stores. Because the liquefaction temperature of hydrogen is 20 K (−253°C), the infrastructure and liquefaction energy requirements are substantial (containers and transfer pipes must be super-insulated). On the other hand, transfer times are low (currently 3 min to charge a passenger car). The energy density is still 4–5 times lower than for conventional fuels (see Table 30.1). The liquefaction process requires very clean hydrogen, as well as several cycles of compression, liquid nitrogen cooling, and expansion.

Metal hydride storage. Hydrogen diffused into appropriate metal alloys can achieve storage at volume densities over two times that of liquid hydrogen. However, the mass storage densities are still less than 10% of those of conventional fuels (Table 30.1), making this concept doubtful for mobile applications, despite the positive aspects of near loss-free storage at ambient pressures (0–6 MPa) and transfer accomplished by adding or withdrawing modest amounts of heat (plus high safety in operation), according to

$$Me + \tfrac{1}{2}\, xH_2 \leftrightarrow MeH_x, \tag{32.22}$$

where the hydride may be body-centred cubic lattice structures with about 6 $\times 10^{28}$ atoms per m^3 (such as $LaNi_5H_6$, $FeTiH_2$). The highest density currently achieved is for metal alloys absorbing two hydrogen atoms per metal atom (Toyota, 1996). The lattice absorption cycle also performs a cleaning of the gas, because impurities in the hydrogen gas are too large to enter the lattice.

Methanol storage. Some prototype hydrogen-fuelled vehicles have used methanol as storage, even if the desired form is hydrogen (because the car uses a hydrogen fuel cell to generate electricity for its electric motor; Daimler-Chrysler-Ballard, 1998). The motivation for this design has been to exploit the simplicity of methanol storage and filling infrastructure. In the long run, transformation of hydrogen to methanol and back seems too inefficient, and it is likely that the methanol concept will be combined with methanol fuel cells (cf. Chapter 19), while hydrogen-fuelled vehicles must find simpler storage alternatives.

Graphite nanofibre stores. Current development of nanofibres has suggested wide engineering possibilities, regarding both electric and structural adaptation, including the storage of foreign atoms inside "balls" or "tubes" of large carbon structures (Zhang *et al.*, 1998). Suggestions have been made that hydrogen may be stored in nanotubes in quantities exceeding those of metal hydrides, and at a lower weight penalty (Service, 1998), but experiments seem to rule out such possibilities (Zhou *et al.*, 2004): Hydrogen seems to attach to the inside of nanotubes only in a monolayer, implying that densities are lower than for attachment on any suitable plane surface.

Regeneration of power from hydrogen
Retrieval of energy from stored hydrogen may be by conventional low-efficiency combustion in Otto engines or gas turbines, or it may be through fuel cells at a considerably higher efficiency, as described in Chapter 19 and in Sørensen (2005).

BATTERY STORAGE

Batteries may be described as fuel cells where the fuels are stored inside the cell rather than outside it. Historically, batteries were the first controlled source of electricity, with important designs being developed in the early 19th century by Galvani, Volta and Daniell, before Grove's discovery of the fuel cell and Planté's construction of the lead–acid battery. Today, available batteries use a wide range of electrode materials and electrolytes, but despite considerable development efforts aimed at the electric utility sector, battery storage is still in practice restricted to small-scale use (consumer electronics, motor cars, etc.).

Efforts are being made to find systems of better performance than the long-employed lead–acid batteries, which are restricted by a low energy-density (see Table 30.1) and a limited life. Alkaline batteries such as nickel–cadmium cells, proposed around 1900 but first commercialised during the 1970s, are the second largest market for use in consumer electronics equipment and recently for electric vehicles. Despite high cost, lithium-ion batteries have made a rapid impact since their introduction in 1991 (see below). They allow charge topping (i.e. charging before complete discharge) and have a high energy-density, suitable for small-scale portable electronic equipment. Uses at larger scale seem to be forthcoming.

Rechargeable batteries are called accumulators or *secondary batteries*, whereas use-once-only piles are termed *primary batteries*. Table 33.1 gives some important characteristics of various battery types. It is seen that the research goals set in 1977 for high-power batteries have been reached in commercially available products, but not quite the goals for high-energy density cells. One reason for the continued high market share of lead–acid batteries is the perfection of the technology that has taken place over the last decades.

An electrochemical storage battery has properties determined by cell voltage, current and time constants. The two electrodes delivering or receiving power to or from the outside are called e_n and e_p (*negative* and *positive electrode*). The conventional names *anode* and *cathode* are confusing in the case of rechargeable batteries. Within the battery, ions are transported between the negative and positive electrodes through an *electrolyte*. This as well as the electrodes may be solid, liquid or, in theory, gaseous. The electromotive force E_0 is the difference between the electrode potentials for an open external circuit,

$$E_o = E_{ep} - E_{en},$$ (33.1)

Type	Elec-tro-lyte	Energy effici-ency (%)	Energy density (Wh kg^{-1})	Power-densities		Cycle life (cycles)	Operating tempera-tures (°C)
				Peak (W kg^{-1})	Sustained (W kg^{-1})		
Commercial:							
Lead–acid	H_2SO_4	75	20–35	120	25	200–2000	−20 to 60
Nickel–cadmium	KOH	60	40–60	300	140	500-2000	−40 to 60
Ni-metal-hydride	KOH	50	60–80	440	220	<3000	10 to 50
Lithium-ion	$LiPF_6$	70	100–200	720	360	500–2000	−20 to 60
Under development:							
Sodium–sulphur	β-Al_2O_3	70	120	240	120	2000	300 to 400
Lithium-sulphide	AlN	75	130	200	140	200	430 to 500
Zinc–chlorine	$ZnCl_2$	65	120	100			0
Lithium–polymer	Li-β-Alu	70	200			>1200	−20 to 60
1977 goal cells:							
High energy		65	265		55-100	2500	
High power		70	60	280	140	1000	

Table 33.1. Characteristics of selected batteries (Jensen and Sørensen, 1984; Cultu, 1989a; Scrosati, 1995; Buchmann, 1998) and comparison with 1977 development goals (Weiner, 1977).

where it is customary to measure all potentials relative to some reference state. The description of the open cell behaviour uses standard steady-state chemical reaction kinetics. However, when current flows to or from the cell, equilibrium thermodynamics is no longer applicable, and the cell voltage V_c is often parametrised in the form

$$V_c = E_0 - \eta\, IR,$$ (33.2)

where I is the current at a given time, R is the internal resistance of the cell, and η is a "polarisation factor" receiving contributions from the possibly very complex processes taking place in the transition layer separating each electrode from the electrolyte. Figure 33.1 illustrates in a highly schematic form the different potential levels across the cell for open and closed external circuit (cf. e.g. Bockris and Reddy, 1973).

The lead–acid battery

In the electrolyte (aqueous solution of sulphuric acid) of a lead–acid battery, three reactions are at work,

$$H_2O \leftrightarrow H^+ + OH^-,$$
$$H_2SO_4 \leftrightarrow 2H^+ + SO_4^{2-}, \qquad (33.3)$$
$$H_2SO_4 \leftrightarrow H^+ + HSO_4^-,$$

and at the (lead and lead oxide) electrodes, the reactions are

negative electrode: $Pb + SO_4^{2-} \leftrightarrow PbSO_4 + 2e^-$,

positive electrode: $PbO_2 + SO_4^{2-} + 4H^+ + 2e^- \leftrightarrow PbSO_4 + 2H_2O.$ (33.4)

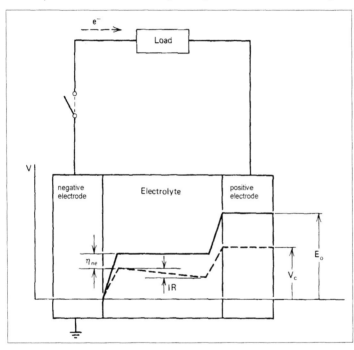

Figure 33.1. Potential distribution through an electrochemical cell: solid line, open external circuit; dashed line, load connected (Jensen and Sørensen, 1984).

The electrolyte reactions involve ionisation of water and either single or double ionisation of sulphuric acid. At both electrodes, lead sulphate is formed, from lead oxide at the positive electrode and from lead itself at the negative electrode. Developments have included sealed casing, thin-tube electrode structure and electrolyte circulation. As a result, the internal resistance has been reduced, and the battery has become practically maintenance-free throughout its life. The energy density of the lead–acid battery increases with temperature and decreases with discharge rate (by about 25% when going from 10 to 1 h discharge, and by about 75% when going from 1 h to 5 min discharge; cf. Jensen and Sørensen, 1984). The figures given in

Table 33.1 correspond to an average discharge rate and an 80% depth of discharge.

While flat-plate electrode grid designs are still in use for automobile starter batteries, tubular-plate designs have a highly increased cycle life and are used for electric vehicles and increasingly for other purposes. The claimed life is about 30 years, according to enhanced test cycling. Charging procedures for lead–acid batteries influence battery life.

Alkaline electrolyte batteries
Among the alkaline electrolyte batteries, nickel–cadmium batteries have been used since about 1910, based upon Jungner's investigations during the 1890s. Their advantage is a long lifetime (up to about 2000 cycles) and with careful use a nearly constant performance, independent of discharge rate and age (Jensen and Sørensen, 1984). However, they do not allow drip charging and easily drop to low capacity if proper procedures are not followed. During the period 1970–90, they experienced an expanding penetration in applications for powering consumer products, such as video cameras, cellular phones, portable players and portable computers, but have now lost most of these markets to the more expensive lithium-ion batteries.

Iron–nickel oxide batteries, which were used extensively in the early part of the 20th century, in electric motorcars, are inferior as regards cell efficiency and peaking capability, owing to low cell voltage and high internal resistance, which also increases the tendency for self-discharge. Alternatives such as nickel–zinc batteries are hampered by low cycle life.

The overall reaction may be summarised as

$$2NiOOH + 2H_2O + Cd \leftrightarrow 2Ni(OH)_2 + Cd(OH)_2. \qquad (33.5)$$

The range of cycle lives indicated in Table 33.1 reflects the sensitivity of NiCd batteries to proper maintenance, including frequent deep discharge. For some applications, it is not practical to have to run the battery down to zero output before recharging.

An alternative considered is nickel–metal hydride batteries, which exhibit a slightly higher energy density but so far with a lower cycle life.

High-temperature batteries
Research on high-temperature batteries for electric utility use (cf. Table 33.1) has been ongoing during several decades, without decisive breakthroughs. Their advantage would be fast, reversible chemistry, allowing for high current density without excess heat generation. Drawbacks include serious corrosion problems, which have persisted and curtailed the otherwise promising development of, for example, the sodium–sulphur battery. This battery has molten electrodes and a solid electrolyte, usually of tubular shape and made from ceramic beta-alumina materials. Similar containment problems have faced zinc–chlorine and lithium–sulphur batteries.

Lithium-ion batteries
Lithium metal electrode batteries attracted attention several decades ago (Murphy and Christian, 1979), owing to the potentially very high energy-density. However, explosion risks stood in the way of commercial applications up until the development of the current lithium-ion concept by Sony in Japan (Nagaura, 1990). Its electrode materials are $LiCoO_2$ and Li_xC_6 (carbon or graphite), respectively, with an electrolyte of lithium hexafluorophosphate dissolved in a mixture of ethylene carbonate and dimenthyl carbonate (Scrosati, 1995). The battery is built from alternating layers of electrode materials, between which the Li-ions oscillate cyclically (Fig. 33.2). The cell potential is high, 3.6 or 7.2 V. Li-ion batteries do not accept overcharge, and a safety device is usually integrated in the design to provide automatic venting in case of emergency. Owing to its high power-density (by weight or volume) and easy charging (topping up is possible without loss of performance), the resulting concept has rapidly penetrated to the high-end portable equipment sector. The safety record is excellent, justifying abandoning the lithium metal electrode despite some loss of power. The remaining environmental concern is mainly due to use of the very toxic cobalt, which requires an extremely high degree of recycling to be acceptable.

Ongoing research aims at bringing the price down and avoiding the use of toxic cobalt, while maintaining the high level of safety and possibly improving performance. The preferred concept uses an all solid-state structure with lithium–beta–alumina forming a layered electrolyte and $LiMn_2O_4$ or $LiMnO_2$ as the positive electrode material (Armstrong and Bruce, 1996), selected from the family of intercalation reactions (Ceder *et al.*, 1998),

$$Li_{x1}MO_2 + (x_2-x_1)Li \rightarrow Li_{x2}MO_2.$$

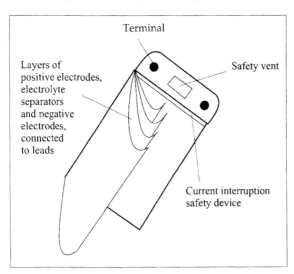

Terminal

Layers of
positive electrodes,
electrolyte
separators
and negative
electrodes,
connected
to leads

Safety vent

Current interruption
safety device

Figure 33.2. Layout of commercial lithium-ion rechargeable battery.

The aim is to reduce the cost of these lithium–polymer batteries to about 20% of the Li-ion battery costs, opening up use not just in small-scale portable equipment but also in electric vehicles. Currently, the ionic conductivity of solid polymer materials is too low to allow ambient temperature operation, but operating temperatures of around 80°C should be acceptable for non-portable applications (Tarascon and Armand, 2001).

Mandatory recycling of batteries has already been introduced in several countries and is clearly a must for some of the recently emerged concepts.

If batteries ever get to the point of being suitable for use in the grid-based electricity supply sector, it is likely to be for short-term and medium-term storage. Batteries have already been extensively incorporated into off-grid systems, such as rural power systems (whether based upon renewable energy or diesel sets) and stand-alone systems such as security back-up facilities. Utility battery systems could serve as load-levelling devices and emergency units, as well as back-up units in connection with variable primary supplies from variable renewable sources.

The small-scale battery options currently extremely popular in the consumer sector carry one very promising possibility, namely, that of creating a de-coupling between production and load in the sector of electricity supply, which traditionally has been strongly focused upon a rigid expectation that any demand will be instantaneously met. If many of the household appliances become battery operated, the stringent requirements placed upon the electricity system to meet peak loads may become less of a burden. A considerable off-peak demand for charging rechargeable batteries could provide precisely the load (as seen from the utility end) control desired, whether it be nearly a constant load pattern or a load following the variations in renewable energy supplies. The same goes for batteries serving the transportation sector (electric vehicles). Freedom from having to always be near an electric plug to use power equipment is seen as a substantial consumer benefit, which many people are willing to pay a premium price for.

Reversible fuel cells (flow batteries)
As mentioned above, fuel cells in reverse operation may replace alkaline electrolysis for hydrogen production, and if fuel cells are operated in both directions, they constitute an energy storage facility. Obvious candidates are further developments of current PEM fuel cells with hydrogen or methanol as the main storage medium. Presently, prototypes of hydrogen/oxygen or hydrogen/air reversible PEM fuel cells suffer from the inability of currently used membranes to provide high efficiency both ways. Whereas a 50% efficiency is acceptable for electricity production, only the same 50% efficiency is achieved for hydrogen production, i.e. lower than conventional electrolysis (Proton Energy Systems, 2003). Fuel cells operated in reverse mode has reached near 100% hydrogen production efficiency, using a few, very large

membranes (0.25 m^2), but such cells are not effective for power production (Yamaguchi *et al.*, 2000). In the laboratory, a high efficiency for both hydrogen production (around 90%) and power production (around 50%) has been demonstrated (Ioroi *et al.*, 2002; 2004).

Early development of reversible fuel cells, sometimes called flow batteries and then referred to as redox flow batteries, used the several ionisation stages of vanadium as a basis for stored chemical energy. The reactions at the two electrodes would typically involve

$$V^{4+} \rightarrow V^{5+} \text{ (positive terminal) } and \text{ } V^{3+} \rightarrow V^{2+} \text{ (negative terminal)}.$$

The electrode material would be carbon fibre, and an ion exchange membrane placed between the electrodes would ensure the charge separation. Specific energy storage would be about 100 kJ kg^{-1}. It has recently been proposed (Skyllas-Kazacos, 2003) to replace the positive electrode process by a halide ion reaction such as

$$Br^- + 2Cl^- \leftarrow BrCl^- + 2e^-$$

in order to increase the storage density. Other developments go in the direction of replacing expensive vanadium by cheaper sodium compounds, using reactions such as

$$3NaBr \rightarrow NaBr_3 + 2Na^+ + 2e^-$$
$$2e^- + 2Na^+ + Na_2S_4 \rightarrow 2Na_2S_2$$

and the reverse reactions. A 120-MWh store with a maximum discharge rate of 15 MW, based upon these reactions, has been installed at Little Barford in the UK. A similar plant is under construction by the Tennessee Valley Authority in the United States (Anonymous, 2002). Physically, the fuel cell hall and the cylindrical reactant stores take up about equal spaces. The cost is estimated at 190 euro per kWh stored or 1500 euro per kW rating (Marsh, 2002).

In the longer range, reversible fuel cell stores are likely to be based on hydrogen, as this storage medium offers the further advantage of being useful as a fuel throughout the energy sector. A scenario exploring this possibility in detail may be found in Sørensen (2005).

CHAPTER

34

OTHER STORAGE FORMS

34.1 Direct storage of light

Energy storage by photochemical means is essential for green plants, but attempts to copy their technique have proved difficult, because of the nature of the basic ingredient: the membrane formed by suitably engineered double layers, which prevents recombination of the storable reactants formed by the solar-induced, photochemical process. Artificial membranes with similar properties are difficult to produce at scales beyond that of the laboratory. Also, there are significant losses in the processes, which are acceptable in the natural systems, but which would negatively affect the economy of man-made systems.

The interesting chemical reactions for energy storage applications accepting direct solar radiation as input use a storing reaction and an energy retrieval reaction of the form

$$A + h\nu \rightarrow B,$$
$$B \rightarrow A + \text{useful energy}.$$

An example would be the absorption of one solar radiation quanta for the purpose of fixing atmospheric carbon dioxide to a metal complex, for example, a complex ruthenium compound. By adding water and heating, methanol can be produced,

$$[M]CO_2 + 2H_2O + 713 \text{ kJ mol}^{-1} \rightarrow [M] + CH_3OH + 1.5O_2.$$

The metal complex has been denoted [M] in this schematic reaction equation. The solar radiation is used to recycle the metal compound by reforming the CO_2-containing complex. The reaction products formed by a photochemical reaction are likely to back-react if they are not prevented from doing so. This is because they are necessarily formed at close spatial separation distances, and the reverse reaction is energetically favoured as it is always of a nature similar to the reactions by which it is contemplated to regain the stored energy.

One solution to this problem is to copy the processes in green plants by having the reactants with a preference for recombination form on opposite sides of a membrane. The membranes could be formed by use of surface-active molecules. The artificial systems consist of a carbohydrate chain containing some 5–20 atoms and in one end a molecule that associates easily with water ("hydrophilic group"). A double layer of such cells, with the hydrophilic groups facing in opposite directions, makes a membrane. If it is closed, for example, forming a shell (Fig. 34.1), it is called a micelle.

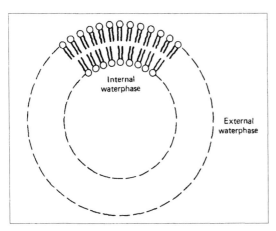

Internal waterphase

External waterphase

Figure 34.1. Double-shell type of micelle. The water-accepting compounds are illustrated by small circles, and the hydrocarbon chain molecules are illustrated by short lines.

Consider now the photochemical reaction bringing A to an excited state A^*, followed by ionisation,

$$A + h\nu \rightarrow A^*$$
$$A^* + B \rightarrow A^+ + B^-.$$

Under normal circumstances, the two ions would have a high probability of recombining, and the storage process would not be very efficient. But if A^+ can be made to form in a negatively charged micelle, the expelled electron would react with B to form B^- outside the micelle, and B^- would not be able to react with A^+. The hope is, in this way, to be able to separate macroscopic quantities of the reactants, which would be equivalent to storing meaningful amounts of energy for later use (see Calvin, 1974), but efficiencies of artificial photosynthetic processes have remained low (Steinberg-Yfrach *et al.*, 1998). Recently, a number of matrix structures have been identified which delay the back-reaction by hours or more, but as yet the retrieval of the stored energy is made correspondingly more difficult. The materials are layered viologen compounds [such as N,N'-dimethyl-4,4'-bipyridinium chloride, methyl viologen (Slama-Schwok *et al.*, 1992), or zirconium phosphate–viologen compounds with Cl and Br inserts (Vermeulen and Thompson, 1992)].

The research on photochemical storage of energy is at a very early stage and certainly is not close to commercialisation. If the research is successful, a new set of storage options will become available. However, it should be stressed that storage cycle efficiencies will not be very high. For photoinduced processes the same limitations exist as for photovoltaic cells, for example, only part of the solar spectrum being useful and losses associated with having to go through definite atomic and molecular energy states. Further losses are involved in transforming the stored chemicals to the energy form demanded.

34.2 Superconducting storage

A magnetic field represents energy stored. When a magnet is charged by a superconducting coil (e.g. a solenoid), the heat losses in the coil may become practically zero, and the system constitutes an electromagnetic energy store with rapid access time for discharging. The maintenance of the coil materials (type II superconductors such as NbTi) near the absolute zero temperature requires cryogenic refrigeration by liquid helium, for example. Owing to the cost of structural support as well as protection against high magnetic flux densities around the plant, underground installation is envisaged. A storage level of 1 GJ is found in the scientific installation from the mid-1970s at the European physics laboratory CERN at Geneva (aimed at preserving magnetic energy across intervals between the accelerator experiments performed at the laboratory). A 100-GJ superconducting store has been constructed in the United States by the Department of Defense, which wants to exploit the fact that superconducting stores can accumulate energy at modest rates but release it during very short times, as required in certain anti-missile defence concepts (Cultu, 1989b). This is still a prototype development. Economic viability is believed to require storage ratings of 10–100 TJ, and one may hope to be able to employ high-temperature superconductors in order to reduce the cooling requirements (however, limited success has been achieved in raising the critical temperature; cf. Locquet *et al.*, 1998).

Only "type II" superconductors are useful for energy storage purposes, since the superconducting property at a given temperature T (below the critical temperature which determines the transition between normal and superconducting phase) disappears if the magnetic field exceeds a critical value, $B_c(T)$ ("magnetic field" will here be taken to mean "magnetic flux density", a quantity that may be measured in units of tesla = V s m^{-2}). Type II superconductors are characterised by high $B_c(T)$, in contrast to type I superconductors.

A magnetic field represents a store of energy, with the energy density w related to the magnetic flux density (sometimes called "magnetic induction") B by

$$w = B^2 / (2\mu_0),$$

where μ_0 (= 1 26 × 10^{-6} henry m^{-1}) is the vacuum permeability. On connecting the superconducting coil to a power source, the induced magnetic field represents a practically loss-free transfer of energy to the magnetic storage. By a suitable switch, the power source may be disconnected and energy could later be retrieved by connecting the coil to a load circuit. Cycle efficiencies will not be 100%, owing to the energy required for refrigerating the storage. Since the magnetic field must be kept below the critical value, increased capacity involves building larger magnets.

Conventional circuit components such as capacitors also represent an energy store, as do ordinary magnets. Such devices may be very useful to smooth out short-term variations in electrical energy flows from source to user, but are unlikely to find practical uses as bulk storage devices.

MINI-PROJECTS AND EXERCISES

1

Find out what kinds of hourly and seasonal variations one can expect for the power production of a photovoltaic array mounted on an inclined rooftop in the area where you are living.

You will need some solar radiation data for your local region. Satellite sources usually only contain data for horizontal surfaces, so it is better to find locally measured data for inclined surfaces suitable for solar panels. If such data are difficult to find, perhaps you can find separate solar radiation data for the direct and scattered parts (or for the direct and for the total radiation, from which you can derive the scattered radiation, at least if you assume it to be identical for any direction within a hemisphere). Such datasets are contained in the so-called reference-year data, which architects and consulting engineers use for design of buildings. They exist for one or more city location in most countries of the world and allow you to calculate the total solar radiation falling on a particular inclined surface at a particular time of the day and the year (translating into a particular direction to the Sun, so that manipulating direct radiation becomes a simple geometric transformation. If you need help to do this, look in Chapter 3 of Sørensen, 2004).

Further, you will have to make assumptions on the functioning of the solar photovoltaic system. Assume as a beginning that it has a fixed efficiency for all the light from the Sun. If you want to go further, you could include wavelength-dependence of the panel efficiency, but in that case you would need to know the frequency composition of the incoming radiation, which is more likely to be available from the satellite data that may be found on the Internet.

2

Discuss the production of heat of the temperature T_L from an amount of heat available, of temperature T above T_L, and access to a large reservoir of temperature T_{ref} below T_L.

Quantities of heat with temperatures T and T_{ref} may be simply mixed in order to obtain heat of the desired temperature.

Alternatively, a thermodynamic cycle may be used between the temperatures T and T_{ref} to produce a certain amount of mechanical or electrical energy, which then is used to power the compressor of a heat pump, lifting heat from T_{ref} to the desired temperature T_L.

As a variant, use a thermodynamic cycle between the temperatures T and T_L, producing less work but on the other hand providing reject heat at the desired temperature. Again, a heat pump is used to provide more heat of temperature T_L on the basis of the work output from the thermodynamic cycle.

What are the simple and the second law efficiencies in the individual cases?

3

Discuss design aspects of a propeller-type wind energy converter for which no possibility of changing the overall pitch angle θ_0 is to be provided.

Consider, for instance, a design similar to the one underlying Figs. 9.3-9.8, but which is required to be self-starting at a wind speed of 2 m s^{-1} and to have its maximum power coefficient at the wind speed of 4 m s^{-1}. Find the tip-speed ratio corresponding to this situation for a suitable overall pitch angle θ_0 and for given assumptions on the internal torque $Q_0(\Omega)$, e.g. that Q_0 is independent of the angular velocity Ω, but proportional to the third power of the blade radius R, with the proportionality factor being given by

$$Q_0 = 10^{-3} \, \pi \rho R^3 u^2_{cut-in},$$

with $u_{cut-in} = 2$ m s^{-1}.

Compare the power produced per square metre swept at a wind speed of 4, 6 and 10 m s^{-1} to that which could be derived from a solar cell array of the same area.

For an actual design of wind energy converters, the blade shape and twist would not be given in advance, and one would, for a device with fixed angular velocity, first decide on this angular velocity (e.g. from generator and gearbox considerations) and then choose a design wind speed and optimise the blade shape $c(r)$ and blade twist $\theta(r)$, in order to obtain a large power coefficient at the design speed and possibly to avoid a rapid decrease in the power coefficient away from the design point.

The blade shape $c(r)$, of course, depends on blade number B, but to lowest order, the performance is independent of changes, which leave the product of $c(r)$ for the individual blades and B unchanged. Name some other considerations that could be of importance in deciding on the blade number.

4

In Section 8.2, a maximum power coefficient C_p far above unity is suggested for certain cases of sail-driven propulsion on friction-free surfaces. Is this not a contradiction, like taking more power out than there is in the wind?

(Hint: C_p is defined relative to the wind power per unit area, which is of immediate relevance for rotors sweeping a fixed area. But what is the area seen by a sail-ship moving with velocity U?).

5

Based, for example, on local data such as those suggested in problem 1, what is the ratio between the daily amount of solar radiation, which on a clear day at different seasons intercepts a fully tracking collector system, and that reaching a flat plate of fixed, south-facing orientation and a certain tilt angle s?

What is the relative importance of east–west tracking alone and solar height tracking alone?

6

Discuss shore-based wave energy converters, e.g. based on letting the wave trains ascend an inclined (and maybe narrowing) ramp, such that the kinetic energy is converted into potential energy of elevation, which may be used to drive a turbine.

The simple solutions for gravity waves in a deep ocean, which were considered in the beginning of Chapter 13, cannot be used directly for waves approaching the shallow coastal waters. However, it may be assumed that the wave period remains unchanged, but that the amplitude a of the wave surface increases in such a way that the total power remains constant as the velocity potential becomes altered due to the boundary condition of no motion at the sea bottom. This assumption implies the neglect of frictional dissipation at the sea floor.

There is a maximum ramp height that will permit a wave to reach the top and cross it into an elevated reservoir. If the turbine used requires a larger height to function well, a narrowing ramp construction may be used to give some fraction of the water mass the desired elevation.

7

Combine biomass productivity data (a global map may be found in B. Sørensen: *Biomass for energy: how much is sustainable?*, which can be downloaded from the download-section at http://energy.ruc.dk; otherwise assume 1-2% of incoming solar radiation converted into biomass) with estimated efficiencies of fuel production by anaerobic fermentation processes (Chapter 22) to give overall conversion efficiencies of bioconversion of solar energy into fuels.

Compare this to other methods of converting solar radiation into mechanical or electrical work, and discuss relative virtues other than conversion efficiency.

8

Estimate the magnitude of and seasonal variation in energy contained in stored food, separately for standing crops on the field, for livestock to be used for food, and for actually stored food (in grain stores, supermarkets, freezers, etc.). Compare food energy stored to the energy value of emergency stores of oil and natural gas, for example, for your own country.

Hint: Data will have to be found at many different places. Some are in Chapter 6 of Sørensen (2004). You could estimate energy stored in supermarkets by your own spot checks of the declarations on typical food products, which should include their energy content.

9

Estimate the potential world production of equivalent crude oil from *Euphorbia* plants, assuming only arid land to be utilised and annual harvest yields of 40 MJ m^{-2}. Areas of different types of land, including arid land, may be found on the Internet.

10

Consider a continuous operation biogas digester for a one-family farm. The digester feedstock is pig slurry (collected daily) and straw (stored). The biogas is used for cooking, for hot water, and for space heating. Estimate the load curves for different seasons and calculate the volume of uncompressed gas storage required, if load is to be met at any time during the year.

11

Consider pure methanol- and pure ethanol-driven cars, and for comparison a gasoline-driven car, weighing 800 kg and going on average 18 km per litre of gasoline and 500 km on a full tank. Calculate the mass penalties for methanol and ethanol fuel tanks, if the same operation range is required. Assume that the fuel usage varies linearly with total mass of the car.

12

Consider a thermal storage of temperature T (say, 200°C) aimed at receiving electric energy through a heat pump (with access to a reservoir of temperature T_{ref}) and delivering electrical energy by means of a thermodynamic engine cycle (available coolant also of temperature T_{ref}). What is the theoretical maximum cycle efficiency, and what can be expected in practice?

13

Discuss energy and power levels for the following "natural hydro energy" concept. Behind a seashore mountain chain there are a number of elevated water reservoirs (e.g. 200 m above sea-level). The sea is characterised by a high evaporation rate (using solar energy), and the prevailing winds carry the evaporated water towards the coast, where a high percentage of the water returns as rain over the mountain region reservoirs (or imagine that precipitation is stimulated by seeding the clouds with some kind of condensation nuclei). Sample evaporation and precipitation rates may be found in meteorological tables or may be extracted from Chapter 2 of Sørensen (2004).

The water is returned from reservoirs to the sea through hydropower plants. In a refined model, storage losses from the reservoirs due to evaporation should be included.

14

On the basis of time sequences of data for an automobile (e.g. your own), try to construct a load-duration curve for the entire period. Based on this, estimate a suitable level of rated power, if the engine should only cover average load, and specify the amount of storage needed for operating a hybrid car with a fuel-based engine rated at the average power needed, a battery store and an electric motor capable of delivering the maximum power required. The electric motor can to a good approximation be considered loss-free.

Hint: A load-duration curve shows the fraction of time during which power demand exceeds a given value as a function of that power value.

15

Construct the load-duration curve for space heating of a dwelling at your geographical location. Assume that this load is to be covered by an energy converter providing constant power year round and that loss-free heat storage is available. Determine the magnitude of the constant converter output that, through use of the storage, will suffice to cover the load at all times. Further, determine the minimum storage size needed. Compare the constant converter output to the range of power provided by currently used heating systems. Use estimated losses for actual storage systems that may be relevant to assess what the required storage capacity would be in case the storage involved realistic losses.

16

A steam turbine power plant with a steam temperature of 700 K and a condenser temperature of 350 K has an efficiency of 0.36 (electric output energy divided by input steam energy), and the turbine efficiency is 20% less than the Carnot efficiency (ε_{max}). Calculate the efficiency of the electric generator.

17

An electric vehicle of mass 1000 kg excluding the battery is designed for a 50-km range in a city driving cycle. The vehicle is equipped with a new 500-kg battery, and the average energy consumption is measured to be 0.3 kWh km^{-1} (from the battery). At a speed of 36 km h^{-1} the required acceleration is 2.0 m s^{-2}, and the power required to overcome the frictional losses is one-third of the total power required. Do the same calculation for an acceleration $a = 1.5$ m s^{-2}.
- Calculate the required average energy density of the new battery.
- Calculate the required power density of the new battery.
- What type of battery could fulfil these requirements (alone or in combination as a hybrid system such as the one considered in problem 14)?

18

Estimate the total hydro resource available for storage applications for a country or region. Characterise the natural reservoirs in terms of water volume and feasible turbine head, and identify the geological formations that could be considered for underground reservoir construction. In this way arrive at rough storage capacities for short- and long-term storage separately. This may be repeated for different regions and results compared.

19

Use (32.7) to express the compressed storage energy (32.6) in terms of the temperature difference $T - T_0$ rather than in terms of the pressure P.

20

Compare the energy needed to compress a certain amount of air by an adiabatic process with that required for an isothermal and an isobaric process.

21

Calculate the shape factor for a thin-rim-type flywheel, as well as mass and volume specific energy density, assuming the material to be steel. Do the same calculation numerically for some of the flywheel types shown in Fig. 31.3 (using data from Tables 31.1 and 31.2) and compare the properties. Discuss priorities for various types of application.

REFERENCES

Achard, P., Lecomte, D., Mayer, D. (1981). Characterization and modelling of test units using salt hydrates, in *Proc. Int. Conf. on Energy Storage*, Vol. 2, pp. 403–410. BHRA Fluid Engineering, Cranfield, UK.

Ackermann, T. (2002). Transmission systems for offshore wind farms. *Renewable Energy World*, **5**, No. 4, 49–61.

Adolfson, W., Mahan, J., Schmid, E., Weinstein, K. (1979). In *Proc. 14th Intersociety Energy Conversion Engineering Conf.*, pp. 452–454. American Chemical Society, Washington, DC.

André, H. (1976). Institute of Electrical and Electronics Engineers, *Transactions* **PAS–95**, 4, 1038–1044.

Andreen, H., Schedin, S. (eds.) (1980). *Den nya energin*, Centrum for Tvärvetenskap/Forlaget Tvartryk, Göteborg.

Angrist, S. (1976). *Direct Energy Conversion*, 3rd ed. Allyn and Bacon, Boston.

Anonymous (1980). *New Scientist*, 11 September, p. 782.

Anonymous (2002). Electricity store. *ABB Review* No. 4, pp. 62–65.

APACE (1982). Technical Information Bulletin PA/111/1, Apace Research Ltd., Hawkesbury, NSW, Australia.

Appropriate Technology Development Organization (1976). *Gobar Gas*, Govt. of Pakistan, Islamabad (undated).

Armstrong, R., Bruce, P. (1996). Synthesis of layered $LiMnO_2$ as electrode for rechargeable lithium batteries. *Nature* **381**, 499–500.

Athey, R. (1976). *Solar Energy* **18**, 143–147.

ATS (2003). Spheral solar technology, website http://www.spheralsolar.com

Baader, W., Dohne, Brenndorfer (1978). *Biogas in Theorie und Praxis*. Landwirtschaftsverlag, Darmstadt.

Bahadori, M. (1977). *Solar Energy Utilization for Developing Countries*, Int. Solar Build. Technology Conf., London, July.

Bakken, K. (1981). System Tepidus, high capacity thermochemical storage/heat pump, in *Proc. Int. Conf. on Energy Storage*, Vol. I, pp. 23–28. BHRA Fluid Engineering, Cranfield, UK.

Ballhausen, C., Gray, H. (1965). *Molecular Orbital Theory*. Benjamin, New York.

Banas, J., Sullivan, W. (1975). Sandia Laboratories Energy Report, SAND75–0530.

Bandel, W. (1981). A review of the possibilities of using alternative fuels in commercial vehicle engines, in *Int. Conf. on Energy Use Management, Berlin, 1981*, Session H–3, Daimler-Benz AG, Stuttgart.

Basore, P. (1991). PC1D v.3 Manual and User Guide. Report 0516/rev. UC–274. Sandia National Laboratory, Albuquerque. Newer versions of the software are available from University of New South Wales, Photovoltaic Centre.

Beale, W. (1976), as quoted in Hughes (1976).

Bechinger, C., *et al.* (1996). *Nature* **383**, 608–610.

Beck, E. (1975). *Science* **189**, 293–294.

Becke, A. (1993). Density-functional thermochemistry. III: The role of exact exchange. *J. Chem. Phys.* **98**, 5648.

BENT SØRENSEN

Beesch, S. (1952). *Ind. Eng. Chem.* **44**, 1677–1682.

Berezin, I., Varfolomeev, S. (1976). *Geliotekhnika* **12**, 60–73.

Berglund, M., Börjesson, P. (2002). Energy efficiency in different types of biogas systems, in *12th European Biomass Conf.*, pp. 219–222. ETA Firenze & WIP Munich.

Bertness, K., Kurtz, S., Friedman, D., Kibbler, A., Kramer, C., Olsson, J. (1994). 29.5% efficient GaInP/GaAs tandem solar cells. *Appl. Phys. Lett.* **65**, 989–991.

Besant, R., Dumont, R., Schoenau, G. (1979). *Solar Age* May, 18–24.

Betz, A. (1959). *Stromunglehre*, G. Brown Verlag, Karlsruhe, BRD.

Biomass Energy Institute (1978). *Biogas Production from Animal Manure*, Winnipeg, Manitoba.

Birck, C., Gormsen, C. (1999). Recent developments in offshore foundation design, in *Proc. European Wind Energy Conference, Nice*, pp. 365–368 (Petersen, Jensen, Rave, Helm, Ehmann, eds.), James & James, London.

Biswas, D. (1977). *Solar Energy* **19**, 99–100.

Blackwell, B., Reis, G. (1974). Sandia Laboratories Report SLA–74–0154, Albuquerque, USA.

Blegaa, S., Hvelplund, F., Jensen, J., Josephsen, L., Linderoth, H., Meyer, N., Balling, N., Sørensen, B. (1976). *Skitse til Alternativ Energiplan for Danmark*, OOA/OVE, Copenhagen (English summary in *Energy Policy*, June 1977, pp. 87–94).

Blomquist, C., Tam, S., Frigo, A. (1979). In *Proc. 14th Intersociety Energy Conversion Engineering Conf.*, pp. 405–413. American Chemical Society, Washington, DC.

Bockris, J., Reddy, A. (1973). *Modern Electrochemistry*. Plenum Press, New York.

Bockris, J., Shrinivasan, S. (1969). *Fuel Cells: Their Electrochemistry*. McGraw-Hill, New York.

Boldt, J. (1978). Solar powered water pump for the rural third world, Danish Technical University, Lab. for Energetics, internal report.

Brehm, N., Mayinger, F. (1989). A contribution to the phenomenon of the transition from deflagration to detonation. VDI–Forschungsheft No. 653/1989, pp. 1–36. (website: http://www.thermo–a.mw.tu–muenchen.de/lehrstuhl/foschung/eder_gerlach.html).

Brown, S. (1998). The automakers' big-time bet on fuel cells. *Fortune Mag.*, 30 March, 12 pages (http://www.pathfinder.com/fortune/1998/980330).

Brüel, P., Schiøler, H., Jensen, J. (1976). Foreløbig redegørelse for solenergi til boligopvarmning, Project report. Copenhagen.

Buchmann, I. (1998). Understanding your batteries in a portable world. Cadex Inc., Canada, http://www.cadex.com/cfm (last accessed 1999).

Buchner, H. (1980). Thermal energy storage using metal hydrides. In *Energy Storage* (J. Silverman, ed.). Pergamon Press, Oxford.

Cacciola, G., Giodano, N., Restuccia, G. (1981). The catalytic reversible (de) hydrogenation of cyclohexane as a means for energy storage and chemical heat pump. In *Proc. Int. Conf. on Energy Storage*, pp. 73–89. BHRA Fluid Engineering, Cranfield, UK.

Callen, H. (1960). *Thermodynamics*, Wiley, New York.

Calvin, M. (1974). *Science*, **184**, 375 – 381.

Calvin, M. (1977). Chemistry, population and resources, in *Proc. 26th Meeting of the Int. Union of Pure Applied Chemistry*, Tokyo. Also Lawrence Berkeley Laboratory Report LBL–6433.

REFERENCES

Casida, M., Casida, K., Jamorski, C., Salahub, D. (1998). *J. Chem. Phys.* **108**, 4439.

Ceder, G., Chiang, Y-M., Sadoway, D., Aydinol, M., Jang, Y.-I. and Huang, B. (1998). Identification of cathode materials for lithium batteries guided by first-principles calculations. *Nature* **392**, 694–696.

Chartier, P., Meriaux, S. (1980). *Recherche* **11**, 766–776.

Cheremisinoff, N., Cheremisinoff, P., Ellerbusch, F. (1980). *Biomass: Applications, Technology, and Production*. Marcel Dekker, New York.

China Reconstructs (1975). December issue, pp. 24– 27 (anonymous).

Christensen, P. (1981). Kemiske Varmelagre, Danish Dept. of Energy, Heat Storage Project Report No. 10, Copenhagen.

Claesson, S. (1974). Jordbrukstekniska Inst., Sweden, Medd. No. 357.

Clarke, F. (1981). Wave energy technology. In *Long-term Energy Sources* (R. Mayer and J. Olsson, eds.), pp. 1269–1303. Pittman Publ., Boston.

Claude, G. (1930). *Mech. Eng.* **52**, 1039–1044.

Clot, A. (1977). *La Recherche* **8**, March, 213–222.

CRES (2002). Wave energy utilization in Europe, Project report for the European Commission, Centre for Renewable Energy Sources, Pikermi.

Cultu, N. (1989a). Energy storage systems in operation. In *Energy Storage Systems* (B. Kilkis, S. Kakac, eds.), pp. 551–574. Kluwer, Dordrecht.

Cultu, N. (1989b). Superconducting magnetic energy storage. In *Energy Storage Systems* (B. Kilkis, S. Kakac, eds.), pp. 551–574. Kluwer, Dordrecht.

Daimler–Chrysler–Ballard (1998). Fuel-cell development programme. Details: http://www.daimler–benz.com/research/specials/necar/necar_e.htm

Damen, K., Faaij, A., Walter, A., Souza, M. (2002). Future prospects for biofuel production in Brazil, in *12th European Biomass Conf.*, Vol. 2, pp. 1166–1169, ETA Firenze & WIP Munich.

Danish Department of Energy (1979). *Sæsonlagring af varme i store vandbassiner*, Heat Storage Project Report No. 2, Copenhagen.

Danish Energy Agency (1992). *Update on Centralized Biogas Plants*. Danish Energy Agency, Copenhagen, 31pp.

Danish Energy Agency (1993). *District Heating in Denmark*, Copenhagen, 58 pp.

d'Arsonval, J. (1881). *Revue Scientifique* **17**, September, 370–372.

Davidson, B., *et al.* (1980). *IEE Proc.* **127**, 345–385.

DEA Wave Program (2002). Final report from Danish Wave Power Committee (in Danish). Rambøll Inc., Virum, website http://www.waveenergy.dk

Deb, S. (1998). Recent developments in high efficiency photovoltaic cells. *Renewable Energy* **15**, 467–472.

Demler, E., Zhang, S.-C. (1998). Quantitative test of a microscopic mechanism of high-temperature superconductivity. *Nature* **396**, 733–735.

deRenzo, D. (1978). *European Technology for Obtaining Energy from Solid Waste*. Noyes Data Corp., Park Ridge, NJ.

Desrosiers, R. (1981). In *Biomass Gasification* (T. Reed, ed.). Noyes Data Corp., Park Ridge, NJ, pp. 119–153.

Dietz, A. (1954). *Diathermarous Materials and Properties of Surfaces* (as quoted by Duffie and Beckman, 1974).

Drewes, P. (2003). Spheral solar – a completely different PV technology. In *Proc. PV in Europe Conf. 2002*, Roma.

Drift, A. van der (2002). An overview of innovative biomass gasification concepts. in *12th European Biomass Conf.*, pp. 381–384. ETA Firenze & WIP Munich.

Duffie, J., Beckman, W. (1974). *Solar Energy Thermal Processes*. Wiley, New York.

Duffie, J., Beckman, W. (1991). *Solar Energy Thermal Processes*, 2nd edition, Wiley, New York.

Dutta, J., Wu, Z., Emeraud, T., Turlot, E, Cornil, E., Schmidt, J., Ricaud, A. (1992). Stability and reliability of amorphous silicon pin/pin encapsulated modules. in *11th EC PV Solar Energy Conference, Montreux* (Guimaraes, L., Palz, W., Reyff, C, Kiess, H. and Helm, P., eds.) pp. 545–548. Harwood Academic Publ., Chur.

Dwayne-Miller, R., *et al.* (2001). High density optical data storage and information retrieval using polymer nanostructures, In *"Proc. 10th Int. Conf. On Unconventional Photoactive Systes*, Diablerets, p. I-27.

EC (1994). *Biofuels* (M. Ruiz-Altisent, ed.), DG XII Report EUR 15647 EN, European Commission.

Eckert, E. (1960). Tidskrift for Landøkonomi (Copenhagen), No. 8.

Einstein, A. (1905). *Ann. der Physik* **17**, 549–560.

El-Hinnawi, E., El-Gohary, F. (1981). In *Renewable Sources of Energy and the Environment* (E. El-Hinnawi and A. Biswas, eds.), pp. 183–219. Tycooly International Press, Dublin

Ellehauge, K. (1981). Solvarmeanlæg til varmt brugsvand, Danish Department of Energy Solar Heat Program, Report No. 16, Copenhagen.

Eltra (2003). Søkabel, Danish power utility webpage http://www.eltra.dk (last assessed 2004).

Energinet (2007). Gaslagre. Danish grid operator webpage http://www.energinet.dk

Erdman, N., *et al.* (2002). The structure and chemistry of the TiO_2-rich surface of SrTiO3(001). *Nature* **419**, 55–58.

Evans, D. (1976). *J. Fluid. Mech.* **77**, 1–25.

Faaij, A., Hamelinck, C. (2002). Long term perspectives for production of fuels from biomass; integrated assessment and R&D priorities, in *12th European Biomass Conf.* vol. 2, pp. 1110—1113. ETA Firenze & WIP Munich.

Fabritz, G. (1954). Wasserkraftmaschinen. In *Hütte Maschinenbau*, Vol. IIA, pp. 865–961. Wilhelm Ernst and Sohn. Berlin.

Fittipaldi, F. (1981). Phase change heat storage. In *Energy Storage and Transportation* (G. Beghi, ed.), pp. 169–182. D. Reidel, Dordrecht, Holland.

Fritsche, H. (1977). In *Proc. 7th Int. Conf. on Amorphous and Liquid Semi-conductors* (W. Spear, ed.), pp. 3–15. University of Edinburgh Pres, Edinburgh.

Fujiwara, L., Nakashima, Y., Goto, T. (1981). *Energy Conversion and Management* **21**, 157–162.

Furbo, S. (1982). Communication No. 116 from Thermal Insulation Laboratory, Technical University, Lyngby, Denmark.

Glaser, P. (1977). *J. Energy* **1**, 75–84.

Glauert, H. (1935). In *Aerodynamic Theory* (W. Durand, ed.), Vol. 4, div. L, pp. 169–360. J. Springer, Berlin.

Gøbel, B., Bentzen, J., Hindsgaul, C., Henriksen, U., Ahrenfeldt, J., Houbak, N., Qvale, B. (2002). High performance gasification with the two.stage gasifier, in *12th European Biomass Conf.*, pp. 289—395. ETA Firenze & WIP Munich.

Golibersuch, D., Bundy, F., Kosky, P., Vakil, H. (1976). Thermal energy storage for utility applications. In *Proc. of the Symp. on Energy Storage* (J. Berkowitz, H. Silverman, eds.). The Electrochemical Society, Inc., Princeton, NJ.

Gramms, L., Polkowski, L., Witzel, S. (1971). *Trans. of the Am. Soc. Agricult. Eng.* **1**, 7.

Granqvist, C., *et al.* (1998). *Solar Energy* **63**, 199-216.

Grasse, W. (1981). *Sunworld* **5**, 68–72.

Green, M. (1994). *Silicon Solar Cells: Advanced Principles and Practice*. PV Special Research Centre, University of New South Wales, Sydney.

Green, M. (2002). Third generation photovoltaics: recent theoretical progress. In *17th European Photovoltaic Solar Energy Conf.*, Munich 2001 (McNelis, Palz, Ossenbrink, Helm, eds.), vol. I, p. 14-17.

Green, M., Zhao, J., Wang, A. (1998). 23% module and other silicon solar cell advances. In *Proc. 2nd World Conf. on PV Energy Conversion*, Vienna (J. Schmid *et al.*, eds.), pp. 1187–1192. JRC European Commission EUR 18656 EN, Luxembourg.

Grimson, J. (1971). *Advanced Fluid Dynamics and Heat Transfer*. McGraw-Hill, London.

Gringarten, A., Wintherspoon, P., Ohnishi, Y. (1975). *J. Geophys. Res.* **80**, 1120–1124.

Grove, W. (1839). On voltaic series and the combination of gases by platinum. *Phil. Mag.* **14**, 127–130.

Hagen, D., Erdman, A., Frohrib, D. (1979). In *Proc. 14th Intersociety Engineering Conf.*, American Chemical Society, Washington, DC, pp. 368–373.

Hagenmuller, P. (1977). *La Recherche* **81**, 756–768.

Hall, C., Swet, C., Temanson, L. (1979). In *Sun II, Proc. Solar Energy Society Conf.*, New Delhi, 1978, pp. 356–359. Pergamon Press, London.

Hamakawa, Y. (1998). A technological evolution from bulk crystalline age to multilayers thin film age in solar photovoltaics. *Renewable Energy* **15**, 22–31.

Hamakawa, Y., Tawada, Y., Okamoto, H. (1981). *Int. J. Solar Energy* **1**, 125.

Hambraeus, G. (ed.) (1975). *Energilagring*. Swedish Academy of Engineering Sciences, Report No. IVA–72, Stockholm.

Hammond, A., Metz, W., Maugh II, T. (1973). *Energy and the Future*. American Assoc. for the Advancement of Science, Washington.

Hammond, A. (1976). *Science* **191**, 1159–1160.

Handley, D., Heggs, P. (1968). *Trans. Econ. & Eng. Rev.* **5**, 7.

Hanneman, R., Vakil, H., Wentorf Jr., R. (1974). Closed loop chemical systems for energy transmission, conversion and storage. In *Proc. 9th Intersociety Energy Conversion Engineering Conf.*. American Society of Mechanical Engineers, New York.

Hara, K., Sayama, K., Ohga, Y., Shinpo, A., Suga, S., Arakawa, H. (2001). A coumarin-derivative dye sensitised nanocrystalline TiO_2 solar cell having a high solar-energy conversion efficiency up to 5.6%. *Chem. Commun.* No. 6, 569-570.

Harth, R., Range, J., Boltendahl, U. (1981). EVA–ADAM System: A method of energy transportation by reversible chemical reactions. In *Energy Storage and Transportation* (G. Beghi, ed.), pp. 358–374. Reidel, Dordrecht, Boston, London.

Hartline, F. (1979). *Science* **206**, 205–206.

Hauch, A., *et al.* (2001). New photoelectrochromic device, website description: http://www.fmf.uni-freiburg.de/~biomed/FSZ/anneke2.htm

Hay, H., Yellot, J. (1972). *Mech. Eng.* **92**, No. 1, 19–23.

Hayes, D. (1977). Worldwatch Paper No. 11. Worldwatch Institute, Washington, DC.

Hein, R. (1974). *Science* **185**, 211–222.

Hermes, J., Lew, V. (1982). Paper CF9/V III/9 in *Proc. UNITAR Conf. on Small Energy*

Resources, Los Angeles 1981, United Nations Inst. for Training and Research, New York.

Heronemus, W. (1975). Proposal to US Nat. Science Found., Report NSF/ RANN/S E/GI–34979/TR/75/4.

Herrick, C. (1982). *Solar Energy* **28**, 99–104.

Hespanhol, I. (1979). *Energia*, **5**, November–December, quoted from El-Hinnawi and El-Gohary (1981).

Hewson, E. (1975). *Bull. Am. Met. Soc.* **56**, 660–675.

Hinze, J. (1975). "Turbulence". McGraw–Hill, New York.

Hirsch, R., Gallagher, J., Lessard, R., Wesselhoft, R. (1982). *Science* **215**, 121–127.

Hoffmann, P. (1998a). ZEVCO unveils fuel cell taxi. *Hydrogen and Fuel Cell Letter*, feature article, August (http://www.mhv.net/~hfcletter/letter).

Hoffmann, P. (1998b). Fuel processors. Record attendance highlight fuel cell seminar in November at Palm Springs. *Hydrogen and Fuel Cell Letter*, feature article, December (http://www.mhv.net/~hfcletter/letter).

Holten, T. van (1976). In *Int. Symp. on Wind Energy Systems*, Cambridge 1976, Paper E3, British Hydromech. Res. Assoc., Cranfield.

Horn, M., *et al.* (1970). *J. Am. Ceram. Soc.* **53**, 124.

Honsberg, C. (2002). A new generalized detailed balance formulation to calculate solar cell efficiency limits. In *17th European Photovoltaic Solar Energy Conf.*, Munich 2001. (McNelis, Palz, Ossenbrink, Helm, eds.), vol. I, p. 3-8.

Hughes, W. (ed.) (1976). *Energy for Rural Development*. US Nat. Acad. Sci., Washington, DC.

Hütte (a Berlin-based association, ed.) (1954). *Des Ingenieurs Taschenbuch*, Vols. I and II. Wilhelm Ernst, Berlin.

Hütter, U. (1976). In *Proc. Workshop on Advanced Wind Energy Systems*, Stockholm, 1974, Swedish Development Board/Vattenfall, Stockholm.

Hütter, U. (1977). *Ann. Rev. Fluid Mech.* **9**, 399–419.

Ichikawa, Y. (1993). Fabrication technology for large-area a–Si solar cells. In *Technical Digest of 7ᵗʰ Int. PV Science and Engineering Conf. Nagoya*, pp. 79–95. Nagoya Institute of Technology.

IEA (1999). Photovoltaic power systems in selected IEA member countries. 3rd survey report of the Power Systems Programme Task 1, International Energy Agency, Paris.

Information (2007). News item based on environmental study. May, Copenhagen.

Ioroi, T., Yasuda, K., Siroma, Z., Fujiwara, N., Miyazaki, Y. (2002). Thin film electrocatalyst layer for unitized regenerative polymer electrolyte fuel cells. *J. Power Sources* **112**, 583-587.

Ioroi, T., Yasuda, K., Miyazaki, Y. (2004). Polymer electrolyte-type unitized regenerative fuel cells. In *15ᵗʰ World Hydrogen Energy Conference*, Yokohama 2004. Paper P09-09. Hydrogen Energy Systems Soc. of Japan (CDROM).

IPCC (1996). *Climate Change 1995: The Science of Climate Change.* Contribution of Working Group I to the Second Assessment Report of the Intergovernmental Panel on Climate Change (Houghton, J.T., Meira Filho, L.G., Callander, B.A., Harris, N., Kattenberg, A., and Maskell, K., eds.). Cambridge University Press, Cambridge, 572 pp.

IPCC (1996b). *Climate Change 1995: Impacts, Adaptation and Mitigation of Climate*

REFERENCES

Change: Scientific–Technical Analysis. Contribution of Working Group II to the Second Assessment Report of the Intergovernmental Panel on Climate Change (Watson, R.T., Zinyowera, M.C., Moss, R.H., Dokken, D.J., eds.). Cambridge University Press, Cambridge, 878 pp.

Isaacs, J., Castel, D., Wick, G. (1976). *Ocean Eng.* **3**, 175–187.

Jamshidi, M., Mohseni, M. (1976). In *System Simulation in Water Resources* (G. VanSteenkiste, ed.), pp. 393–408. North-Holland Publ. Co., Amsterdam.

Jensen, J. (1981). Improving the overall energy efficiency in cities and communities by the introduction of integrated heat, power and transport systems. In *Proc. IEA Inl. New Energy Conservation Technologies Conf.* (J. P. Millhom and E. H. Willis, eds.). Springer-Verlag, Berlin, p. 2981.

Jensen, J., Sørensen, B. (1984). *Fundamentals of Energy Storage.* Wiley, New York, 345 pp. (see p. 217).

Johansson, T. (2002). Energy for sustainable development – a policy agenda for biomass, in *12th European Biomass Conf.*, Vol. 1, pp. 3-6. ETA Firenze, WIP, Munich.

Johnson, J., Hinman, C. (1980). *Science* **208**, 460–463.

Jørgensen, L., Mikkelsen, S., Kristensen, P. (1980). Solvarmeanlæg i Greve, Danish Department of Energy Solar Heat Program, Report No. 6, Copenhagen (with follow up: Report No. 15, 1981).

Joukowski, N. (1906). *Bull. de l'Inst. Aeronaut. Koutchino*, Fasc. I, St. Petersburg.

Juul, J. (1961). In *Proc. UN Conf. on New and Renewable Energy Sources of Energy*, Rome, Vol. 7, paper W/21 (Published 1964 as paper E/Conf. 35 by UN Printing Office, New York).

Kahn, J. (1996). Fuel cell breakthrough doubles performance, reduces cost. Berkeley Lab. Research News, 29. May (http://www.lbl.gov/science–articles/archive/fuel–cells.html).

Kamat, P., Vinodgopal, K. (1998). Environmental photochemistry with semiconductor nanoparticles. Ch. 7 in *Organic and Inorganic Photochemistry* (V. Ramamurthy & K. Schanze, eds.).

Kaye, G., Laby, T. (1959). *Tables of Physical and Chemical Constants*, 12th edn. Longmans, London.

Keenan, J. (1977). *Energy Conversion* **16**, 95–103.

Khaselev, O., Turner, J. (1998). *Science* **280**, 425–427.

Klein, S. (1975). *Solar Energy* **17**, 79–80.

Kohn, W., Sham, L. (1965). Self-consistent equations including exchange and correlation effects. *Phys. Rev.* **140** (1965) A1133.

Koppen, C. van, Fischer, L., Dijkamns, A. (1979). In *Sun II, Proc. Int. Solar Energy Society Conf., New Delhi*, pp. 294–299. Pergamon Press, London.

Korn, J. (ed.) (1972). *Hydrostatic Transmission Systems.* Int. Text Book Co., London.

Kraemer, F. (1981). En model for energiproduktion og økomomi for centrale anlæg til produktion af biogas. Report from Physics Laboratory 3, Danish Technical University, Lyngby.

Krasovec, O., et al. (2001). Nanocrystalline WO_3 layers for photoelectrochromic and energy storage dye sensitised solar cells. In *Proc. 4th Int. Symp. New Materials*, Montréal, pp. 423-425.

Kreider, J. (1979). *Medium and High Temperature Solar Processes.* Academic Press, New York.

Kuemmel, B., Nielsen, S. K., Sørensen, B. (1997). *Life-cycle Analysis of Energy Systems*. Roskilde University Press, Copenhagen, 216 pp.

Kussmaul, K., Deimel, P. (1995). Materialverhalten in H_2-Hochdrucksystemen. *VDI Berichte* **1201**, 87-101.

Kutta, W. (1902). Auftriebkrafte in stromende Flüssigkeiten, *Ill. aeronaut. Mitteilungen*, July.

Ladisch, M., Dyck, K. (1979). *Science* **205**, 898–900.

Ladisch, M., Flickinger, M., Tsao, G. (1979). *Energy* **4**, 263–275.

Lehmann, J. (1981). Air storage gas turbine power plants: a major distribution for energy storage. In *Proc. Int. Conf. Energy Storage*, pp. 327–336. BHRA Fluid Engeneering, Cranfield, UK.

Leichman, J., Scobie, G. (1975). The Development of Wave Power – a Techno-economic Study, part 2, UK Dept. of Industry, and Nat. Eng. Lab., East Kilbride, Scotland.

Li, A., Antizar-Ladislao, B., Khraisheh, M. (2007). Bioconversion of municipal solid waste to glucose for bio-ethanol production. *Bioprocess Biosyst. Eng.* **30**, 189-196.

Li, Chen (1951). *Kho Hsueh Thung Paro* ("Science Correspondent") **2**, No. 3, 266.

Linden, S. van der (2003). The commercial world of energy storage: a review of operating facilities. Presentation for 1st Ann. Conf. Energy Storage Council, Houston, Texas.

Lissaman, P. (1976). In *Proc. Int. Symp. on Wind Energy Systems, Cambridge*. Paper C2: BHRA Fluid Engineering, Cranfield, UK.

Loferski, J. (1956). *J. Appl. Phys.* **27**, 777–784.

Locquet, J.–P., Perret, J., Fompeyrine, J., Mächler, E., Seo, J., Tendeloo, G. (1998). Doubling the critical temperature of $La_{1.9}Sr_{0.1}CuO_4$ using epitaxial strain. *Nature* **394**, 453–456.

Losciale, M. (2002). Technical experiences and conclusions from introduction of biogas as a vehicle fuel in Sweden, in *12th European Biomass Conf.*, Vol. 2, pp. 1124-1127. ETA Firenze & WIP Munich.

Luzzi, A. (1999). IEA Annual Report from Agreement on Production and Utilization of Hydrogen, pp. 35-42.

Ma, W., Saida, T., Lim, C., Aoyama, S., Okamoto, H., Hamakawa, Y. (1995). The utilization of microcrystalline Si and SiC for the efficiency improvement in a–Si solar cells. In *1994 First World Conf. on Photovoltaic Energy Conversion, Kona* pp. 417–420. IEEE, Washington, DC.

Manassen, J., Cahen, D., Hodes, G., Sofer, A. (1976). *Nature* **263**, 97–100.

Marchetti, C. (1973). *Chem. Econ. & Eng. Rev.* **5**, 7.

Mardon, C. (1982). High-rate thermophilic digestion of cellulosic wastes. Paper presented at the 5th Australian Biotechnology Conference, Sydney, August 1982.

Margen, P. (1980). *Sunworld* **4**, 128–134.

Marks, S. (1983). *Solar Energy* **30**, 45–49.

Maron, S., Prutton, C. (1959). *Principles of Physical Chemistry*. Macmillan, New York.

Marsh, G. (2002). RE storage – the missing link. Elsevier Advanced Technology website: http://www.re-focus.net/mar2002_4.html

Masterson, K., Seraphin, B. (1975). *Inter-Laboratory Comparison of the Optical Characteristics of Selective Surfaces for Photo-Thermal Conversion of Solar Energy*, Nat. Sci. Found. (USA) Report NSF/RANN–GI–36731X.

REFERENCES

Masuda, Y. (1971). Paper presented at *Int. Colloquium on Exploitation of the Oceans*, Bordeaux, France.

Mathew, X., *et al.* (2001). In *Proc. 4th Int. Symp. New Materials*, Montréal, 420-421.

Maugh II, T. (1979). *Science* **206**, 436.

Mcbride, J. (1981). Chemical heat pump cycles for energy storage and conversion. In *Proc. Int. Conf. on Energy Storage*, Vol. 2, pp. 29–46. BHRA Fluid Engineering, Cranfield.

McCormick, M. (1976). *Ocean Eng.* **3**, 133–144.

McCoy, E. (1967). *Trans. Am. Soc. Agricultural Engineers* No. 6, 784.

McGowan, J. (1976). *Solar Energy* **18**, 81–92.

McKay, R., Sprankle, R. (1974). In *Proc. Conf. on Res. for the Devt. of Geothermal Energy Resources*, pp. 301–307. Jet Propulsion Lab., California Inst. of Technology, Pasadena, California.

Meibom, P., Svendsen, T., Sørensen, B. (1999). Trading wind in a hydro-dominated power pool system. *Int. J. of Sustainable Development* **2**, 458–483.

Meinel, A., Meinel, M. (1972). *Phys. Today* **25**, No. 2, 44–50.

Meinel, A., Meinel, M. (1976). *Applied Solar Energy*. Addison-Wesley, Reading, MA.

Millner, A. (1979). *Technology Review* November, 32–40.

Minami, E., Kawamoto, H., Saka, S. (2002). Reactivity of lignin in supercritical methanol studied with some lignin model compounds, in *12th European Biomass Conf.*, pp. 785—788. ETA Firenze & WIP Munich.

Mock, J., Tester, J., Wright, P. (1997). Geothermal energy from the earth: its potential impact as an environmentally sustainable resource. *Ann. Rev. Energy Environ.* **22**, 305–356.

Mollison, D., Buneman, O., Salter, S. (1976). *Nature* **263**, 223–226.

Moore, W. (1972). *Physical Chemistry*, 5th edn. Longman Group, London.

Moser, J. (1887). Notitz über Verstärkung photoelektrischer Ströme durch optische Sensibilisierung. *Monatshefte für Chemie und verwandte Teile anderer Wissenschaften* **8**, 373.

Müller, C., Falcou, A., Reckefuss, N., Rojahn, M., Wiederhirn, V., Rudati, P., Frohne, H., Nuyken, O., Becker, H., Meerholz, K. (2003). Multi-colour organic light-emitting displays by solution processing, *Nature* **421**, 829–833.

Murphy, D., Christian, P. (1979). *Science* **205**, 651.

Nagaura, T. (1990). Paper for *3rd Int. Battery Seminar*, Dearfield Beach, FA.

Nakicenovic, N., Grübler, A., Ishitani, H., Johansson, T., Marland, G., Moreira, J. Rogner, H. (1996). *Energy primer.* pp. 75–92 in IPCC (1996).

Nazeeruddin, M., *et al.* (1993). Conversion of light to electricity by cis-X2Bis(2,2'-bipyridyl-4,4'-dicarboxylate) ruthenium-(II) charge transfer sensitizers (X=Cl-, Br-, I-, CN- and SCN-) on nanocrystalline TiO_2 electrodes, *J. Am. Chem. Soc.* **115**, 6382–6390

Nazeeruddin, M., *et al.* (2001). Engineering of efficient panchromatic sensitizers for nanocrystalline TiO_2-based solar cells. *J. Am. Chem. Soc.* **123**, 1613–1620.

Nelson, J., Haque, A., Klug, D., Durran, J. (2001). Trap-limited recombination in dye-sensitised nanocrystalline metal oxide electrodes. *Phys. Rev.* **B 63**, 205321-9.

Nielsen, S., Sørensen, B. (1996). Long-term planning for energy efficiency and renewable energy. Paper presented at Renewable Energy Conference, Cairo April 1996; revised as: Interregional power transmission: a component in planning for renewable energy technologies, *Int. J. Global Energy Issues* **13**, No. 1-3 (2000), 170–180.

Nielsen, S., Sørensen, B. (1998). A fair market scenario for the European energy system. In *Long-Term Integration of Renewable Energy Sources into the European Energy System* (LTI–research group, ed.), pp. 127–186. Physica–Verlag, Heidelberg.

Norman, R. (1974). *Science* **186**, 350–352.

Novozymes (2006). Novozymes and NREL reduce enzyme cost. Announcement posted 2005 on http://www.novozymes.com/htm

Offshore Windenergy Europe (2003). News item dated December 2002, Technical University Delft, posted on the website http://www.offshorewindenergy.org

Ogilvie, T. (1963). *J. Fluid Mech.* **16**, 451–472.

Öhgren, K., Vehmaanpera, J., Siika-Aho, M., Galbe, M., Viikari,L., Zacchi, G. (2007). High temperature enzymatic prehydrolysis prior to simultaneous saccharification and fermentation of steam pretreated corn stover for ethanol production. *Enzyme and Microbial Technology* **40**, 607–613

Olsen, H. (1975). Jordbrugsteknisk Institut Medd. No. 24, Royal Veterinary and Agricultural University, Copenhagen.

Onsager, L. (1931). *Phys. Rev.* **37**, 405–426.

O'Regan, B., Grätzel, M. (1991). *Nature* **353**, 737.

Osterle, J. (1964). *Appl. Sci. Res.*, section A, **12**, 425–434.

Oswald, W. (1973). Progress in Water Technology, *Water Qual. Mgt. Pollut. Contr.* **3**, 153.

Ovshinsky, S. (1978). *New Scientist*, 30 November, 674–677.

Patil, P. (1998). The US DoE fuel cell program. Investing in clean transportation. Paper presented at Fuel Cell Technology Conference, London, September, IQPC Ltd, London.

Pattie, R. (1954). *Nature* **174**, 660.

Pelser, J. (1975). In *Proc. 2nd Workshop on Wind Energy Conversion Systems* (F. Eldridge, ed.), pp. 188–195. The Mitre Corp., McLean, Virginia, Report NSF–RA–N–75-050.

Petersen, J. (1972). Statens Byggeforskningsinstitut (Danish Building Res. Inst.), Notat No. 20.

Pfister, G. and Scher, H. (1977). In *Proc. 7th Int. Conf. on Amorphous and Liquid Semiconductors* (W. Spear, ed.), pp. 197–208. University of Edinburgh Press, Edinburgh.

Pigford, T. (1974). In *Energy, Ecology and the Environment* (R. Wilson and W. Jones, eds.) pp. 343–349. Academic Press, New York.

Pines, D. (1994). Understanding high-temperature superconductivity, a progress report. *Physica B* **199–200**, 300–309.

Pippard, J. (1966). *The Elements of Classical Thermodynamics*. Cambridge University Press, Cambridge, UK.

Popel, F. (1970). *Landtechnische Forschung*, Heft 5, BRD.

Pöpperling, R., Schwenk, W., Venkateswarlu, J. (1982). Abschätzung der Korrosionsgefärdung von Behältern und Rohrleitungen aus Stahl für Speicherung von Wasserstoff und wasserstofhältigen Gasen unter hohen Drücken. VDI Zeitschriften Reihe 5, No. 62.

Post, R., Post, S. (1973). *Sci. Am.*, December, 17–23.

Prengle, H., Sun, C.-H. (1976). *Solar Energy* **18**, 561–567.

Proton Energy Systems (2003). Unigen. Website http://www.protonenergy.com (last accessed 2003).

Rabenhorst, D. (1976). In *Wind Energy Conversion Systems* (J. Savino, ed.). US Nat. Sci. Found., Report NSF/RA/W–73–006.

Rabl, A., Nielsen, C. (1975). *Solar Energy* **17**, 1–12.

Rajeswaran, G., *et al.* (2000). In *SID 00 Digest*, No. 40 (4 pp)

Rambøll Inc. (1999). Wave energy conditions in Danish North Sea, Report (in Danish) from Danish Energy Agency Project 51191/97-0014, Rambøll Inc., Virum.

Reed, T. (ed.) (1981). *Biomass Gasification*, Noyes Data Corp., Park Ridge, NJ.

Reeves R., Lom, E., Meredith, R. (1982). *Stable Hydrated Ethanol Distillate Blends in Diesels*. Apace Res. Ltd., Hawkesbury, NSW, Australia.

Ren, Y., Zhang, Z., Gao, E., Fang, S., Cai, S. (2001). A dye-sensitized nanoporous TiO_2 photochemical cell with novel gel network polymer electrolyte, *J. Appl. Electrochemistry* **31**, 445-447.

Ricaud, A. (1999). Economic evaluation of hybrid solar systems in private houses and commercial buildings. In *Photovoltaic/Thermal Solar Systems, IEA Solar Heating & Cooling/Photovoltaic Power Systems Programmes, Amersfoort*, 13 pp., Ecofys, Utrecht.

Rijnberg, E., Kroon, J., Wienke, J., Hinsch, A., Roosmalen, J., Sinke, W., Scholtens, B., Vries, J., Koster, C., Duchateau, A., Maes, I., Hendrickx, H. (1998). Long-term stability of nanocrystalline dye-sensitized solar cells, in *2nd World Conference on PV Solar Energy Conversion, Vienna* (Schmid, J., Ossenbrink, H., Helm, P., Ehman, H. and Dunlop, E, eds.), pp- 47–52. European Commission, Luxembourg.

Robinson, J. (ed.) (1980). *Fuels from Biomass*. Noyes Data Corp., Park Ridge, NJ.

Roseen, R. (1978). *Central Solar Heat Station in Studsvik*, AB Atomenergi Report ET–78/77, Studsvik.

Rubins, E., Bear, F. (1942). *Soil Sci.* **54**, 411.

Russell, F., Chew, S. (1981). In *Proc. Int. Conf. on Energy Storage*, pp. 373–384. BHRA Fluid Engineering, Cranfield.

Sakai, H. (1993). Status of amorphous silicon solar cell technology in Japan. In *Technical Digest of 7th Int. PV Science and Engineering Conf. Nagoya*, pp. 169–172. Nagoya Institute of Technology.

Salter, S. (1974). *Nature* **249**, 720–724.

Sariciftci, N., Smilowitz, L., Heeger, A., Wudl, F. (1992). *Science* **258**, 1474.

Savonius, S. (1931). *Mech. Eng.* **53**, 333–338.

Schlieben, E. (1975). In *Proc. 1975 Flywheel Technology Symp., Berkeley, CA*, pp. 40–52. Report ERDA 76.

Schulten, R., Decken, C. van der, Kugeler, K., Barnert, H. (1974). Chemical Latent Heat for Transport of Nuclear Energy over Long Distances. In *Proc. British Nuclear Energy Society Int. Conf., The High Temperature Reactor and Process Applications*. BNES, London.

Scrosati, B. (1995). Challenge of portable power. *Nature* **373**, 557–558.

Service, R. (1998). Superstrong nanotubes show they are smart, too. *Science* **281**, 940–942.

Sforza, P. (1976). In *Proc. Int. Symp. on Wind Energy Systems, Cambridge, 1976*, paper E1. British Hydromech. Res. Assoc., Cranfield.

Shaheen, S., *et al.* (2001). 2.5% efficient organic plastic solar cells. *Appl. Phys. Lett.* **78**, 841–843.

Shelton, J. (1975). *Solar Energy* **17**, 137–143.

Shigley, J. (1972). *Mechanical Engineering Design*, 2nd ed., McGraw-Hill, New York.

Shirland, F. (1966). *Advanced Energy Conversion* **6**, 201–221.

Shklover, V., Ovchinnikov, Y., Braginsky, L., Zakeeruddin, S., Grätzel, M. (1998). Structure of organic/inorganic interface in assembled materials comprising molecular components. Crystal structure of the sensitizer bis[4,4'-carboxy-2,2'-bipyridine)(thiocyana-to)]ruthenium(II). *Chem. Mater.* **10**, 2533–2541.

Shockley, W. (1950). *Electrons and Holes in Semiconductors*. Van Nostrand, New York.

Sjoblom, C.-A. (1981). Heat storage in phase transitions of solid electrolytes, in *Proc. of the 16th Intersociety Energy Conversion Engineering Conf.*, Vol. I, paper 8 19441. The American Society of Mechanical Engineers, New York.

Skyllas-Kazacos, M. (2003). Novel vanadium chloride/polyhalide redox flow battery. *J. Power Sources*, **124**, 299–302.

Slama-Schwok, A., Ottolenghi, M., Avnir, D. (1992). Long-lived photoinduced charge separation in a redox system trapped in a sol–gel glass. *Nature* **355**, 240–242.

Slotta, L. (1976). In *Workshop on Wave and Salinity Gradient Energy Conversion* (R. Cohen and M. McCormick, eds.), Paper H, US ERDA Report No. C00–2946–1.

SMAB (1978). *Metanol sam drivmedel*, Annual Report, Svensk Metanol-utveckling AB, Stockholm.

Sørensen, B. (1981). A combined wind and hydro power system. *Energy Policy* March, 51–55.

Sørensen, B. (1986). A study of wind-diesel/gas combination systems. Energy Authority of New South Wales, EA86/17. Sydney (82 pp).

Sørensen, B. (1994). Model optimization of photovoltaic cells. *Solar Energy Materials and Solar Cells* **34**, 133–140.

Sørensen, B. (1996). Does wind energy utilization have regional or global climate impacts? In *1996 European Union Wind Energy Conference, Göteborg* (Zervos, A., Ehmann, H. and Helm, P., eds.), pp. 191–194. H. Stephens & Ass., UK.

Sørensen, B. (1999). Wave energy. In *Encyclopedia of Desalination and Water Resources*, UNESCO project. EOLSS Publ., Oxford, by subscription at website http://www.desware.net

Sørensen, B. (2000a). Role of hydrogen and fuel cells in renewable energy systems. In *Renewable energy: the energy for the 21st century*, Proc. World Renewable Energy Conference VI, Reading, Vol. 3, pp. 1469–1474. Pergamon, Amsterdam.

Sørensen, B. (2000b). PV power and heat production: an added value, in *16th European Photovoltaic Solar Energy Conference*, vol. 2, pp. 1848-1851. (H. Scheer *et al.*, eds), James & James, London.

Sørensen, B. (2002a). Modelling of hybrid PV-thermal systems, in *Proc. 17th European Photovoltaic Solar Energy Conference*, Munich 2001, Vol. 3, pp. 2531-2538. WIP and ETA, Florence.

Sørensen, B. (2002b). Biomass for energy: how much is sustainable?, in *12th European Biomass Conf.*, Amsterdam 2002, Vol. 2, pp. 1394-1397. WIP-Munich and ETA, Florence.

Sørensen, B. (2003a). Understanding photoelectrochemical solar cells. In *PV in Europe*, Rome 2002, pp. 3-8. WIP-Munich and ETA, Florence.

REFERENCES

Sørensen, B. (2003b). Progress in nanostructured photoelectrochemical solar cells. In *Third World Conference on Photovoltaic Energy Conversion, Osaka 2003*, paper 1\1O-B13-04, CDROM.

Sørensen, B. (2004). *Renewable Energy*, 3rd ed., Elsevier Academic Press, 956 pp., Boston; previous editions 1979 and 2000.

Sørensen, B. (2004b). Surface reactions in photoelectrochemical cells. In *Proc. 19th European PV solar energy conference, Paris*. Paper 1AV.2.13, CDROM. WIP Munich and ETA Florence.

Sørensen, B. (2005). *Hydrogen and Fuel Cells*. Elsevier Academic Press, 464 pp. Boston.

Sørensen, B. (2006). Appraisal of bio-hydrogen production schemes. In *Proc. 16th World Hydrogen Energy Conf., Lyon*, Paper S07-109, #109, IHEA CDROM, Sevanova, France.

Sørensen, B., Meibom, P. (2000). A global renewable energy scenario. *Int. J. Global Energy Issues* **13**, No. 1-3, 196-276.

Sørensen, B., Petersen, A., Juhl, C., Ravn, H., Søndergren, C., Simonsen, P., Jørgensen, K., Nielsen, L., Larsen, H., Morthorst, P., Schleisner, L., Sørensen, F., Petersen, T. (2001). Project report to Danish Energy Agency (in Danish): Scenarier for samlet udnyttelse af brint som energibærer i Danmarks fremtidige energisystem, *IMFUFA Texts* No. 390, 226 pp., Roskilde University

Spear, W., Le Comber, P. (1975). *Solid State Commun.* **17**, 1193–1196.

Squires, A. (1974). *Ambio* **3**, 2–14.

Stafford, D., Hawkes, D., Horton, R. (1981). *Methane Production from Waste Organic Matter*, CRC Press, Boca Raton, FL.

Stambolis, C. (1976). In *Proc. UNESCO/WMO Solar Energy Symposium, Geneva 1976*, Paper ENG.S/Doc. 3, WMO Paper No. 477 (1977).

Steinberg-Yfrach, G., Rigaud, J.–L., Durantini, E., Moore, A., Gust, D., Moore, T. (1998). Light–driven production of ATP catalysed by F_0F_1-ATP synthase in an artificial photosynthetic membrane. *Nature* **392**, 479–482.

Stewart, D., McLeod, R. (1980). *New Zealand Journal of Agriculture* September, 9–24.

Stewart, G., Hawker, J., Nix, H., Rawlins, W., Williams, L. (1982). *The Potential for Production of Hydrocarbon Fuels from Crops in Australia*. Commonwealth Scientific and Industrial Research Organization, Melbourne.

STI (2002). Sustainable Technology International, Queanbeyan, NSW Australia; Website: http://www.sta.com.au

Stock, M., Carr, A., Blakers, A. (1996). Texturing of polycrystalline silicon, *Solar Energy Materials and Solar Cells* **40**, 33.

Street, R., Mott, N. (1975). *Phys. Rev. Lett.* **39**, 1293–1295.

Strickland, J. (1975). Sandia Laboratories Energy Report SAND75-0431, Albuquerque, New Mexico.

Svendsen, S. (1980). Effektivitetsprøvning af solfangere, Danish Technical University, Laboratory for Heat Insulation, Communication No. 107, Lyngby.

Tabor, H. (1967). *Solar Energy* **7**, 189.

Tafdrup, S. (1993). Environmental impact of biogas productiom from Danish centralized plants. Paper presented at IEA Bioenergy Environmental Impact Seminar, Elsinore, 1993.

Taiganides, E. (1974). *Agricult. Eng. (Am. Soc. Agric. Eng.)* **55**, No. 4.

Takahashi, K. (1998). Development of fuel cell electric vehicles. Paper presented at

Fuel cell technology conference, London, September, IQPC Ltd., London.

Takamoto, T., Ikeda, E., Kurita, H., Yamaguchi, M. (1997). Over 30% efficient In-GaP/GaAs tandem solar cells with InGaP tunnel junction. In *14th European PV Solar Energy Conf.* pp. 970–975. HS Stephens & Assoc., Bedford.

Tang, C. (2001). Organic light emitting diodes, In *Proc. 10th Int. Conf. On Unconventional Photoactive Systems*, Diablerets, p. I-23

Tarascon, J., Armand, M. (2001). Issues and challenges facing rechargeable lithium batteries. *Nature*, 414, 359-367.

Taylor, G., *et al.* (1993). Full scale measurements in wind turbine arrays, In *EC Wind Energy Conference, Travemünde*, pp. 755–758. H. Stephens & Ass., Felmersham.

Telkes, M. (1952). *Ind. Eng. Chem.* 44, 1308.

Telkes, M. (1976). In *Critical Materials Problems in Energy Production* (C. Stein, ed.). Academic Press. New York and London.

Templin, R. (1976). In *Proc. Workshop on Advanced Wind Energy Systems, Stockholm 1974* (O. Ljungstrom, ed.). Swedish Development Board/Vattenfall, Stockholm.

Thorpe, T. (2001). Current status and development in wave energy. In *Proc. Conf. Marine Renewable Energies*, pp. 103-110, Institute of Marine Engineers, UK.

Tideman, J. and Hawker, J. (1981). *Search* 12, 364–365.

Toland, R. (1975). In *Proc. 1975 Flywheel Technology Symp.*, pp. 243–256. Berkeley, CA, Report ERDA 76–85.

Toyota (1996). High-performance hydrogen-absorbing alloy (last accessed 1999), http://www.toyota.co.jp/e/november_96/electric_island/press.html

Trenkowitz, G. (1969). *Die Värmepumpe*, Verein Deutscher Ingenieurs Berichte Nr. 136.

Trinidade, S. (1980). Energy Crops – the Case of Brazil, in *Int. Conf. on Energy from Biomass*, Brighton, UK, 1980, Centro de Tecnologia Promon, Rio de Janeiro

Trombe, F. (1973). Centre Nationale de Récherche Scientifique, Report No. B–1–73–100.

Tsang, C., Lippmann, M., Wintherspoon, P. (1979). In *Sun II, Proc. Solar Energy Society Conf., New Delhi 1978*, pp. 349–355. Pergamon Press, London.

Tsubomura, H., Matsumura, M., Nomura, Y., Amamiya, T. (1976). Dye sensitised zinc oxide: aqueous electrolyte: platinum photocell. *Nature* 261, 402–403

Tuttle, J., Ward, J., Duda, A., Berens, A., Contreras, M., Ramanathan, K., Tennant, A., Keane, J., Cole, E., Emery, K., Noufi, R. (1996). The performance of Cu(In,Ga)Se$_2$-based solar cells in conventional and concentrator applications. *Proc. Material Research Society Symposium* 426, 143.

UN (1981). *World Energy Supplies*, United Nations, New York.

US DoE (1979). *Peat Prospectus*, United States Department of Energy, Washington, DC.

US EPA (1980). US Environmental Protection Agency Report EPA–600/7–80–040 (D. deAngelis *et al.*), Washington, DC; also earlier reports EPA–600/2–76–056 and EPA–600/7–77–091.

Vermeulen, L., Thompson, M. (1992). Stable photoinduced separation in layered viologen compounds. *Nature* 358, 656–658.

Vindeløv, S. (1994). Research activities in wave energy, *Sustainable Energy News*, No. 7 (December), 12–13.

Vohra, K. (1982). Rural and urban energy scenario of the developing countries and

REFERENCES

related health assessment. In *Proc. Int. Symp. on Health Impacts of Different Sources of Energy, Nashville, 1981*, pp. 79–96. Int. Atomic Energy Agency, Vienna, Paper No. IAEA–SM–254/102.

Wagner, U., Geiger, B., Schaefer, H. (1998), Energy life cycle analysis of hydrogen systems. *Int. J. Hydrogen Energy* **23**, 1–6.

Wan, E., Simmins, J., Nguyen, T. (1981). In *Biomass Gasification* (T. Reed, ed.), pp. 351-385. Noyes Data Corp., Park Ridge, NJ

Wang, J., Wang, D., Smith, K., Hermes, J. (1982). In *The Future of Small Energy Resources*, pp. 465–472. McGraw-Hill, New York.

Ward, R. (1982). *Solar Energy* **29**, 83–86.

Weber, O. (1975). *Brown Boveri Mitt.* **62**, No. 7/8, 332–337.

WEC (1991). *District heating/combined heat and power*. World Energy Council, London.

Wehausen, J., Laitone, E. (1960). In *Handbuch der Physik* (S. Flügge, ed.), Vol. 9 (Stromungsmechanik III), pp. 446–778. Springer-Verlag, Berlin.

Weiner, S. (1977). The Sodium–Sulphur Battery: Problems and Promises. Chapter 12 in *Solid State Chemistry of Energy Conversion and Storage* (J. Goodenough and M. Whittingham, eds.), Advances in Chemistry Series 163, American Chemical Society, Washington, DC.

Weinstein, J., Leitz, F. (1976). *Science* **191**, 557–559.

Wenham, S., Robinson, S., Dai, X., Zhao, J., Wang, A., Tang, Y., Ebong, A., Hornsberg, C., Green, M. (1995). Rear surface effects in high efficiency silicon solar cells. In *1994 First World Conf. on Photovoltaic Energy Conversion, Kona*, Vol. 2, pp. 1278–1282. IEEE, Washington, DC.

West, C. (1974). *Fluidyne Heat Engine*, Harwell Report AERE–R6775, U.K. Atomic Energy Agency.

Wilbur, P., Mancini, T. (1976). *Solar Energy* **18**, 569–576.

Wilson, R., Lissaman, P. (1974). *Applied Aerodynamics of Wind Power Machines*. Oregon State University, Report No. NSF–RA–N–74–113.

Winsberg, S. (1981). *Sunworld* **5**, 122–125.

Wise, D. (1981). *Solar Energy* **27**, 159–178.

Wittenberg, L., Harris, M. (1979). In *Proc 14th Intersociety Energy Conversion Engineering Conf.* pp. 49–52. American Chemical Society, Washington, DC.

Wizelius, T. (1998). Potential for offshore transmission. *Windpower Monthly* December, p. 25.

Wolf, M. (1963). *Proc. Inst. Electr. Electron. Eng.* **51**, 674–693.

Wolf, M. (1971). *Energy Conversion* **11**, 63–73.

Wolfbauer, G. (1999). The electrochemistry of dye sensitised solar cells, their sensitisers and their redox shuttles, Ph.D. thesis, Monash University, Victoria, Australia.

Wrighton, M., Ellis, A., Kaiser, S. (1977). Conversion of visible light to electrical energy: stable cadmium selenide photoelectrodes in aqueous electrolytes. In *Solid State Chemistry of Energy Conversion and Storage* (J. B. Goodenough and M. S. Whittingher, eds.), pp. 71–92. Advances in Chemistry Series 163. American Chemical Society.

Wulff, H. (1966). *The Traditional Crafts of Persia*. MIT Press, Cambridge, MA.

Wurster, R. (1997). PEM fuel cells in stationary and mobile applications. Paper for Biel Conference (http://www.hyweb.de/knowledge).

Wysocki, J., Rappaport, P. (1960). *J. Appl. Phys.* **31**, 571–578.

Yamaguchi, M., Horiguchi, M., Nakanori, T., Shinohara, T., Nagayama, K., Yasuda, J. (2000). Development of large-scale water electrolyzer using solid polymer electrolyte in WE-NET, in *Hydrogen energy progress XIII* (Mao and Veziroglu, eds.), Vol. 1, pp- 274-281. IAHE Beijing.

Yamamoto, K., Yoshimi, M., Tawada, Y., Okamoto, Y., Nakajima, A. (1999). Cost effective and high performance thin film Si solar cell towards the 21st century, in *Technical digest of the international PVSEC-11, Sapporo* (Saitoh, T., ed.), pp. 225-228. Tokyo Univ. of Agriculture and Technology, Tokyo.

Yartym, J., *et al.* (2001). In *Proc. 4th Int. Symp. New Materials*, Montréal, pp. 417–419.

Yazawa, Y., Tamura, K., Watahiki, S., Kitatani, T., Minemura, J., Warabisako, T. (1996). GaInP single-junction and GaInP/GaAs two-junction thin-film solar cell structures by epitaxial lift-off. In *Technical Digest of 9th PV Science & Eng. Conf., Miyazaki*, p. 865. Tokyo Inst. of Technology.

Yoneda, N., Ito, S., Hagiwara, S. (1980). Study of energy storage for long term using chemical reactions, 3rd Int. Solar Forum, Hamburg, Germany, June 24–27.

Yu, G., Gao, J., Hummelen, J., Wudl, F., Heeger, A. (1995). Polymer photovoltaic cells. *Science* **270**, 1789–1791.

Zener, C., Fetkovich, J. (1975). *Science* **189**, 294–295.

Zhang, Y., Suenaga, K., Colliex, C., Iijima, S. (1998). Coaxial nanocables: silicon carbide and silicon oxide sheathed with boron nitride and carbon. *Science* **281**, 973–975.

Zhao, J., Wang, A., Altermatt, P., Wenham, S., Green, M. (1995). 24% efficient solar cells. In *1994 First World Conf. on Photovoltaic Energy Conversion, Kona*, Vol. 2, pp. 1477–1480. IEEE, Washington, DC.

Zhao, J., Wang, A., Green, M. (1998). 19.8% efficient multicrystalline silicon solar cells with "honeycomb" textured front surface. In *Proc. 2nd World Conf. on PV Energy Conversion, Vienna* (J. Schmid *et al.*, eds.), pp. 1681–1684. JRC European Commission EUR 18656 EN, Luxembourg.

Zhou, L., Zhou, Y., Sun, Y. (2004). A comparative study of hydrogen absorption on superactivated carbon versus carbon nanotubes. *Int. J. Hydrogen Energy* **29**, 475-479.

Zittel, W., Wurster, R. (1996). *Hydrogen in the energy sector.* Ludwig-Bölkow-ST Report: http://www.hyweb.de/knowledge/w–i–energiew–eng

Zlatev, Z., Thomsen, P. (1976). Numerical Inst. at Danish Technical University, Internal Reports No. 76–9/10, Lyngby, Denmark.

SUBJECT INDEX

A

Absorber, see Collector, Solar energy conversion
Absorptance, 140-141
Absorption,
  of light in solar cell, 103-108
  of ocean waves, 90
Absorption cooling cycle, 163-164
Acceptor level, 98
Airfoil theory, 38-42
Alternating current transmission, 228-229
Amorphous solar cells, 116-118
Anaerobic process, 196-200
Anode losses, 172-173
Aquifer storage, 276, 281-182
Aquifers, 30
Axial interference factor, 37

B

Batteries, 169-172, 262, 288-294
Bernoulli's theorem, 10
Betz limit, 38, 67-68
Bias, of semiconductor junction, 103
Bio-energy, see Biological energy conversion and storage
Biogas, 195-205
Biological energy conversion and storage, 195-213
  efficiency of conversion, 188, 204, 210, 213, 221
  gaseous fuel production, 195
  heat production, 188
  liquid fuel production, 207
Biomass energy, see Biological energy conversion
Blade-element theory of wind converters, 43-48
Boltzmann distribution, 102
Booster mirrors, 155
Brayton cycle, 14-15
Brine screw, 30
Burning, 186

C

Carnot cycle, 5-6
  efficiency of, 6
Cathode losses, 171-172
Characteristic of solar cell, 104 107
Chemical potential, 7
Coefficient of performance (COP), 15
Cogeneration, 22
Collector, solar, 137 , see also Solar energy conversion
Combined cycle, see Cogeneration
Composting, 191
Compressed air storage, 262, 275
Concentrating solar collectors, 120
Concentration ratio of solar collector, 120-126
Conduction band, 96
Conductor, 98
Coning, 56
Convective energy transport, 145
Conversion of energy, see Energy conversion
Cooling cycle, 15, 163
Cross-wind converters, 64
Currents,
  conversion of energy in, see Turbines
Cusp collectors, 155
Cyclic processes, 5, 14

D

Darcy, 282
Darrieus rotor, 64
Dialytic battery, 182
Diesel cycle, 14
Direct current transmission, 228
Direct energy conversion, see Energy conversion processes, principles of
Dissipation of energy, 7 , see also Frictional dissipation, Attenuation, etc.
District heating, 225 , see also Community-size heating systems
Donor level, 98
Doping of semiconductor material, 98
Drag forces, 38-42
Driven cell, 169

Drying, by sun, 188
Ducted rotors, 69 , see also Turbines

E
Effective mass of electron in semiconductor, 100
Efficiency of energy conversion, see Energy conversion processes, efficiency of
Electric field, 11
Electricity production,
  geothermal, 30
  hydro- and tidal power, 76
  photovoltaic, 94
  quality of wind-produced electricity, 74
  solar thermal, 22, 159
  wave energy based, 83
  wind turbine/asynchronous generator, 55
  wind turbine/synchronous d.c. generator, 53
  with use of ocean thermal gradients, 31
  with use of salinity gradients, 182
Electric transmission, see Energy transmission
Electrochemical energy conversion, 169
Electrolysis, 178, 284
Electrolyte resistance, 172
Emission current of electrons, 20
Emittance, 141
Energy bands in solids, 98
Energy conservation, see Energy conversion processes, efficiency of
Energy conversion processes, see Solar energy conversion, Wind energy conversion, Wave energy conversion, etc.
  efficiency of, 9, 77, 89, 153, 157, 160
  principles of, 3
Energy forms, 4
Energy storage, 234, 261
  cold, 163
  cycle efficiency of, 262
  energy density in, 262
  heat, 138, 149-150, 234-259
  high-quality, 261-298

latent heat, 250
  sensible heat, 234, 262
Energy transmission, 223-233
  as heat, 224
  by conducting lines, 228
  by microwaves, 232
  by superconducting lines, 230-231
Engines, see Thermodynamic engines
Enthalpy, 5
Entropy, 4
Ericsson cycle, 14, 22
Ethanol production, 209
Extensive variables, 8

F
Faradays constant, 170, 284
Fermentation, 195
Fermi energy, 17, 98
Fermi-Dirac distribution, 17
First law efficiency, 11
First law of thermodynamics, 11
Flat-plate solar collector, 140
Flywheel storage, 262, 267
Focusing solar collectors, 120, 155
Fossil fuels,
  present usage, 186
  recoverable reserves, 186
Foundations, off-shore wind, 62-63
Francis turbine, 77
Free energy, 7
Free stream turbine, 35
Fresnel formula, 143
Fresnel reflectors and lenses, 156
Fuel cells, 169, 174
  efficiency, 176
  vehicle use, 177
Fuel production from biological material, 195
Fuels, energy density of, 262

G
Gearbox exchange ratio, 50
Geothermal energy, 30
  conversion of, 30
Gibbs potential, 7
Glauber salt, 252
Gray surface, 140-142
Gray-body emitters, 140

Grid, electric, see Transmission of
power

H
Heat exchanger, 29, 160
Heat flow in pipes, 225-227
Heat of fusion storage, 250
Heat pipe, 20, 227
Heat pump, 26
Heat storage, see Energy storage, heat
Heat transfer coefficient, 29, 225
Heating systems,
  biological, 191-194
  direct, 188-191
  geothermal, 30
  heat pump, 26
  solar, 149-156
  wind, 74-75
Helmholtz' potential, 8
Hetero-junction, 107-108
Hole description of electron vacancies,
  98
Hopping, of electrons,
Hot water production, see Heating sys-
  tems, hot water
Hybridisation of atomic orbitals, 95
Hydrogen fuel use, 174-178
Hydrogen production, 283-287
Hydrogen storage, 577 -579
Hydropower, 76
  pumped storage, 261
  turbine types and efficiency, see Turbines

I
Impurities in solids, 99
Induced current, 81
Insulation materials, heat, 240
Intensive variables, 7
Irreversible thermodynamics, 7-9

K
Kutta-Joukowski theorem, 39

L
Lattice structure, 97
Lift forces, 40
Lorentz force, 81

M
Magnetic storage, 297
Magnetohydrodynamic (MHD) con-
  verter, 81
Manure, composting of, 191
Metabolic heat, of livestock, 193
Metal hydride storage, 259, 286-287
Methane production, 194
Methanol, use as a fuel, 176, 262, 287
Mobility of electrons, 106
Monocrystalline solar cells, 112
Multicrystalline solar cells, 114

N
Nagler turbine, 77
Nanofibre storage, 287
Natural gas reserves, 186
Network, electrical, see Transmission
  of power

O
Ocean thermal gradients, electricity
  generation from 31-32
Ohmic losses, 228-229
Ohm's law, 9
Oil reserves, 186
Onsager relations, 8
Optical systems, 120
Orbital theory of molecules, 94
Organic solar cells, 127
Otto cycle, 14-15

P
Panemone, see Cross-wind converter
Parabolic solar collectors, 154
Particles,
  in buildings, 190-191
Pelton turbine, 77
Permeability, electric in vacuum, 298
Permeability, soil, 282
Photoelectrochemical cells, 127
Photolysis, 179, 295-297
Photovoltaic conversion, 94
  efficiency, 111-113,
  temperature dependence, 111
Pipeline transport, 225, 232
Pitch angle, 47

p-n junction, 99
Power coefficient, of wind energy converter, 38, 47, 67, 69
Primary battery, 169
Propeller-type converters, 43, 76
Pumped hydro storage, 262
Pumping devices,
solar, 165-166
wind-driven, 75
Pyrolysis, 216-220

Q
Quantum efficiency, of solar cell, 108

R
Rankine cycle, 14, 164
Recombination lifetime, 106
Recycling of materials, 204
Refrigeration cycle, 15
Regenerative fuel cell, 169
Resistance, electric, 228
Reversible fuel cell, 169, 285, 293
Reynolds number, 40
Rock storage, 248

S
Sail-ships, energy conversion by, 38-41
Salinity differences, electricity from, 179
Salt dome storage of gas, 283, 286
Salt storage of heat, 245-247
Satellite solar power, 125-126
Savonius rotor, see Cross-wind converter
Second law efficiency, 12, 35
Secondary battery, 169
Seebeck coefficient, 18
Selective surface, 141
Self-starting wind turbines, 67, 52
Semiconductor theory, 94
Sensible heat, 235
Short-circuit current of solar cell, 105
Shottky junction, 108
Soil, energy storage in 244
Solar cell, 103, 109
modules, 119
optical subsystem, 120
concentrators, 120

Solar energy conversion, 93
concentrating collectors, 120, 153
distillation, 167
electricity production, 159
flat-plate collector, 140, 161
focusing systems, 156
heat production, 137
passive heating and cooling, 137-139
photo-thermoelectric converters, 160-161
photovoltaic converters, 109-119
pumps, 165-166
refrigerators, 164
tracking systems, 153
work delivered by hot-air engine, 22
Solar pond, 139
Solar still, 167
Stacked solar cells, 115
Standard reversible potential, 171
Steady-state flow, 11 (non-steady flow, 54)
Stirling cycle, 14, 165
Storage, stored energy, see Energy storage
Stratification of water storage temperatures, 243, 245
Stream flow converters, see Turbines
Streamtube model, 43, 57
Superconducting storage, 262, 297
Superconducting transmission lines, 230
Surface recombination velocity, 105-107

T
Tangential interference factor, 45
Tensile strength, 268
Thermal equilibrium, 102
Thermionic generator, 19
Thermocouple, 17
Thermodynamic engines, 13-16, 22-25
Thermodynamic equilibrium, 8
Thermodynamic laws, 8-12
Thermodynamic theory, 7
Thermoelectric generator, 17
Thin-film solar cells, 118
Tidal energy, 79
Tip-loss factor, 48
Tip-speed ratio, 48

Tip-vane, 71
Torque, 11
Tracking systems of solar collectors, 153
Transmission,
  of heat, 225
  of power, 228
Transmission-absorption product, 143
Transmittance, 142
Transport of energy, 224
Tuning of wave energy converter, 89
Turbines, 35, 76
Twist, of airfoil, 49

U
Utility system, electric, 228

V
Valence band, 98
Voltage factor, 109-110
Vorticity, 48, 73

W
Wake, 38
  of wind turbines, restauration of wind profile in, 59-62
Waste heat, utilisation of, 22
Waste, utilisation of energy in, 188-194
Water storage, 235
Wave energy conversion, 83
  oscillating vane converter, 90
  pneumatic converter, 86
Wind, restoration of kinetic energy in 61

Wind energy conversion, 38, 43
  arrays of converters, 61-62
  augmenting devices, 68
  blade-element theory, 43
  cross-wind converters, 64
  delta wing and artificial tornado concepts, 72-73
  ducted rotors, 69
  efficiency, 38
  electricity production, 74
  fuel production, 74
  heat production, 74
  height scaling, 63
  momentum theory of, 35
  multiple streamtube model of Darrieus rotor, 65
  non-uniform wind conditions, 56
  off-shore foundation, 62
  propeller-type converters, 43
  pumping, 74
  streamtube models, 43, 64
  tip-vane rotors, 71
  wind field behind converter, 60
Wind parks, see Wind energy conversion, arrays of converters
Wind power, wind turbine, see Wind energy conversion
Wing theory, 38-42
Wood energy, 188, 262

Y
Yawing of wind energy device, 51-55
Yeast fermentation, 197-199

Printed and bound by CPI Group (UK) Ltd, Croydon, CR0 4YY

08/05/2025

01864879-0001